Electricity at the World's Fair of 1893 Columbian Exposition:
Illustrated Enlarged Special Edition

by John Barrett

Originally published in 1894

Cover Design by Mark Bussler

www.CGRpublishing.com

ELECTRICITY

AT THE

COLUMBIAN EXPOSITION

INCLUDING

AN ACCOUNT OF THE EXHIBITS

IN THE

ELECTRICITY BUILDING, THE POWER PLANT IN MACHINERY HALL, THE ARC AND INCANDESCENT LIGHTING OF THE GROUNDS AND BUILDINGS, THE TRANSMISSION OF POWER FOR THE OPERATION OF MACHINERY, THE SUBWAY AND CONDUITS, ELECTRIC FOUNTAINS, SEARCH LIGHTS, INTRAMURAL ROAD, ELECTRIC LAUNCHES, ELECTRIC ELEVATORS, LIGHTING OF THE FERRIS WHEEL, CORONA LIGHTING OF THE MANUFACTURES BUILDING, ETC.

By J. P. BARRETT

CHIEF OF DEPARTMENT

CHICAGO

1894

"Tipped with golden domes, touched with the pomp of Asia, in the midst of the White City, beside the gleaming waters of Lake Michigan, looking upon the rippling lagoon and the dazzling fountains of the ideal Venice that in the heart of America is the radiant shell of the Columbian World's Fair, one of the exhalations of that wonderful frozen dream, whose exquisite hues and airs and lines are a picture in which genius has been prodigal, and where are gathered the glories and mysteries of human achievement, rises the Electrical Building, stored with the most marvellous of the marvels of the age.

The Fair, considered as an electrical exposition only, would be well worthy the attention of the world. Look from a distance at night, upon the broad spaces it fills, and the majestic sweep of the searching lights, and it is as if the earth and sky were transformed by the immeasurable wands of colossal magicians; and the superb dome of the structure, that is the central glory of the display, is glowing as if bound with wreaths of stars. It is electricity! When the whole casket is illuminated, the cornices of the palaces of the White City are defined with celestial fire. The waters that are at play leap and flash with it. There are borders of lamps around the làgoon. The spectacle is more resplendent than the capitals of Europe ever saw when ablaze with festivals to celebrate triumphant peace or victorious war."

Murat Halsted in "Cosmopolitan."

DEDICATION.

TO THE EXHIBITORS IN THE DEPARTMENT OF ELECTRICITY,

THE MEN WHO GAVE THEIR TIME, MONEY AND INTELLECT TO CROWN THE SUCCESS OF THE COLUMBIAN EXPOSITION, AND TO WHOM POSTERITY OWES THE LARGEST MEASURE OF PRAISE FOR THE GRANDEST ACHIEVEMENTS OF THE NINETEENTH CENTURY, THIS MODEST WORK IS AFFECTIONATELY DEDICATED, BY ONE WHO BELIEVES THAT ELECTRICITY IS THE ACME OF THE ARTS AND INDUSTRIES, AND WHO KNOWS THAT THE MEN OF TO-DAY HAVE WOVEN THE LAUREL WREATH THAT RESTS UPON HER BROW.

THE AUTHOR.

PREFACE.

This volume is not set before the public with any feeling of pride in its authorship. It nearly resembles a compilation. But it is with much elation that I sum up the great features of the greatest of all expositions, and classify them in a book called Electricity.

It is not a little thing to say that electricity, which so many people are pleased to call an infant, supplied to the exposition all its light, both service and decorative, outside and interior; all the power for the operation of exhibits in the various buildings and about the grounds; the sole means of transportation from place to place—if we except the roller chairs; and the means of transportation upon the waterways of the exposition. And it is even more gratifying to know that no such service in any of these various directions was ever given before on a like occasion. Moreover, the outside decorations, such as the cornice lighting of the buildings, the border lighting of the lagoons, the crown of the Administration building, the Electrical fountains, the great search lights, and the circles of light around the Ferris wheel, all these together made a jewel setting to the Fair that brought exclamations of wonder and admiration from all visitors.

Electricity had won its way well to the front in the forward march of the sciences, arts and industries, even before the Columbian Exposition became a fact, but that occasion was worth a decade to the young and growing industry. It gave an opportunity to the workers in its ranks to show what was being done; it dissolved much of the mystery that had pervaded its domain; it brought electricity to the people in the light of a servant not as an awful master; and finally it created an impression of stability and soundness among the thinking and progressive element of the people that will mean wider commercial development in the near future. I may be pardoned for adding that not a small part of the good to the industry will come because of the fact that the people were taken for the time into the confidence of the craft and were taught to know genuineness from charlatanism, good from bad, honest practice from dishonest, and were made to feel moreover that electricity was not a great mystery, but rather one of the exact sciences, as easy calculable as mathematics or physics.

For the rest, I may say that I have made no effort to even name all the electrical exhibits that were at the exposition. I have chosen only those that seemed to me to be worthy of consideration. I have praised where I felt that praise was due, and I have dismissed with bare facts some things that might be well left to individual judgment. I have been especially careful of statements, and I believe I may urge accuracy in excuse for an absence of technical finish, or the evidences of elaborate and scholarly execution. I have made no effort to brighten the book with a view to popularizing it, as I found that such a course would lessen its value for the electrical people, while not accomplishing the end desired on behalf of the general public.

Those who read the book will do so for the information it contains; they may take the various chapter introductions as merely expressive of the opinions of one individual, and not intended to be in any sense oracular.

I have felt that the matter in the body of the book ought to be preserved—hence its publication.

THE AUTHOR.

TABLE OF CONTENTS.

CHAPTER VII.

CHAPTER VIII.

CHAPTER IX.

CHAPTER X.

CHAPTER XI.

CHAPTER XII.

CHAPTER XIII.

CHAPTER XIV.

CHAPTER XV.

TABLE OF CONTENTS.

CHAPTER I.

INTRODUCTORY.

IN a work of this character, a history of electricity would be manifestly inopportune, excepting as items of historical interest may arise in connection with the development and progress of the various branches of the art. Since we are to discuss electricity in its relation to the Columbian Exposition, however, it will not be out of place, and may be of interest, to digress long enough to observe the beginning and growth in importance of electricity in its bearing upon expositions in the past.

As early as 1867, at the Paris Exposition of that year, an attempt was made to entertain visitors at the exposition grounds and in the buildings in the evening. Gas and oil lamps were used lavishly for the purpose, music and theatricals were supplied of satisfactory character and quality, restaurants and cafés were kept open, and the exposition generally was given as gay and festive an air as possible. The prodigal expenditure of time, money and labor were without avail however, and the effort to force attendance after dark was a signal failure, solely because of the insufficient light, and the refusal of the people to be entertained in the dark. People promenaded the Champ de Mars and attended the concerts given in the salons of the International Club; but they would not be amused at the exposition proper. It was this failure that caused the authorities in charge of the exposition of 1878 to decline to keep open in the evenings. Many appliances for lighting by electricity were shown in the French section, but as a commercial factor in illumination electricity was unknown. Five years before, at the Vienna Exposition of 1873, the Gramme machine, a Paris product, was exhibited, and in 1878 it was again shown in a modified and vastly improved form. In 1878 also, M. Gramme exhibited his alternate current machine for use with the Jablochkoff candles, by means of which a number of lights could be operated from one machine. Of course dynamo and electro-magnetic machines for illuminating purposes had been exhibited prior to 1878. In the London Universal Exposition of 1862, Nollett's Alliance machine and the Holmes machine were shown. It was also at this exposition that Dubosq and Serrin exhibited their arc lamps.

In Paris in 1867 two magnetic electric machines were also shown, the Alliance in the French section, and that invented by Mr. Ladd, by the British. It was here too that the first dynamo electric machine was exhibited by Dr. Wilhelm Siemens, who described it before the Berlin Academy on January 17th of that year. Dr. Siemens' invention marks really the beginning of the use of electricity commercially, since without the dynamo for the transformation of mechanical into electrical energy the inventions for lighting and for the transmission of power were absolutely of no avail. Between 1878 and 1881 so many improvements had been made in the utilization of the electrical current for lighting purposes, and so great was the public interest in the subject, that the ever memorable electrical exposition of 1881 was held at the Palais de l'Industrie. It was then that incandescent lighting was first shown on anything like a large scale, and formed with the telephone the chief element of success for that exposition. Arc lighting was also indulged in commercially, 1,350 horse power being produced to operate 1,383 arc lamps, or an aggregate lighting capacity of 500,000 candles, a larger amount of light than had previously been used to light the city of Paris.

This exposition was the first which was successfully kept open in the evening. The electric syndicate of Paris undertook the installation and operation of the service on payment by the administration of a bonus of 10,000 francs and 50 centimes for each evening visitor. This form of contract clearly shows that the administration gravely doubted the expediency of the attempt, and that the electrical people were obliged to bear the burden of any failure that might come of it. That the experiment was successful may be judged by the following figures: Money taken in at the doors mornings and afternoons, 297,600 francs; that taken in the evenings only, 377,696 francs. The success of this undertaking decided the question of keeping open future expositions.

While the commercial history of electricity was thus being made in Europe, Americans not being an exposition people, were content to invent methods of procedure, new apparatus, and to push the work of popularizing electricity for illumination and power purposes. At the Centennial Exposition, however, there was exhibited the telephone, and half a dozen arc lights. At that time Mr. Brush was deep in his experiments with special arc lighting dynamos, and Mr. Edison was on the eve of the perfection of the incandescent lamp.

The first considerable exposition in this country at which electricity played a part of consequence was the "Southern Exposition" at Louisville, held for 100 days, beginning August 1, 1883. The building in which that exposition was held contained 677,400 square feet, with an average interior height of 40 feet. In its lighting, the Edison Company, which had the contract, employed 3,998 c.p. light, though the contract called for a maximum of 4,600 lamps. The gas companies were allowed to compete in the bidding for this contract of lighting, and their figures called for one six-foot burner for every 45 square feet, against one 16-candle power electric lamp for every 175 square feet. The gas companies computed that it would require between 7,000 and 12,000 jets. That is the last exposition at which gas was at all considered in the plans for illumination.

At the Cotton Centennial, held at New Orleans in the Winter of 1884-85, bids for illumination specified electricity as the luminant, and the contracts were awarded to four companies—the Edison Company, the Louisiana Electric Company, the Fort Wayne Jenney Company, and the Brush Electric Company. The art gallery was lighted by 4,800 16 c.p. lamps, the main building employed 800 arc lights of 2,000 c.p., the U. S. Government building had 300 arcs. The grounds were lighted by the tower system of out-door lighting. There were five towers, each containing fifty arcs of 2,000 c.p. In all there were at that exposition 1,500 arc lamps, and 5,000 incandescent lamps. A note made in the records of that exposition, that will be of interest now, states that one of the companies operated an arc lamp of 120,000 candle power. In view of the two hundred million candle power lights of the Columbian Exposition, the light of 1884 shines with the comparative intensity of a tallow dip.

A number of expositions of electrical industries held in Brussels and Munich, and that held under the auspices of the Franklin Institute in Philadelphia in 1884, created a most favorable impression in the minds of the administration of the Paris Exposition of 1889, and in consequence the electricians of Paris and other places were early called into the Councils of the Administration, the graceful treatment of them being in no small degree due to the efforts of M. Berger, the Director-General, himself an electrician and a man of advanced views. It will not be profitable to rehearse the preliminaries that led up to the execution of the well organized plans for the entertainment of the public at that time, by means of brilliant illumination, and the provision of such novelties as the grand illuminated fountain and other attractions of hardly less interest; suffice it to append a resumé of the splendid results. The public or official electric lighting of the Paris Exposition, exclusive of the illuminated fountains, was as follows: 48 arc lights, of 60 amperes; 86 arc lights, of 25 amperes; 911 arc lights, of 8 amperes; 66 incandescent lights, 500 c.p.;

415 incandescent lights, 16 c.p.; 2,525 incandescent lights, 10 c.p.; 5,830 incandescent lights, 4 c.p.; giving a total of 1,045 arc lamps, and 8,837 incandescent lamps. For private lighting, the International Syndicate of Electricians had installed, 1,005 incandescent lights, 10 c.p.; 1,047 incandescent lights, 16 c.p.; 886 incandescent lights, 7 c.p.; 12 incandescent lights, 50 c.p.; 4 incandescent lights, 150 c.p.; 6 incandescent lights, 250 c.p.; total, 3,860 incandescent lamps; 382 arc lamps, 500 c.p.; 27 arc lamps, 1,000 c.p.; 2 arc lamps, 3,000 c.p.; total, 411 arc lamps. Public lighting (including illuminated fountains): 1,093 arc lights, and 8,837 incandescent lights; total of electric lighting of exhibition, 1,716 arc lights, and 12,847 incandescent lights; total c.p. incandescent lamps, 87,410 (official lighting); total c.p. arc lamps, 2,822,000 (official lighting); total c.p. incandescent lamps, 26,204 (private lighting); total c.p. arc lamps, 201,000 (private lighting). Total, 3,136,614.

Passing the Paris Exposition, we are brought at once to our own great exposition, where electricity seems undoubtedly to have made a stride almost suddenly, that placed it at least abreast of the sister sciences, arts and industries, and demonstrated at once that for all time it was to take precedence of, when not ministering to, all the other commercial industries. A few brief years only, not centuries, were required to develop so vast a factor in human progress, and before we proceed further, a few pictures of the remote past may be entertaining.

What we know as electricity was first discovered by the Phœnicians, and a little later by the Greeks. Always looking to the poetical and the mythical however, for their causes and effects, these historical peoples formed theories upon their discoveries that were not only characteristic, but make so pretty a story that even the whir and buzz of our own machinery should not drive their memory away. The Phœnicians have transmitted to us in their romantic language the story that the pieces of Amber sometimes washed up by the waves of the ocean were the petrified tears of maidens, who, disappointed in love, had cast themselves into the arms of Mother Ocean and had after years returned like Galatea to their original source. These maidens' tears were highly prized and were symbols of purity. They were precious articles of barter, worth more than pure gold. They were worn as amulets by the rich and their presence about one's person insured immunity from evils of soul and body. The Greek philosopher Thales, who discovered the peculiar properties of amber about 500 years before Christ, thought he had called back to its petrified body the soul of a virgin, who in life had been so lovely and pure, so perfect that even inanimate objects were attracted to her ; and with prophetic mind, almost with inspiration, the old Greek predicted that the "Amber soul" would henceforth live in beneficent guardianship over the world.

Electricity Building at Night.

CHAPTER II.

INCANDESCENT LIGHTING.

ALL the lighting of the Columbian Exposition, decorative as well as service, and the operation of all machinery, was done by electricity. It was the first time in the history of expositions that this was the case.

At Paris in 1889, there were 10,000 incandescent lamps ranging in candle power from four to fifty, creating an aggregate of 200,000 c.p. in incandescent light. At the Columbian Exposition, in the Electricity Building alone there were 15,000 incandescent lamps of an aggregate light capacity of 240,000 c.p.

In the Electrical Building great variety of lighting was done, and plans of decoration by means of incandescent lighting were even lavish. These special features may best be enumerated under the exhibit headings. Outside the Electrical Building, the incandescent lighting was executed under a single contract by the Westinghouse Electric & Manufacturing Company. There were provided for in this contract, 90,000 16-c.p. lights. All the various systems of generators, transmission apparatus, and lamps were given severe test in operative comparison, and the special advantages of each were developed in a most convincing degree. The great flexibility of the alternating current system of incandescent lighting was strikingly illustrated by the enormous amount of such lighting that was done at the World's Columbian Exposition, both for service and decorative purposes. Comparatively little of this lighting was done by the direct current system, although some new systems such as the three-wire feeder system and the five-wire equalizing system were shown to great advantage in several instances. A number of new forms of the incandescent lamp, some of which were a radical departure from the common type, and some of very great candle power, were shown. For the direct current system the use of the three-wire feeder plan was a decided advance on the old two-wire system, not alone in its economy in the matter of copper conductors but in the maintenance of a proper voltage on the whole system, which was practically impossible when the lines were of any great extent with the two-wire system. For similar reasons the five-wire system was also a great advance on the old methods. These systems were shown in detail in Electricity Building and elsewhere and with them some beautiful decorative effects were obtained.

SERVICE LIGHTING OF THE EXPOSITION.

The plans sent to the various electric lighting companies, with the specifications dated February 22, 1892, were seventy-five in number, and showed the general design and construction of the buildings, the lights not being shown in them as the exhibit aisles were not then laid out. The following is a list of the buildings sent out with the specifications and the number of lights estimated for each: Administration Building, 5,000 lights; Manufactures Building, 10,000; Mines Building, 3,000; Machinery Hall, 4,000; Agricultural Building, 4,000; Fine Arts Building, 9,626; Woman's Building, 3,000; Horticultural Building, 800; Transportation Building, 2,500; Grand Basin, 7,080. The specifications also called for lights as follows for which no plans could be obtained at the time: Gallery of Fine Arts, Annexes, 6,616 lights; Music Hall and Casino, 3,500; States, Government, and Midway, 26,500.

Administration Building, Plant No. 1, Interior	3253	lights
Administration Building, Plant No. 1, Exterior	1824	"
Manufactures Building, Plant No. 12, Cornice	490	"
Machinery Hall, Plant No. 10, East Cornice	400	"
Machinery Hall, Plant No. 9, North Cornice	545	"
Agricultural Building, Plant No. 9, North Cornice	511	"
Agricultural Building, Plant No. 8, West Cornice	275	"
Woman's Building, Plant No. 1, 1st floor	1549	"
Woman's Building, Plant No. 2, 2nd floor	1095	"
Women's Building, Plant No. 3, Attic	304	"
Gallery Fine Arts Building, Plant No. 1, 1st floor	8172	"
Gallery Fine Arts Building, Plant No. 2, 2nd floor	1198	"
Mines Building, Plant No. 9, Cornice	196	"
Main Basin, Plant	2600	"

It was further required of the bidder to figure for service connections for:

U. S. Government Building	5000	lights
Fisheries Building	500	"
Service Building	1500	"

Making a total of 92,622 lights.

April 2, 1892, the bids for the above lights were opened and rejected, and new specifications sent out May 5, 1892, with the following changes: Machinery Hall, 1,000 lights; Music Hall and Casino, 4368; Main Basin, (Decorative) 6212; Fisheries Building, 500. The total number of lights called for in these specifications was 89,622. The bids were opened May

16, 1892, and the contract awarded to the Westinghouse Electric & Manfacturing Co., at $5.25 per 16-c.p. lamp. The construction work was started on underground work the middle of July, 1892. The inside work was started the latter part of the same month, the Connecticut State Building being the first wired. When the contracts were made for the incandescent and arc lighting, no plans were out for the Railway Terminal Station, or the Choral Hall. The Wooded Island had been reserved for an exhibit of outside incandescent lighting, and as this had not been accepted by any electrical company, it was necessary to include it in a contract for lighting the Railway Terminal Station and Choral Hall, which was made on July 25, 1892, with the Siemens & Halske Company, of Berlin, Germany. This contract called for approximately 3,800 16-c.p. incandescent lights, and approximately 163 2000-c.p. arc lights. Construction work in the Railway Terminal Station started about the middle of April, 1893, and the incandescent lights were put in service April 29th, 567 16-c.p. lights being used. The construction work on the Choral Building was started April 20, 1893, and finished July 28, 1893, although a few of the lights were turned on June 20th, the others being turned on as the wiring progressed. The building was wired for 968 16-c.p. lights, 708 25-c.p. lamps, and 64 50-c.p. lamps. The Wooded Island was lighted by 130 25-c.p. lamps, operated on the five-wire system. The duct on the Wooded Island was installed under an arrangement with the Siemens-Halske Company, the total number of feet installed being about 17,000. A small special foundation was designed to receive the incandescent lamp posts and 101 of these were set at an average distance of thirty feet apart in main paths. The incandescent lamp posts used had an ornamental triangular base eighteen inches high, in which was fitted a wooden column about 5 ft. 6 in. long. This was surmounted by a circular cap to which were attached three ornaments. The lamp stood perpendicularly on top of cap and was covered by an ornamental colored glass globe.

In addition to the general lighting of the grounds and buildings by arc and incandescent lighting, it was necessary to provide for arc lights for exhibitors and concessionaires. Accordingly, a contract was made with the Helios Company of Philadelphia, for 261 alternating current arc lamps to be installed and operated on incandescent circuits. Later, an additional order was placed for 328 Helios lamps, making a total of 589 purchased from this company. Of these, 346 were installed by the New York Insulated Wire Company, the balance by the Exposition Company. The general distribution of the Helios Arc Lamps is shown in table with the distribution of incandescent lights, installed as follows:

	INCANDESCENT 16 C.P.	HELIOS.
Exposition, offices, etc.	40,864	189
Exposition, decorative	5,096	14
Government and States	4,439	20
Foreign Governments	442	36
Exhibitors and Concessionaires in buildings	7,468	36
Concessionaires in Park	5,416	21
Concessionaires in Midway	1,947	253
Total	65,872	533

Eight alternating current arc lamps were wired and installed to give the desired effect upon the painting of the Upper Dome of Administration Building, under the contract with the Westinghouse Electric & Manufacturing Co.

The Westinghouse Company laid 146,739 feet of duct, for which the Exposition Company did the excavating and filling in. There were 242,441 feet of wire installed for primary service, of which 77,462 feet were Grimshaw wire used in the Subway, and 164,979 feet of Waring cable used in the ducts. These figures are the distances one way, or the length of one conductor, the total number of feet of conductor being twice the amount above mentioned. The sizes of the conductors ranged from No. 0 to No. 6 B. & S. gauge. There were 327 Converter Pits put in position, and 385 Converters installed with a total capacity of 66,760 16-c.p. lights.

On the first of May, 1893, the Westinghouse Electric & Manufacturing Company had ten 10,000, and two 4,000 light machines, and 25 primary circuits ready for operation. On the first of September, there were 33 circuits connected and in service on Exposition lighting, and one additional auxiliary or emergency circuit from Machinery Hall to the Midway, to be used in case of accident to either of the other circuits supplying light to the Midway. All circuits, except one, to the south grounds, were run entirely underground. The cables, from the switchboard to converters, were duplex Waring cables, except in the Main Subway, where Grimshaw rubber covered wire was run on insulators. All converters were placed in fire and water proof pits outside the buildings, and the secondary wires were led into the buildings in vitrified tile ducts. The largest converter used had a capacity of 200 lights, and nearly all were of that size. Every converter on the grounds had its own independent secondary circuit so that no trouble upon a secondary or inside circuit would ever put out more than 200 lights. The secondary wiring was controlled in all cases by switches and cutouts located at points where the wires entered the buildings, and the circuits were in general distributed from asbestos lined boxes in which the cutouts and switches were bunched at centers of distribution. The inside wiring was done entirely with the best grade of Grimshaw rubber and taped wire, and the wires were run almost entirely in standard moulding, or in interior conduit. The lamps were of 105 volts, all the new "Stopper" lamp of the Westinghouse Company. The general standard of the Exposition lighting for lamps suspended at the ordinary height from the floor was 40 square feet of floor space per 16-c.p. lamp, but inasmuch as the plant contained an almost infinite variety of incandescent lighting, the intensity of lighting varied greatly, ranging from 14 9-10 square feet in the large gallery of the Fine Arts, and 9 3-10 square feet per 16-c.p. lamp for the smallest gallery, to 3,662 square feet per 16-c.p. lamp for the system of patrol lighting under the floor of the Manufactures Building.

The most novel service lighting was that of the tanks in the Fisheries Building, the aquaria being lighted only by invisible lights shining through the water of the tanks, and hidden behind. The most brilliant was the lighting of the gallery of the Fine Arts, where the lamps were placed in reflecting screens around all sides of each picture gallery, the lights being only eight inches from socket to socket for nearly two miles of screens. The finest lighting was undoubtedly that of the Administration Building. The lighting of the interior was uniform, and in conjunction with the exterior decorative lighting, formed probably the most difficult and beautiful piece of incandescent lighting ever executed. In the incandescent lighting of the Exposition it was the aim throughout to avoid display lighting, but to secure sufficient and uniform illumination, and where the lights were placed for decorative effects, to place them so as to be inconspicuous by day, and bring out at night the decoration and lines of the buildings. Wall sockets were largely used, and stiff pendant fixtures used but in one building, simple clusters from flexible cord being almost universally used. This method of construction proved very satisfactory and had the advantage of being easily and quickly installed. It is easily maintained, and what is important to an exposition, the position of the lights can be easily and quickly changed without disfiguring or damaging the ceilings.

INCANDESCENT LIGHTING IN THE ELECTRICAL BUILDING.

All lighting in the Electricity Building was done pursuant to special rules quoted hereafter. Those exhibitors operating their own plants in Machinery Hall were allowed certain freedom in the way of incandescent lighting within their own exhibition spaces and contracts were made with other exhibitors providing for the lighting of their spaces from the service lighting of the former class on terms stated in the rules. The result was that the large companies were elaborate, even extravagant in their incan-

Mural Decoration, South Wall, Electricity Building.

descent lighting, and those who were present will remember that the building in the evening was almost a blaze of light. The General Electric Company, occupying space in the center of the building, operated 10,000 incandescent lights of an average intensity of 16 c.p. The Westinghouse Electric & Manufacturing Company operated in decorative incandescent lighting, 5,000 16-c.p. lamps: The Western Electric Company, including the Scenic Theater, operated 5,500 16-c.p. lamps; The Brush Electric Company operated 1,500 incandescent lamps; The Siemens & Halske Company operated for themselves and for other German exhibitors within the confines of the German section, about 1,500 lights of 16 c.p.; The Fort Wayne Electric Company operated 250 16-c.p. lamps; the Eddy Electric Manufacturing Company, the C. & C. Motor Company, the Mather Electric Company, and the Jenney Electric Motor Company operated an aggregate of 2,000 lights within the confines of their own exhibits, and the smaller exhibitors who had to rely for their service on other companies, operated an aggregate of 3,000 incandescent 16-c.p. lamps, making a total of 28,750 incandescent lamps of 16-c.p. in the building or a grand total of 460,000 c.p. This amount of lighting added to the 1,344,000 c.p. of arc lights aggregated 1,804,000 c.p. of lighting, which if distributed evenly over the Electrical Building would give approximately six c.p. per square foot of floor space—by far a greater amount of lighting than has ever before been confined for lighting purposes within a like area.

A part of this lighting was supplied by means of heavy cables conducting current at 110 volts from Machinery Hall and applied directly to the lamps by the direct current system. A part of it was supplied by means of 500-volt power mains run from the Machinery Hall plants to 500-volt motors in Electricity Building, these in turn operating direct current 110-volt dynamos from which the light was taken. In the case of the Westinghouse Company the current was brought in by means of lead-covered cables at 2,000 volts alternating current, and was transformed down after reaching the building to 105 volts, this being the voltage of the Westinghouse alternating current lamp.

THE WESTINGHOUSE ELECTRIC & MANUFACTURING COMPANY.

In its contract work with the Exposition, and also in its incandescent lamp exhibit, the Westinghouse people used exclusively the Sawyer-Mann lamp. The new form of lamp has a ground glass stopper which allows the renewing of the filament and the usual vacuum is replaced by an attenuated atmosphere of nitrogen. The possibility of re-using parts of the lamp, cheapens it to a great degree. In this lamp iron leading in wires are substituted for platinum, a great saving in first cost. The company have aimed rather at high efficiency than long life, the comparatively low cost of the lamp reducing in a measure the importance of longevity.

In addition to the marvelous exhibit of incandescent lighting made in connection with its contract with the Exposition Company, the Westinghouse people did much to render the Electrical Building attractive, especially to night visitors. The first view of the interior of the building had by visitors entering from the north, was the great mural decoration on the south wall above the gallery. The decoration consisted in the name of the company, modest in size of letters, made of incandescent lamps in colors, about fifteen lamps to each letter, the name extending in a half circle above the balance of the decoration. Below this was a mammoth bust of Columbus painted in oil upon a canvas sixteen feet high, the painting being emphasized and enriched by the presence of 900 16-c.p. lamps so arranged as to bring the features and the bust into prominence. Below were the name Columbus and the dates 1492–1892, in incandescent 16-c.p. lamps. As a whole the decoration was one calculated to charm the beholder, and was certainly an embellishment to the building. The company used for an office in the Electrical Building a beautiful little pavilion of Gothic architecture in white and gold. The outlines of the pavilion were emphasized by a border of incandescent lamps.

THE WESTERN ELECTRIC COMPANY.

By all odds the most attractive and popular exhibit in the Electrical Building, and one of the most interesting at the Exposition, was that made by the Western Electric Company. This result was due largely to the fact that the greatest possible use was made of incandescent lighting, and the decorative effects produced were of a character to attract the public, without in any degree detracting from the scientific and dignified aspect.

Scenic Theatre.—The most interesting and unique of these was the Scenic Theatre, located in the extreme southeast corner of the building. It was desired by the company when the plans were first formed for making an exhibit at the World's Fair, to devise means by which its apparatus employed on theater stages could be fitly exhibited, and after several plans had been discussed it was finally decided to re-produce in as realistic a manner as possible the natural transitions of light during a period of twenty-four consecutive hours of night and day. The scenery employed to produce these effects represented the Swiss Alps. The stage opening was ten feet wide and nine feet high. The front portion of the scene represented a village consisting of several houses and a church, which were connected by a bridge, with a castle on the opposite side. Under the center of the bridge was an arch through which the rain and mountain streams of real water flowed to a lake below, over a cascade about twelve inches high. In the rear of the picture were the foot hills, and in the extreme distance were the snow-covered peaks of the Alps extending their stony crags far up into the sky. The audience room of the theater seated about 175 people and was tinted a salmon red. Large revolving fans were used to give free ventilation through the room. The lights in the room were extinguished at the opening of each performance and music from a piano began at the rise of the curtain; revealing a Swiss landscape in the evening light. The village and bridge are lighted by miniature lamps placed on posts twelve inches high, which are exact representations of the larger posts manufactured by the Western Electric Company. Midnight approaches and the lights in the houses as well as the street lamps over the bridge are extinguished, and the darkness of the scene suggests that the midnight hour has been reached. As morning dawns a tinge of soft gray light creeps gradually over the scene, and the eastern sky is tinted with a glow of red. The red tint deepens in tone until the entire sky and peaks of the mountains catch the light, which melts away through the golden and yellow tints, until finally the sun bursts over the scene with a white light, revealing clearly the mountains, lake and the cascade below, as well as the village scene. Soon after daybreak a procession starts over the bridge toward the castle, consisting in all of seventy-five people. The procession is headed by a military band, and is followed by a troop of cavalry and two troops of militia. The procession disappears through the large arch leading to the castle. The sun passes from east to west, the shadows changing accordingly, until the middle of the afternoon is reached, when the sky begins to be overcast, and it is evident to the audience that a storm is approaching. Darkness once more creeps into the picture, and sheet and chain lightning play upon the sky beyond the mountains. After the storm has ceased the sun gradually emerges from the clouds and a rainbow is seen arching across from one side of the sky to the other. This rainbow is seen exactly the same as in nature. The base of the bow is first seen, which gradually creeps toward the zenith and finally forms a complete arch. It dies away in the same manner. After this storm has passed away peasants are seen passing over the bridge towards the village, with umbrellas over their heads, followed by other peasants and a load of hay. Gradually night comes on and the soft mellow tints of evening are seen to creep over the landscape, followed by the orange, red, and purple of evening, until the sun sets in the western sky. As the darkness deepens the stars are seen to emerge one by one, until finally the whole heavens are studded with stars, and the curtain falls upon the evening scene in the moonlight. All these effects are produced by automatic machinery, requiring the attendance of but one man. The number of people who viewed this miniature theater exceeded 3,000 per day. The demand for seats was so great that it was finally decided to give consecutive performances from 10 o'clock in the morning until 9 o'clock in the evening, in order to accommodate all who might desire to see these effects. During all the changes of lights, music from a piano was arranged to correspond with the different scenes, and to interpret as much as possible the times of day. The circuit used was 110 volts pressure, which was furnished by means of a motor generator, similar to that described for lighting the column. The scenery was designed and painted by Sosman and Landis of Chicago.

Another of the attractive exhibits was a column of light two feet in diameter and over twenty feet high, encircled with red, white and blue lamps, from the top of which extended four radial zigzag arms, typical of lightning. At the ends of two of these arms were large revolving spheres forty-two inches in diameter, studded with 300 lamps each of 16-c.p., and revolving at a high rate of speed. At the ends of the other two arms were conical clusters of lamps of different colors. Periodically, throughout the day and night, bands of light three feet wide rushed from the base of this column to the summit and then out upon the arms into the balls and cones, producing the most weird and wonderful effect. The balls were at one instant filled with red light which suddenly changed to white and again to blue. This effect was constantly repeated. Over 2,600 lamps of 16-c.p. each were used in this exhibit alone. The controlling switch was placed beneath the floor and automatically produced all these various changes. This switch consisted of 137 circuits which were connected and disconnected by means of a revolving cylinder, upon which were placed cams so arranged as to make contact at the proper time.

In the Egyptian Temple, in the main pavilion, of this company, there were more than 1,200 16-c.p. lights, not one of which was in sight, the lighting being done by indirect rays entirely. The office of the company was located in the northwest corner of space four. It was of Grecian style of architecture. The interior was formed into two rooms, the front one being a drawing room and the rear one a private office. The illumination of the company's office was novel in the extreme, the front room being the green room and the rear one the yellow room. The lighting was entirely concealed behind a frieze of paper over which was placed a light cloth of the same tint as the room. The paper was cut into a beautiful design, back of which were placed the lamps, thus giving an appearance of an illuminated wall, and bringing out the design in a highly artistic manner. The effect was novel and original. In the front portion of the building was placed an elegant painting on glass nine feet high and six feet wide, entitled "Fairy Queen," the lighting of which showed the evenness with which light may be distributed over a surface of this size. On the walls were illuminated pictures of various buildings lighted by the Western Electric system, one of the Ashland Block, Marshall Field's wholesale store, Masonic Temple, the whaleship Progress, the Leiter Building, and the Ferris wheel, all of Chicago, and the factories of the Western Electric Company, Chicago and New York.

A unique and to the public puzzling exhibit, was that of writing machines. The exhibit was on a large cabinet, upon the top of which were the letters "W.E.Co." in 16-c.p. lamps. Behind these lamps was a pointer operated by means of a controlling mechanism placed inside the cabinet. This arm commenced at the beginning of the letter "W," passed across the cabinet tracing each letter in rapid succession and apparently lighting the lamps as it passed, after which it retraced its course and extinguished the lamps in the same manner in which they were lighted. One hundred and fifty-six lamps were used in this exhibit. Near this exhibit was placed another exhibit which spelled the words "Western Electric Co." by moving a lighted lamp through the air. Commencing at the letter "W" it wrote in script letters. The mechanism which controled this arm was also placed in a small cabinet, and was in continual operation day and evening. Near the center of the south end of the exhibit was placed another writing machine operating a pen and writing the letters "Western Electric Co." on a card. This was enclosed in a case, partly of glass so as to disclose the working parts and arrangement of the case necessary to cause the pen to trace the letters.

THE GENERAL ELECTRIC COMPANY.

The central circular space of the Electrical Building around which were grouped the several other exhibits of the General Electric Company, was occupied by the Tower of Light, intended as a tribute to the Edison Lamp and the Edison System of Incandescent Lighting. The Tower, which was over eighty-two feet in height, arose from the circular pavilion thirty-two feet in diameter, occupied by the Phœnix Glass Company with a beautiful display of glassware for electrical purposes. The base of this tower was faced with mirrors and between these and the circular row of columns supporting the roof of the pavilion the Phœnix exhibit was artistically arranged. Above the pavilion was a dome from which arose the tall cylindrical shaft. The dome had alternately long and short tibs of 5-c.p. lamps in various colors. The shaft, ten feet in diameter at the base, of pure white, was relieved by lines of dark molding arranged in

Edison Tower.—Electricity Building.

geometrical designs and studded with 10,000 6-c.p. lamps in colors. Surmounting the shaft was an ornamental cap with lamps in pretty settings, and on each of whose four sides were the words "Edison Light" in miniature lamps. Above the whole was a mammoth incandescent lamp built up of about 30,000 cut glass prisms, lighted up within by a number of small incandescent lamps, bringing out the prismatic colors most beautifully. By means of a commutating device within the base of the tower, the lights were thrown on and off in a great variety of charming combinations, with pleasing effect. Above the tower and crowning lamp on a background of black, were spelled out in miniature lamps of dazzling whiteness the words "General Electric Company," showing by whom it was designed, erected and operated. The incandescent lamp exhibit was the first complete commercial exhibit ever assembled, there being fifty-two cases containing over 2,500 lamps, no two of which were alike, being in many colors and in candle power ranging from ½-c.p. to 300-c.p. There were cases showing fibrous vegetable growths experimented on by Mr. Edison in his search for the best substance from which to make filaments; cases showing filaments in all stages of manufacture, before and after carbonization; cases showing steps in the manufacture of lamps; cases showing experimental and historical lamps (among which are two of the first ever made by Mr. Edison), and two cases showing lamps made in 1880 and which have been in daily service for twelve years with small currents. An interesting feature in connection with the lamp exhibit of this company was the automatic lamp circuit control. On a large stand was a board with lamps in series on a 1,000-volt alternating circuit, connected to a bank board at the station switchboard, showing the company's system of incandescent street lighting. Should a lamp in the series break, it was instantly and automatically short-circuited, and the fact indicated by a meter on the station switchboard, when another lamp was thrown in circuit in its stead, the switching in of a lamp or the use of a reactive coil being two methods of station control. At the north end of the General Electric Company's space was an immense fan decorated with over 1,300 lamps of 1-c.p. each, beautifully arranged to show the outlines and ribs of the fan, and the decorations in colors. The current was switched on and off in such a way as to give the effect of opening and closing the fan. There was also a case containing in bound volumes the testimony before the United States Courts on the Incandescent Lamp Filament Case, in which the patent was sustained and Edison declared the inventor of the Commercial Incandescent Lamp.

THE BRUSH ELECTRIC COMPANY.

The special decorative lamp exhibit of this company, consisted of the system of lighting in their pavilion or office booth. The interior of the booth was a rotunda and dome, lighted from the sides by lamps placed behind a molding out of view from the floor. The dome was pink tinted and threw the light about the room evenly, and in such a mellow form as to create an optical illusion, the eye failing to appreciate the full force of the light until a paper or fine printed matter was looked at. There it could be at once seen that while the room seemed to be dark, it was really well lighted. In the way of decorative lighting this company also showed two large flags, one the American, the other the green flag of Ireland with the harp in the center. The flags were each sixteen feet long by ten feet wide. They were painted on canvas board backed and filled in with about 800 16-c.p. lamps. The lights were operated by means of an automatic controller in such a way as to throw in and out consecutive rows of lamps across the flags from the staff to the outer border, giving the impression of waves. The flags attracted much attention and were an ornament to the building.

OTTO RAUDA.

This exhibit made in connection with the Ansonia Electric Company consisted of lamps mounted on plain stands, a lamp mounted on a gas burner, which could be screwed to any gas fixture, a pendant lamp with a large plated reflector and a sub-aquatic lamp, which was designed to burn under water. The Electric Light Scarf Pin consisted of a small incandescent lamp mounted on a scarf pin, with silk covered conducting wires, giving a brilliant light whenever the interrupter, attached to the conducting wires is closed. The current was supplied by a two cell pocket battery whose principal qualities are safety from damage by leak, easy access for opening, refilling and recharging and easy renewal of wearing parts.

LIGHTING OF THE FERRIS WHEEL.

It was not originally contemplated to light the wheel, the intention being merely to light the offices, border and fences and illuminate the supporting pillars. Later, however, during the progress of wiring, it was thought better to extend the plant and illuminate the wheel also. All of the lamps on the iron work were set as nearly as possible eighteen inches apart and were all placed upon porcelain bases which in turn were screwed to a wooden block, this being bolted to the iron structure. All of these blocks, also all of the unglazed porcelain surfaces were thoroughly painted with P and B insulating compound. The illumination of the wheel proper consisted of two outer circles, one on each side, 250 feet in diameter, carrying 360 lamps in twelve circuits; each circuit being protected by a double pole safety cut-out of porcelain. Within these were two circles of 170 feet diameter carrying 200 lamps each and connected to the outer circle by twelve radial arms—six on each side. In addition to these there was a large star near the axle on each side and a spiral of lamps around the axle, together using 210 lamps. It will thus be seen that there had to be taken off from the two collector rings current sufficient for 1,540 16-c.p. Edison incandescent lamps at 110 volt. These two rings were made of solid copper fifty-six inches in diameter, five-eighths of an inch thick and three inches wide, each placed upon an insulating collar. The brushes used allowed the wheel to revolve in either direction for the purpose of taking up the wear on the gears. An elliptical form was therefore selected one and one-fourth inches wide built up of fifty sheets of No. 36 copper wound first in a ring and screwed and soldered to the brush holders. Connection from the two 0000 mains was made to the brush holder by two compression rings, which slipped over the stud inside of the block which held the brush holders. The lamps were placed at an angle of 45 degrees just within the outer rim of the wheel as from there they were entirely invisible from the carriages and still cleared the supporting towers which were but two inches from the iron rim and could be seen when one was directly in front of the wheel. The lamp was attached to the porcelain base which in turn was fastened to the circular wood block which was secured to the ironwork by screwing on to the projection of the bolts that held the angle plates together. All of the wire on the ironwork was of the best quality "Okonite" or "Simplex" protected at angles of the iron, and other places where abrasion was likely to occur, by insulating conduit or tape. All joints were carefully made, soldered, wiped and thoroughly taped with Okonite and smeared with P and B compound. The whole line tested free from ground upon completion and remained so with the exception of short repair intervals. All of the special fittings were made and all wiring done by the Western Electric Company of Chicago. In addition to the wheel, the two supporting pillars composed of four stanchions each, were illuminated from the ground up, culminating at the axle in the star cluster before mentioned. In addition, the connecting arches at the bottom had each two rows of lamps and three braces at equal intervals up the pillar, and were lighted, giving the whole a more solid appearance. The same care was taken in wiring these supports as in the wheel proper.

The fences, garden and offices were also illuminated, adding very much to the effect when one was near the wheel. This swells the total number of lamps to 2,900 16-c.p. incandescent lamps. The central station supplying current for these lamps, was placed outside of the ground at a distance of 700 feet and was carried by several Okonite cables as follows: four 000, two 00, one 0 from the generators to the terminal house, from which it was distributed over offices, fences and pillars and current for the wheel was carried by two 0000 cables up one of the pillars to the collector rings. These seven cables were bunched together and run in a box made of rough white pine boards one inch by six inches and returned in an adjacent similar box. These boxes were not painted or protected in any manner from the damp swampy soil through which they were laid and run throughout their length within twelve inches of the steam pipe supplying the driving engines. This pipe was covered with mineral wool and an out-covering of cement.

The machines supplying this current were two Edison machines giving 200 amperes at 110 volts each and two Western Electric machines giving 400 amperes at 110 volts each. Each pair of machines was belted in tandem with a double expansion 125-h.p. Dick & Church engine, the Edison machines being run at a speed of 1,300 and Western Electric at 900 revolutions per minute. On the switchboard each machine was supplied independently with an ammeter starting box and switch. Each pair was also equalized by bus-bars at the fuse blocks and furnished with a voltmeter. The instruments in connection with each pair of machines being of same manufacture as machine. The fuses were all placed upon slate bases at the back of board and well ventilated cables were connected directly to the fuse blocks and entered conduit immediately under the switchboard. An extra switch was placed before the completion of the wheel wiring so that the Western Electric machines might be used to light the pillars in the event of a mishap to the General Electric machines installed for that purpose. The Western Electric machines were used for the illumination of the wheel and those of the General Electric Company to light the offices, grounds and pillars supporting the wheel.

There were three 250-h.p. Heine boilers delivering steam at 100 lbs., pressure. Two of these were in constant use and supplied motive power for the wheel, the other being used for the operation of the lighting plant.

The fuel used was oil taken from the same mains that supplied the Exposition power plant.

The whole electric plant gave entire satisfaction and probably served better to advertise and make the wheel an evening attraction than anything else, and illustrated one more feature of this gigantic Exposition in which electricity played a most important part as an agent of ornamentation as well as a motive power and means of illumination.

ALLGEMEINE ELECTRICITATS GESELLSCHAFT.

Incandescent Lamps.—The manufacture of incandescent lamps is a special feature of this company's business. This branch of manufacture embraces all modern advances in technical science and in point of durability and economy fulfils every requirement. To suit their manifold uses the incandescent lamps are manufactured in different forms (pear shaped, candle shaped, tube shaped, etc.,) and for every voltage required. The annual production amounts to 1½ millions of lamps. On a special board the manufacture of an incandescent lamp was shown in its various processes. The sockets most in demand were placed together on one board.

Stage Lighting.—This firm showed a most modern and perfect system of stage and auditorium lighting for theaters. All the various effects of light in this complicated form of plant are rendered possible by the use of a specially devised stage regulator. This contains all the mechanism required for producing the different degrees of light and effects of colors by means of white and colored lamps installed on the stage and in the theater, and is the means at the same time of throwing into and out of circuit all the lamps in use. Notwithstanding the variety of its functions, the safety and simplicity of the apparatus when at work and the facility with which it can be examined and manipulated are in no way lessened thereby. Even on the largest stages one man is able to produce the quickest and most complicated changes of light and colors, and to ensure that natural succession of the same which is of the greatest importance for the success of the piece. The arrangements for changing the intensity of light consist of eight regulating levers, conducting their corresponding currents to sliding-contacts between the rheôstats; from here the current flows in full strength or weakened, as required, through the "color-switches" to the white or colored lamps. Every complete stage-lighting device possesses two levers for controlling the rheostats, a mechanism for connecting the colors with three contact plates for white, red and green, and a mechanism for flashing. Several of these systems may be put together, being united into one device, according to the size of the stage. The lamps are fixed by strong brass pieces, each group of stage lamps being mounted on strong separate wooden ledges. According to the three colors of the lamps (white, red and green) three circuits are arranged possessing a common return wire leading to a special switch-board. For better utilization of the light the lamps are shaded by white reflectors. Against damage from outside they are protected by this plate-like reflector, and in front by wire net-work. The intermittent lighting is effected by flexible, well-insulated cables, joined to the mains as required. To slowly darken a very small number of lamps a "portable stage regulator"

Painted Window Lighted by Incandescent Lamps.
Western Electric Company.

is used. For lighting effects arc lamps (System Lautenschlager) and a portable incandescent lighting system are applied. A parabolic reflector that can be turned is fixed to a pillar and contains a so-called "Focus-keeping lamp." For feeding these lamps a small accumulator battery of 2 amperes at 15 volts is used.

INCANDESCENT LAMP SOCKETS.—The sockets showed important improvements. The keyholders might be turned only in one direction. By fastening the socket with a bayonet coupling the inconvenient separation of the socket parts from the base plate was to be avoided. Of special pattern were the so-called spring sockets, by which overscrewing of the lamps in the sockets was avoided.

THE INTERNATIONAL LIGHTING COMPANY, PARIS.

Luminous buoys, and floating lights, in model and natural sizes, were the exhibit of this company. The lights were of the Cordon system, clear and steady, of high candle power, and required no watching.

PROF. MOSES G. FARMER'S HISTORICAL LAMP EXHIBIT.

There was an old mantelpiece set up near the Egyptian Temple of the Western Electric Company which was taken out of the Farmer home in Salem, Massachusetts. The principal object of interest in this mantelpiece centered around two small incandescent lamps placed upon it in exactly the same manner as Mr. Farmer employed in 1859. These lamps consisted of small standards six inches long, one end of which was fastened to the mantelpiece, and to the other end of which was fastened a small platinum wire, No. 28, by means of small brass pincers. Over this filament was a glass cylinder with a brass cap. These lamps were used by Mr. Farmer at his house in 1859, and when measured with a calorimeter were found to be 32 c.p. each. Prof. Farmer has said that the commercial failure of his incandescent lamp was due entirely to the absence of an electric generator, and that he would have succeeded had the dynamo been in existence at the time.

LIGHTING FIXTURES.

THE PHŒNIX GLASS COMPANY, PITTSBURGH.

Directly in the center of the Electricity Building, occupying the circle that divided the avenue leading to every entrance, stood the ivory-colored pavilion of the Phœnix Glass Company. Ever rich in its classic, quiet elegance, in the evening hours when a blaze of golden glory streamed from every point, this exhibit flashed and scintillated as though adorned with gems of the Orient. The dome of the Phœnix Glass Company's pavilion was supported on graceful columns rising from the floor, that was reached by ascending three low steps, the floor forming a resting place for thousands of weary sight seers; suspended from the ceiling by golden chains, or placed on mirrored shelving, were samples of rich and daintily designed glass for electric and gas illumination; not a great mass of cheap glassware of obsolete pattern, inartistically arranged, but selections to indicate the latest advance in etching and tintings, in cuttings and colorings, from the lowest priced globe to the $200 cut-glass, pineapple design, for a richly carved newel-post, and the $250 hand-painted pendant. There were delicate tints in sapphire and topaz, in cream and ruby, orange and citron, silver etchings in rococo, arabesque and empire designs, and etchings with delicate tinted edges, Venetian threads in tints; star cut balls of the world-famed Phœnix cutting, thirty-six inches in diameter, said to be the largest cutting in the world, and Phœnix diamond-cut pineapples twenty inches in diameter. For the wealthy, the architects of the millionaires and decorative specialists have found out where they can secure decorative glass so perfect that it has been likened to crystallized drops of water. The exhibit was another visible illustration of the material benefits following the utilization of electricity; for when gas was the principal illumination, the artist was hampered by the upward, outward flow of the flame that limited the forms and shapes of globes and shades. But with the incoming of electricity, and the rapid spread of the incandescent bulb, the artists of the Phœnix Glass Company cast aside all thought of mere utility, and crystallized their most airy fancies into gems of skilled handicraft.

CHAPTER III.

ARC LIGHTING.

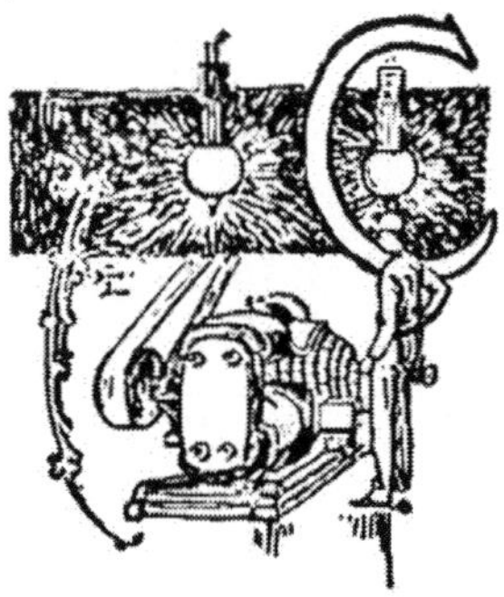

CONSIDERED alone from the standpoint of quantity arc lighting is by all odds the best and most economical artificial light. In the illumination of the grounds of the World's Columbian Exposition various types of arc lights were used with signal effect, and the fact was conclusively demonstrated that for public and private use the arc light is the most satisfactory substitute for natural light. Several systems employing both the direct and alternating current were employed, and illustrated the possibility of the use of the arc light on any circuit, no matter what its voltage. As a rule service lamps were used in series on each circuit. Lamps for such use were fitted with regulating device consisting of two solenoids, one in series with the circuit and the other in shunt. In some cases a single solenoid is used, double wound and in opposite directions in series and shunt. This device acts as follows: When the lamp is not in use the carbons touch each other, in which case the current flows through without giving light. The resistance, however, is slight and the current flows readily through the series winding of the solenoid, the plunger of which is moved to actuate a pawl which lifts up the carbon holder and establishes an arc between the two carbons. When the carbons become too widely separated the resistance becomes great enough to cause the current to be deflected through the shunt winding of the solenoid, causing an opposite movement of the plunger, which allows the carbons to approach to the proper distance. It has not been found feasible to make arc lamps of a higher voltage than about 45 to 50 volts, and this fact has militated against their use on incandescent circuits for the reason that the customary voltage on such circuits is 110 volts. By using two arc lights in series, this series being in parallel on the incandescent circuit, it has been found necessary to introduce an artificial resistance of from 10 to 20 volts, which are wasted. For special uses arc lamps with an artificial resistance contained in the lamp have been devised which work very successfully on incandescent circuits. These lamps are so designed that whether a lamp burns or not it still offers a resistance equal to one-half of the voltage of the circuit; for, in using lamps in this way both lamps must burn or the voltage must be divided between them, otherwise one would be speedily burned out. This same principle applies to direct currents of constant but high potential, as on railway circuits of 500 volts. In this case ten lamps are burned in series with very successful results.

For use on alternating circuits it has been found that the light is diffused too much. In the direct current type of lamp the upper or positive carbon has formed on its end a crater which results in throwing most of the light produced in a downward direction where it is most wanted. In the alternating current type of lamp both carbons burn as points, the result of which is that much of the light is wasted in an upward direction. To overcome this defect a small reflector is used on the upper carbon which intercepts much of this waste of light and casts it downward in the proper direction. Another defect of the alternating type of arc light was the buzzing noise which seemed to be inseparably connected with it. This defect has been overcome by the use of but little iron in the controlling mechanism, and that of the softest character and laminated. Such lamps as these were very successfully used on alternating current circuits in the Terminal Station, the yards of same and the approach to the elevated railway.

In order that the special systems of arc lighting of the various companies might be given practical demonstration under circumstances that would be of benefit to intending users and visitors, the Electrical Building was divided as to its service lighting between the General Electric Company, the Western Electric Company, the Brush Company, the Fort Wayne Company and the Excelsior Company. The German firm of Siemens & Halske supplied by their special system the lighting for the German section of the building. The circuits of each company were entirely independent of all other circuits, so that a failure on the part of any set of lamps could not be properly chargeable to failure on the part of any other company's apparatus.

Much ingenious and effective special lighting was done by means of arc lamps, as well as decorative lighting, the special features of these to be described under Decorative Arc Lighting to follow.

Search and marine lights were special features, most interesting and attractive. The great lights established on the corners of the Manufactures Building, on the Colonnade between the Agricultural and Machinery Buildings, the smaller lights using tinted slides, located about the Court of Honor, and the lights upon the Battle Ship, served in a remarkable degree to accentuate the general night illumination, and undoubtedly played no small part in attracting the crowds that visited the Exposition in the evenings. The buildings about the Court of Honor, the MacMonnies allegorical fountain, the electric fountains, and the pleasure boats that traversed the lagoons, owed much of their effectiveness at night to the small search lights located, one on the Casino, one on the Music Hall of the Peristyle, one on the northwest corner of the Agricultural Building, and the large lights on the Manufactures Building. All these lights had colored slides attached so that they could be thrown in tints to heighten the beauty of the electric fountains. As a matter of fact the men operating the search lights were drilled to play the colors in harmony with the colors in the electric fountains, and the effect of this arrangement can be much more readily imagined than described.

ARC LIGHTING OF THE WORLD'S COLUMBIAN EXPOSITION.

In July, 1891, a corps of draughtsmen was employed to prepare plans for the arc lighting of the various buildings of the Exposition. The buildings were taken up in the following order: Mines, Agriculture, Horticulture, Manufactures, Machinery, Fisheries and Woman's Building. The lights were laid out at a distance of 20 feet apart over the entire floor area, there being no plans showing aisles at that time. These plans were presented to the electrical committee, were approved by this committee and duplicate tracings made for reserve. According to these plans it would have required 6000 arc lights exclusive of the out-of-door lighting. Later on these plans were changed to provide for aisle lighting only. The original layout of the grounds required 1650 lights. This left so few for the buildings that it was decided to cut this figure, and accordingly the number was cut to 1434, which was the amount of outside lighting operated nightly during the Exposition. The construction work for the permanent plant was begun in June, 1892. Owing to an unsatisfactory arrangement with the contractors it was considered more economical to take up the work with the Exposition employees, and in that way the remainder of the underground work which connected the Machinery Hall with all parts of the park and buildings, also the subway, was completed. Points and grades for foundations and manholes were given by the Surveys and Grades Department from maps made from the standard atlas, showing all locations and connections. There were 21 of these plans made to a standard scale of 40 feet to the inch. Later smaller maps were made to render the work of inspection and installation easier. These were 100 feet to the inch and proved to be the most convenient plans in use. The underground work was completed June 8th, 1893, with 1087 standard foundations in the grounds, 307,837 feet of duct were placed in the ground and 117 points prepared for posts on piers, docks, etc. The method of doing this work was to cut a hole in the planking

Night View Agricultural Building, From Electricity Building.

and set a vitrified tube in the hole, cementing it in. The posts were set over this tube. There were also 1559 manholes, single and double set, and covers attached. Designs were prepared providing for a standard one-light post, two and three-light cluster posts and a special combination post for one arc and two incandescent lights. Special posts were designed to go on railings south of the Art Gallery and on the three main canal bridges. The design for this one-light post closely resembled that of the Western Electric Company's Columbian Arc Post. The contract for supplying these posts was let to R. B. Ayres on September 7, 1892, providing for 1100 standard posts and 112 cluster and special posts. The first shipment was received November 15, 1892, and the contract completed May 19, 1893, Mr. Ayres having delivered 1212 posts and erected 1196 posts complete. The balance were held in reserve to replace broken posts. Before the posts were placed on foundations the elevation of each foundation was carefully checked by the Surveys and Grades Department, and the corrections necessary were made to a subgrade one inch below finished grade of walk. In all, the elevations of 945 foundations were corrected before posts were erected. After the elevations of the foundations were corrected, foundations were aligned so as to bring the posts into alignment when erected.

On July 15, 1892, a contract was made with the Safety Insulated Wire & Cable Company for supplying the cable needed for all underground work, exclusive of the subway. The cables supplied were two kinds, lead-covered No. 6, extra heavy pure rubber insulation, single tape between lead and rubber, also No. 6 single tape, being the same wire without the lead cover. The work of drawing in cable was commenced January 20, 1893, south of the Manufactures building. A great deal of difficulty was experienced because of frozen duct, necessitating thawing out with steam. For this purpose three steam rollers and eleven portable boilers were used for a period of seven weeks, there being at one time forty-five miles of duct frozen. On May 1st there were 700,300 feet of cable in the ducts, and on June 19th the entire work was completed. The total of underground cable was 773,460 feet.

For convenience arc lighting will be divided into two general heads, "Outside" and "Inside Lighting."

OUTSIDE ARC LIGHTING.

As originally projected, the outside arc lighting contemplated the use of 1650 lights, of which 500 lamps were to burn all night. This was made on a total arc light capacity of 6000 lamps. When the final capacity of the plant was fixed at 4500 lights, the exterior lighting was cut to 1500, of which 10 per cent. were designed for patrol service. The avenues of distribution for the outside lighting were to be the subway for the section of the park north of Machinery Hall, and the Intramural Railway structure for the west half of the park and the south portion. For the arc circuits in the subway there were drawn in 270, 242 feet of No. 6 rubber-covered braided wire, starting with 84 wires at Machinery Hall. After making several plans showing the arrangement of wires to be strung on the Intramural it was decided objectionable to place them there, owing to the great number of wires necessary, and the solution settled upon was a duct trunk line starting at the Machinery Annex connecting with the end of the subway and running north across the railway terminal tracks along the fence to the north end of the California building. The trunk line was divided into two parts, starting from Machinery Hall with fifteen ducts. The number of ducts was decreased as the number of wires diminished towards the end

ARC POST, BRIDGE BETWEEN AGRICULTURAL AND MACHINERY HALLS.

of the line. In the main trunk line special manhole boxes were designed. The inside measurements of these boxes were 36x30x22 inches deep, giving ample room for making splices and pulling in. The ducts were placed at one end of the box, giving room for a man to stand while working, without injury to the wires. These boxes were placed about 175 feet apart, the exact distance depending upon the length of the wire as it came from the reel, so as to reduce cutting into the cable to a minimum. The main trunk line duct, inclusive of arc and incandescent, aggregated 121,096 feet of duct, and there were in it 136,673 feet of No. 6 safety taped wire for arc purposes. In addition to the main trunk line and subway, two other trunk lines were constructed, one southwest of Machinery Hall, containing eight ducts carrying 39 wires, supplying Agriculture, Leather, Forestry, and Anthropological buildings, and all overhead circuits throughout the south grounds; the other line connected up to the elevated road, and had a riser pole at the Electrical and Mechanical office building; a third trunk line of seven ducts connected the Agricultural building with the elevated road and contained 23 wires supplying the Agricultural building and outside circuits on the Casino pier and dock. The foundations for posts as first designed provided for a concrete base with cast iron manhole box attached. This was decided to be too expensive for the short time the work was to be in use, and another design was constructed and used, which consisted of four piles and a crib work on top, and a wooden manhole box. The foundation consisted of a crib work 6x6 inches hard pine timber firmly bolted together. At each foundation was placed a manhole box 14 inches to 23 inches deep, with an iron cover. The foundations were set an average distance of 75 feet apart although a minimum distance of 25 feet occurred north of the Art Building. The aim in laying out this work was to preserve straight lines to conform with the general plan of the grounds. In each foundation connecting with the manhole box was placed a ¼ curved vitrified tube, 4 inches inside diameter. This was firmly cemented in at the foundation and insured a dry channel for the wires leading up to the poles. The manholes were spaced to give a pull of cable of not more than 130 feet. In several cases to get around the foundations and where several ducts led to one manhole, a double manhole was employed. Inside measurements of this were 32x14x21 inches deep. This manhole was provided with two standard iron covers. The wooden pump log, or duct, in use was 4x4 inches with a hole 2¼ inches in diameter accommodating four wires. It was in six foot lengths jointed with a plain dry thimble joint. There were 307,837 feet used. It was placed in a trench of an average depth of 18 inches, laid in the ground with corners up, the better to resist a downward pressure of the settling dirt on top. It was decided to install overhead work for that portion of the grounds south of Machinery Hall and west of the Anthropological building, also for lighting the entire fence line from Fifty-sixth street and the lake, to the southeast corner of the grounds. For this lighting there were used 115 poles, forty feet high, and 62,300 feet of rubber braided wire. When this overhead work was planned it was proposed to light the courts of the stock sheds with arcs but this plan was abandoned and incandescent lighting substituted.

When the subject of circuits was considered, the first plan adopted was to indicate the course of the circuits and to plot them in colored inks on tracings; changes in circuits were so frequent that to change the tracings each time was not practicable; it was finally decided to represent the circuits on a board with colored thread; the lamps were represented by brass tacks. This plan was found very satisfactory and was the method by which all circuits were planned. In designating outside circuits the letters of the alphabet were used; patrol circuits were distinguished by the use of the letter "P" preceding the letter such as "P.A.," etc. In order to have these circuits convenient for ready reference copies of each were prepared, and proved very convenient for use outside. The same method was used for preparing copies of the inside circuits. In designating the locations of foundations and manholes the following plan was adopted: each building and locality was given a letter of the alphabet as Manufactures E, Midway X; and manholes were covered, numbered in series, as "E 143," the series letter denoting the general location of the manhole or foundation, and the number indicating the exact location. This was found to be a most satisfactory manner of locating the work. The same form was used in locating the posts.

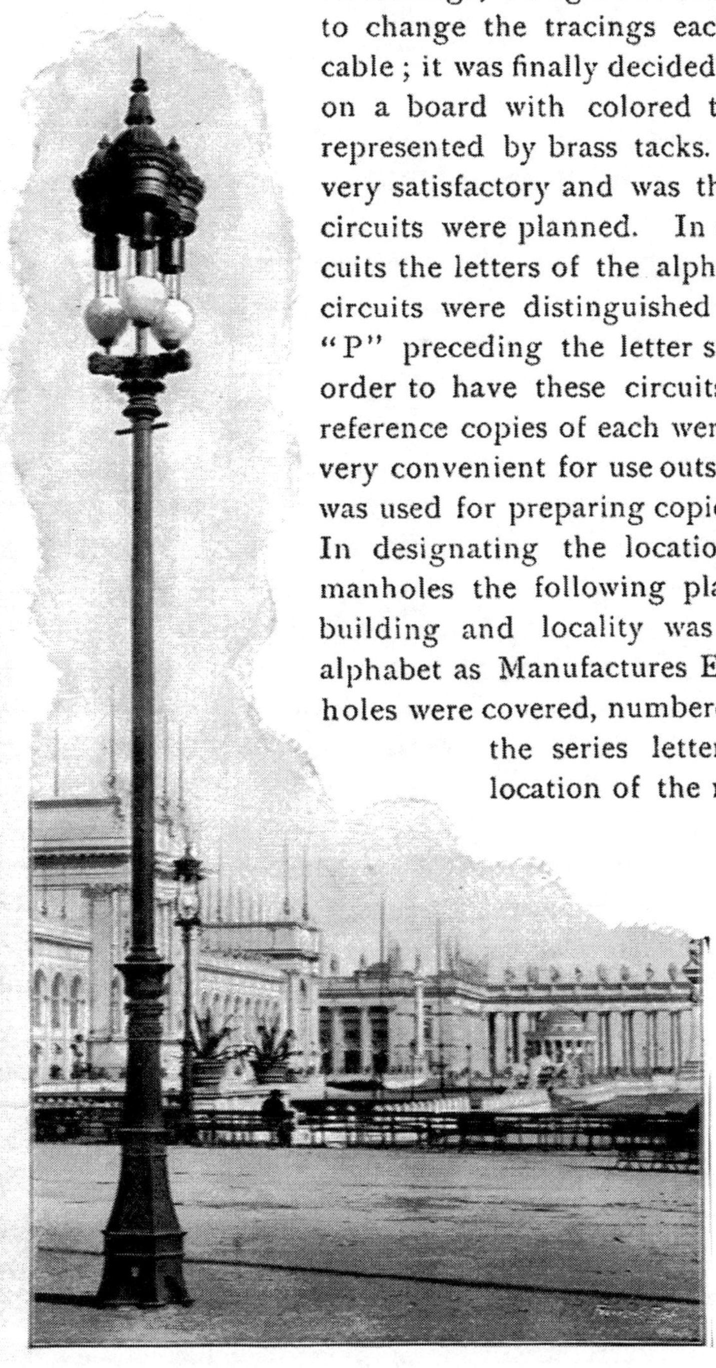

THREE-LAMP ARC POST, EXPOSITION GROUNDS.

In series with patrol arc circuits were a system of range lights for the lagoon bridges. These lamps burned on 9 6-10 amperes of current, were of 30-c.p., two being used under each bridge, one red and one white, 36 in all for this purpose. In addition to these lights, on certain circuits were the combination posts before mentioned, each with two incandescent lamps and one arc. These were located on the main canal bridges and on the railing south of the Art Gallery. There were twenty-eight of these incandescent lights on the circuits mentioned and thirty-two lights on the Art Gallery railings, making fifty-six in all on the four circuits. On the circuit operating the lights on the Casino pier was placed the Government buoy line of thirteen 100-c.p. lamps lighting the lake channel to the city.

CHANGES IN THE MIDWAY.—Owing to uncertainty, for a long time, of the layout at the east end of the Midway and the construction of viaducts spanning the center walk, together with many other changes in this locality, the arc layout was subjected to numerous changes. The original scheme was to place posts on either side of the center walk at intervals of seventy-five feet and on the two side alleys at intervals of 150 feet. When the viaducts were planned it was proposed to operate one evening and one patrol light under each viaduct and to place three lights on special posts secured to the viaduct rail. This was modified to the extent that no lights were suspended under Woodlawn and Madison avenues, illumination at these points being provided by special iron poles, five on each viaduct. These iron poles were five inch steam pipe, fifteen feet long and fitted at the top with regular pole arms tapped to fit the pipes. A regular pole hood covered the whole. This made a very neat and inexpensive pole which fully answered the demand. The lighting of these viaducts consisted of three patrol lights and two evening lights to each viaduct. The Stoney Island avenue viaduct was lighted underneath by two suspended evening lights and by a fence pole line of six lights. For lighting the subway passage under the Illinois Central Railway, four arc lights were planned to be suspended from the structure. These were afterward abandoned and incandescent lighting substituted.

Patrol lights in all buildings were placed on circuits which included outside lighting except in two cases, one in Manufactures and one in Agriculture, which were called patrol circuits entirely in the buildings named.

INSIDE ARC LIGHTING.

The early plans for lighting the buildings provided for uniform distribution of light over the entire lower area of the buildings, but owing to the increased number of lamps required by this method it was decided to light the aisles only; the latter method was employed exclusively except in the center nave of the Manufactures building, where five coronas hold-

View of Electric Fountain.—Switchboard and Reflectors.

ing 414 lamps were suspended 140 feet from the floor, giving a uniform illumination over the entire area. An average illumination of one lamp of 2000-c.p. nominal, to 1544 square feet was allowed for all main buildings. The total space illuminated exclusive of small buildings was 4,877,965 square feet. The systems contracted for by the Exposition were the Brush, Wood, Standard, Thomson-Houston, Western Electric and Excelsior. The allotment of those systems to the different large buildings is shown in the following table :

Manufactures : patrol lighting, Brush ; evening lighting, Thomson-Houston. Machinery Hall : patrol lighting, Brush ; evening, Standard, Wood, and Thomson-Houston. Machinery Hall Annex : patrol, Standard and General Electric. Mines and Mining : patrol and evening, Brush. Illinois State building : patrol, Brush : evening, Wood. Fisheries : patrol, Brush ; evening, Thomson-Houston. Transportation : patrol, Western Electric ; evening, Western Electric. Anthropological : patrol, Standard ; evening, Western Electric and Standard. Horticulture : patrol, Brush ; evening, Excelsior. Agriculture : patrol, Standard ; evening, Standard. Leather : patrol, Standard ; evening, Standard. Forestry : patrol, Standard ; evening, Standard.

Different schemes for suspending lamps were all considered, including those used at the Paris Exposition in '89. It was carefully estimated that by the use of raising and lowering devices, 700,000 feet of wire would be saved, which at the market price represented a probable outlay of $35,000. These reasons determined the construction which was adopted and which was the first extensive use of this style of suspension. No trouble was experienced, as the work was done in a most careful and substantial manner. The hangers as used afforded perfect insulation between the lamp and the ground and were not in the least unsightly, the entire construction being hardly noticed.

The following is a report of the wiring and installation of the arc lights in each building of the Fair.

Transportation.—The work installed provided for 317 lights, 26 pole lights or patrol, and 291 evening lights. The illumination of the main building with lamps about 32 feet apart was 1296 square feet to the lamp.

Agriculture.—In this building it was first proposed to suspend four coronas, two of 25 and two 30 arc lights, each in the main wells, and also to use aisle lighting. This project was afterwards abandoned and the regular aisle scheme was followed. Wiring of the building provided for 416 evening lights, and 50 patrol lights. There were 50,728 feet of wire used. The average distance apart was 35 feet, giving an illumination of 1684 square feet per light.

MANUFACTURES.—The lighting of this building differed materially from that of any other large building ever erected. The plan of the building was peculiar and for that reason was well adapted to the solution of problems in lighting large buildings. There was a center nave or open space of 433,000 square feet area, and around this on all sides was another set of roofs covering the main floor and gallery. The plans as originally laid out called for 2400 lamps, but these were later abandoned and another scheme considered. Some difficulty was met with. In the lighting of the side naves of the building the construction was easy; the regular scheme of aisle lighting was followed as in other buildings, but the lighting of the center nave, spanned as it was by arches 202 feet 9 inches high with a span of 368 feet, became more difficult. It was too high to drop suspension wires economically and the amount of open overhead wire would interfere with the erection of exhibits. It then remained to solve a problem of lighting this building in a satisfactory manner. In July, 1892, Mr. L. Stieringer, Consulting Electrical Engineer, submitted a lighting plan for the building, providing for a system of aisle lighting surrounding the center nave, while the lighting of the central nave was an entirely original plan. It provided for the suspension of five electroliers or coronas at the height of 140 feet above the floor of the building. These coronas were to be of two sizes—one at the exact center of the building, being 75 feet in diameter and carrying 102 lamps, 98 evening and 4 patrol, and the others, 60 feet in diameter each, four in number, located two on each side of the center of the building over the main aisle, and containing 78 lamps each, 72 evening and 6 patrol. These coronas were constructed of light angle iron and were suspended by cables from the roof. The method of suspending the lamps was ingenious. They were suspended from light iron arms, and to avoid the objectionable dead weight of counter poisers, each lamp balanced its neighbor; thus no extra weight was carried by the corona. For trimming purposes each corona was constructed so as to provide a rail foot walk entirely around it. To get to this walk from the roof a ladder was provided. While burning, the lamps were suspended below the level of the bottom of the coronas. In order to trim the lamps the trimmer raised one lamp to reaching distance, and its neighbor was thereby lowered. In replacing the lamp to operating distance the level of both lamps then became the same. This method of suspension proved most satisfactory, the 414 lights illuminating 434,000 feet of floor area, giving a pleasing, soft and mellow effect. The illumination of this portion of the building was 1048 square feet per light. The side floors and galleries were planned for 782 lamps, 68 patrol and 714 evening. With these a tolerably satisfactory lighting was obtained, the division being 1687 square feet per light, with light space 30 feet apart on the main floor, and 1828 square feet per light, with light space 40 feet apart in the galleries. Seventy of the lights were used for loggia lighting, 70 feet apart. The total amount of wire used was 137,280 feet.

HORTICULTURE.—The plan executed for this building called for 186 evening and 30 patrol lights placed 29 feet apart and giving an illumination of 951 square feet per light. The total of wire required was 15,590 feet.

MINES.—This building was planned for 174 evening lights and 12 patrol lights. This gave an illumination of 1964 square feet per light. About 11,000 feet of wire were used. The average distance between lamps was 35 feet.

FISHERIES.—When first considered it was proposed to light the entire Fisheries building with arcs but a question arose as to whether these lights would be preferable in the aquarium department. Experiments with arc and incandescent lights were convincing that incandescent lights would be preferable so that the arc lighting was confined to the main building of the angling department. There were 49 lights installed of which 9 were patrol. About 5540 feet of wire were used. The illumination was 1910 feet per light.

MACHINERY HALL.—The original plans for Machinery Hall provided for 622 evening lights and 45 patrol. When the scheme was changed to aisle lighting it was first proposed to light this building with a margin of outside circuits both evening and patrol. Later it was decided to light it with full aisle circuits. The total number of lights installed in the building, including pump house, boiler house, and annex adjoining, was 326 evening lights (of which 26 were in the loggia) and 137 patrol lights. The average distance between lamps was 25 feet and the illumination, 1505 square feet per lamp. About 33,600 feet of wire were used in the building.

ILLINOIS STATE BUILDING.—This building was wired for 78 lamps, 60 evening and 18 patrol. There were about 6340 feet of wire used. An illumination of 898 square feet per lamp was obtained. The plans called for 986 incandescent lights and the building was accordingly wired for that number, but only 303 were installed.

FORESTRY.—This building was provided with 59 lights of which nine were patrol. The illumination was 1483 feet per light.

LEATHER.—There were installed 79 lights, 69 evening and 10 patrol. The illumination was 1712 square feet per light. Wire required, 6482 feet.

ANTHROPOLOGICAL.—Seventy-seven evening lights and 10 patrol lights were placed. The lamps were spaced 37 feet apart and gave a maximum illumination of 1743 square feet per light. The amount of wire used was 3940 feet.

ARC LIGHTING.

In addition to 2796 lights in operation on circuits feeding exclusively in the buildings, there were 268 lights for patrol service in buildings on outside circuits, thus raising the total number of inside lights to 3064 arcs.

GENERAL SUMMARY.

There were operated from Machinery Hall a total of 89 prominent circuits of which 33 were outside and 56 for building lighting. In addition to these there was a temporary circuit for lighting the Stock Pavilion. In the power plant the total number of arc machines devoted to the use of the Exposition was 90 as shown in the following table; these machines were driven by 20 engines, aggregating 4660 h.p.; Brush, 12 machines; Wood, 13 machines; Standard, 22 machines; Thomson-Houston, 27 machines; Western Electric, 10 machines; Excelsior, 6 machines.

ARC LIGHTING OF ELECTRICITY BUILDING.

In all plans of Electrical equipment, this building was excepted from the rules and contracts that operated in other parts of the Exposition. Pursuant to an agreement between the Exposition Company and the larger electrical exhibitors to be dilated upon hereafter, the latter agreed to contribute all of the service arc lighting in the building, the agreement being essentially as follows: The building was divided according to the proportion of lighting agreed to be furnished by each company, and within the bounds of each of these points of division, the lights of only one company were used. The Exposition Company furnished the fuel at the plants in Machinery Hall, also met certain other conditions provided for in the regular contracts with the companies under which their plants were originally installed, some of these conditions being the furnishing of foundations for generators, steam connections from the boilers to the engines, cartage of machines, the setting up of machines, etc. The several companies maintained and operated at their own expense the machines necessary for the service of the Electricity building, furnished and installed all cables and wiring between the generators in Machinery Hall and the lamps in Electricity building, also furnished all lamps and attendance, carbons, globes, etc. The building operated 422 lights for evening service up to 10:30 o'clock, this lighting being divided as follows: General Electric Company, 105; Brush Electric Company, 75; Fort Wayne Electric Company, 80; Siemens & Halske Company, 40; Excelsior Electric Company, 22; Western Electric Company, 100, total 422.

The lamps of the Fort Wayne, Brush, Siemens & Halske, and Excelsior Companies were all installed on pendants hung from the ceiling or on brackets hung from the iron trusses over the main aisles, all of the lamps of these companies being under or on the galleries or over the main aisles in reach of the iron trusses. The Western Electric Company and the General Electric Company operated part of their lamps on pendants, but both of these companies also operated additional lamps upon highly ornamental posts in the main floor, away from the galleries or under the points of high roof at the two extreme ends of the building on the galleries and at the transepts. The posts used by the General Electric Company were imported, highly ornamental, cast iron posts with wrought iron trimmings or trimmings of bronze and steel. The Western Electric Company's posts were those known commercially as the Columbian Arc Post, being fourteen feet high, of cast iron beautifully ornamental in character, with the switching device inside the post near enough to the ground to be easily in reach of the operator. The posts furnished for the occasion by the Western Electric Company were painted in ivory white with gold trimmings, and were a donation to the Exposition for the use of the Department. All of the arc lights in the building were 2000 c.p.

On behalf of every company participating in this service, it must be said that throughout the period of the Exposition the service was maintained in a manner highly satisfactory to the Department and most creditable to the companies.

In addition to this service arc lighting all of the contributing companies operated lights within the bounds of their own exhibits to an elaborate extent for ornament and for general lighting effect in the building. A few lights were furnished the smaller exhibitors on their application, the total number of this extra exhibit lighting in the spaces of the large and the small companies being 250 arc lamps, or a total of 700 arc lights of 2000 c.p. in the building, making a grand total of 1,344,000 c.p. in arc lights in the building.

THE BRUSH ELECTRIC COMPANY.

The Brush Arc Lamps.—The Brush Electric Company exhibited many styles of arc lamps, to one of which especial attention is called as durable to a remarkable degree and capable of giving a good light, when neglected and wrongly handled to an extent that would entirely unfit most any other kind of lamp for use. There is no spring about it. There are two electrical connections made of coiled wire, to cause a positive and simple contact, but no springs at all. The tension of a spring changes very materially with the temperature, and more so with use. As the adjustment and feeding of an arc lamp is a very delicate matter, the doing away with springs, common in most makes, is a marked improvement. It has heretofore been considered a very important point to have the rod slip down through the clamp very gradually, or in shop terms, a "sneak feed."

Interior View of Electric Fountain.—Changing the Colors.

This has been one of the strong points in the old Brush Lamp. The new lamp does not release the rod gradually, but allows it to drop suddenly when the clutch reaches the releasing point. The "sneak feed," or gradual downward slipping of the rod through the clutch or clamp, is good practice when the rod is clean and perfectly true, but experience has demonstrated that the rod in use becomes sometimes dirty and sticky. By a peculiar construction of the clamp in the new lamp it is unnecessary to have a true or clean rod. This clamp grips the rod tightly, and after lowering it for a certain distance lets go, and takes hold again higher up. Therefore, as far as the correct feeding of the lamp is concerned, the rod might be dirty, rough or even threaded along its entire length. This point obviates the difficulties heretofore encountered arising from wear on the rod. The electric action which takes place as the carbons approach each other causes the lamp to again lift the rod, having taken hold a little higher up. It has been found that a sudden feed with the proper mechanism maintains a much quieter and steadier arc, with closer regulation than is possible with a "sneak feed" lamp. There is no dash-pot, glycerine, wax or mercury about the new lamp. The cut-out is mechanical and has no adjustment whatever. It is positive in operation and cannot get out of order. All of the carbon holders are made adjustable, so that they will take either $\frac{7}{16}$, $\frac{1}{2}$ or $\frac{5}{8}$ carbons, and hold carbons central. The fine wire is connected to the main magnet spools by coil springs, which are very convenient in connecting and disconnecting. The lamp is easily taken apart and put together, without any soldering whatever. The rods in both the single and double lamps can be easily removed from the lamp while it is hanging in position. All the parts of both the single and double lamps are, as far as possible, similar—clutches, carbon-holders, hanger-boards, carbon rods, rod protectors and spark arresters. The only tool necessary to change the position of the feeding point to take up the wear on the rod, is a screw driver. This is accomplished by moving the clutch trip up or down. A new method of dropping the globe for trimming was shown. This is accomplished in two seconds' time by simply loosening a thumb nut under the globe holder. A rod running up the inside between the lower carbon holders acts as a guide, and an adjustable nut on end of same as a stop. Trimmers will appreciate the saving of time by this device. It is applied only to the double lamps. There were 652 of these double and 352 single lamps used at the Exposition.

DECORATIVE ARC LIGHTING.—The Brush Company used for their own decoration a continuous row of 2,000-c.p. lamps around their space. The lamps were twelve and one-half feet apart, and added in no small degree to the brilliancy of the building. They were hung either by cords from the gallery floor or on extension pulleys out from the gallery, over the main aisle.

THE WESTERN ELECTRIC COMPANY.

In arc lighting, apart from the contract work done by this company, there were in the Electrical Building 100 2000-c.p. lamps devoted to the service of lighting various portions of the building, and 30 were placed on Columbia lamp posts manufactured by the Western Electric Company, located in the galleries and main floor. Eight of these lamps were used for patrol service. Thirty-four arc lamps were used to illuminate signs, which are described elsewhere. The remaining lamps were used for lighting various portions of the Western Electric Company's exhibit. In a portion of the space was a bronze lamp-rack, upon which were suspended, ready for operation, all varieties of lamps manufactured by the Western Electric Company, consisting of double and single lamps with wide and narrow frames, the single carbon lamps being designed to burn either seven or fourteen hours, as desired. Also arc lamps for burning on incandescent circuits with perfect uniformity and freedom from hissing, and give a nominal 2000-c.p. light. These latter lamps are provided with cut-outs, and are what are termed gravity feed lamps, the rack and pinion usually employed in lamps of this kind being dispensed with. There are also shown three Krueger lamps for theater use, one being equipped with a reflector, another with an olivet hood provided with colors, and a third with lenses for concentrating and intensifying light. These lamps are designed to burn at any angle from

THE BRUSH LAMP.

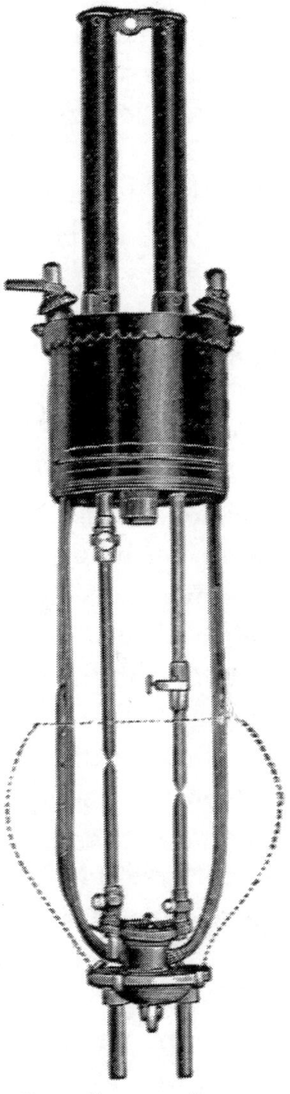

THE BRUSH LAMP.

vertical to horizontal, and are peculiarly adapted for following moving figures or objects on theater stages. They burn singly or in pairs up on a 110-volt circuit.

DECORATIVE ARC LIGHTING.—Directly over the south end of the Egyptian Temple was a large sign made in various ruby jewels, and illuminated behind by twenty arc lamps of nominally 2000-c.p. each, producing scintillating effects which attracted the eyes of visitors the moment they entered the doors of the building. Over the aisle dividing two spaces was a large glass sign reading "Western Electric Company, Chicago, New York, London, Antwerp, Paris and Berlin." This sign, 22 feet long, 8 feet high and 4 feet wide, reading on both sides. The interior of the sign was lighted with fourteen arc lamps. The letters were formed of different colors of glass crushed into small pieces, and the lamps swinging continually caused a scintillating effect to be produced resembling diamonds, rubies, sapphires and emeralds, making a very attractive display.

GENERAL ELECTRIC COMPANY.

THOMSON-RICE ARC LAMPS.—The arc lamps exhibited are constructed in various forms. The lamps are made to burn single or double carbons, the former being used where it is desired to operate the lamps for a period of about six to eight hours, and the latter where the lamps are to burn from twelve to sixteen hours. The new lamps are more ornamental than the older lamps, being fitted with a brass case instead of the plain black enameled case which is used in the Standard type, and a brass crown has been added for the top of the globe. The lamps burn with quietness and entire freedom from hissing, are self-adjusting, and do not require expert attention.

TUNNEL LAMPS.—For mining operations, underground work, and places where lamps have to be frequently handled, a special lamp was shown. These lamps are substantially constructed and capable of standing much hard usage. The side rods are of heavy gas-pipe, rigidly bolted to the frame at the top and bottom. They are made either full length, for six or eight hours' burning, or short length, burning about one-half that time.

SHORT ARC LAMPS.—In many cases, especially in low-studded mills, it has been found desirable to use a lamp which will not hang so low or require so long a carbon as the Standard lamp. To meet this demand, single and double short arc lamps were shown, which possess all the good features of the others with the added advantage of short length, being but twenty-seven inches long. A single lamp of this type will burn about four hours, and a double lamp about eight hours. They are made both plain and ornamental.

THE RIBBON FEED ARC LAMP.—A new arc lamp has recently been perfected for indoor lighting on either arc or incandescent circuits. It is but 27¼ inches from top to bottom, 6 inches wide, and weighs 21 pounds without the globe. The feeding and regulation are perfect, the arc being always the same length, perfectly steady, brilliant, and free from hissing. The entire mechanism is carefully protected from dust and dirt. On arc circuits the lamps are run in series and regulate and cut off automatically. On incandescent circuits they are run in multiple or two in series, a small rheostat or bank of incandescent lamps being used as a controlling device. There were forty-two arc lamps on Walworth poles surrounding the spaces occupied by the General Electric Company, and illuminating the center of the Electrical building. Suspended from highly ornamented brackets manufactured by the Fixture Department of the General Electric Company were sixteen constant potential arc lamps of 3000-c.p. each, connected on the "three-wire" system, which also were used for service lighting around the exhibits. In addition to the lamps described above there were shown twenty-two arc lamps of special finish, showing all the different styles of the company's make.

THE FORT WAYNE ELECTRIC COMPANY.

Twenty-seven different types of "Wood" Arc Lamp, for all kinds of current, were shown by this company. Every part was interchangeable, the winding of the coils simply being changed in the different forms of lamp to suit different currents.

ALTERNATING CURRENT ARC LAMP.—This Lamp is essentially a gravity feed lamp, regulation being effected by means of a solenoid, a U-shaped laminated core being used. This core slips into two coils of copper wire, through which the whole current of ten amperes passes, and the current in these coils weakened through the arc lengthening out, the U-shaped core retires from the coils, thus allowing the lamp to feed. The magnet is connected with the main frame of the gear carrier, which is pivoted about the center to two uprights, serving two purposes: first, as a medium of securing the rocker, and, secondly, the last one of which meshes with the carbon rod, at one end, the other extreme end engaging with a detent spring, which—unless the lamp is at its proper feeding point—retains the gears, and prevents any motion in the carbon rods. In the double lamp an automatic cut-over is arranged on the main spindle, whereby the rod carrying the burnt-out carbon is disengaged and the second rod brought into service. A combination rheostat hanger-board and fuse-box, made in porcelain, accompanies each lamp, which it is found necessary in practice to use. The Standard single and double gear-feed lamps have an H-shaped core, two legs of which enter the main coils on the top, and the other two, the shunts situated on the bottom of the lamp. The feed and escapement in this lamp are identical with the alternating lamp. This lamp is fitted with a relighter, two absolute cut-outs—one hand and one automatic, and when the carbons are entirely consumed, another double contact automatic cut-out short circuits and cuts out the lamp entirely.

Wood Arc Lamp.

Constant Current Clutch Lamp.—This lamp has an H-shaped core, connected with the main rocker, and the main and shunt coils are placed relatively in the same position as in the gear-feed lamp. When the current is established the main coils pull up the H-shaped core and the rocker attached to it, and the clutch mechanism, being connected to this rocker, seizes the rod and strikes the arc. As the arc lengthens, the rocker as well as the H-shaped core, lowers down until one arm of the clutch trips on the regular tripping piece, and allows the rod to move slightly, thus adjusting the arc without any alteration of the light. This lamp, like the gear-feed lamp, is fitted with automatic and absolute cut-outs, and can be fitted either a straight bow frame or short frame for low ceilings.

DECORATIVE ARC LIGHTING.

This company was given the wall area at the north end of the Electrical building in which to make a mural decoration in arc lights. A large and handsomely painted sign was placed high up against the wall. At a distance of twenty-three feet out from the wall there was a row of twelve 2000 candle lights of the Wood type in tinted globes, swung from the iron arch-way, and dropped below the arches to a distance of twenty feet. The lights served to vastly improve the appearance of that end of the building, and proved quite attractive to visitors.

Interior View of Electric Fountain.—Signal Slides.

THE STANDARD ELECTRIC COMPANY.

Fifty Standard arc lamps were suspended from four arches following the aisle lines on three sides of the exhibit. Other lamps were shown suspended from mast arms and from an automatic lamp hanger and cut-out. Special provision was made by means of lamps suspended at proper distance from the floor, for illustrating the correctness of feed mechanism by means of a portable voltmeter attached to the lamp, which was kept in plain view at all times. Five different styles of lamps were exhibited; single service lamp for inside use, double service or fourteen-hour lamp for indoor burning—single service; weather-proof pattern for out-door or street use—double service; all night weather-proof lamp for municipal lighting, and the ornamental, adapted for any of the different purposes mentioned. The Standard arc lamp possesses features of merit which may be summarized as follows: a reliable feeding mechanism; an automatic relighting cut-out of unique design; the absence of contacts in the bushings or at fixed points upon the rod; the adaptability of the single service lamp to all-night service by means of the Standard elliptical carbons; and the improved globe holder extension which facilitates rapid trimming of the lamp and yet insures perfect safety to the globe. The lamp is independent in feed, always maintaining an arc of the same voltage no matter how varying the current may be in volume. This guarantees a silent arc and a steady white light free from fluctuations. The current is carried to the rod by means of a copper brush bent in the shape of an elongated horseshoe. This brush grips the rod on both sides equally, the pressure being governed by an adjusting spring. This brush is also mounted upon a fulcrum, which permits it to move freely upon the rod and on this account offers no friction whatever to the rod when the arc is formed. The clutches are all insulated by mica, and no current can possibly reach the carbon rod through them. The weather-proof lamp requires no weather-hood and no additional hanging board. All of the contact points of the lamp are encased within a weather-proof cap immediately above the case of the lamp. This device is so constructed as to permit the lamp being removed from the line quickly without breaking the circuit. The weather-proof cap is provided with a suitable cut-out which is amply protected from the weather. The elliptical carbon for all-night burning requires but one mechanism, one carbon rod and one pair carbons. The exhibit of the Standard Electric Company in the power plant in Machinery Hall consisted of 22 50-light series arc dynamos with 1100 arc lamps in daily service throughout the grounds and buildings.

To afford visitors and interested parties special opportunity of carefully investigating the perfection of feed in the Standard arc lamp, and to further illustrate the desirable and important features of the double service arc lamp which employs an elliptical carbon for all-night service, a hard wood cabinet was provided, inside of which an arc lamp was kept in operation, having its arc reflected and magnified some twelve or fifteen times and projected on a screen at the farther end of the cabinet. By means of properly devised apertures and openings, visitors could watch the operation of the arc upon the carbons and the closeness of feed. This object lesson proved an attractive and instructive feature of the exhibit.

GENERAL INCANDESCENT ARC LIGHT COMPANY.

The exhibition of this company consisted of various kinds of arc-lamps for both direct and alternating current, classified as standard, ornamental, railway, chain, and Bijou lamps. There were seventy lamps exhibited, ranging from the plainest Standard lamp, which is itself very sightly in appearance, to the beautiful and comparatively small so-called Bijou lamp, which latter style is best illustrated by a 6-light electrolier which carries six of these lamps, and is not only capable of lighting an interior most successfully, but at the same time furnishes a centerpiece of great beauty. All the different styles and sizes of lamps were shown in various beautiful effects from comparatively simple to the most elaborate triumphs of the metal worker's art, and finished in polished brass, rich gilt, old brass, imitation steel, wrought iron, etc. They were shown singly and in pairs, suspended from brackets and pendants and on all kinds of fixtures. The Standard arc lamp of the past has been the 10-ampere so-called 2000-c.p. lamp; this company makes its lamps of any style to consume a current of from two amperes up. It has been found that the 4-ampere lamp giving about 600 c.p. lends itself to the most perfect distribution of lighting, and that with this size of lamp almost any kind of interior can be more successfully and economically lighted than by any other method. Marked changes and improvements have likewise been developed in facility of handling and trimming the lamps. This point is one which the importance need not be enlarged on to any user of arc lights.

The two principal types are Rack Feed and the Chain lamp, and the points of difference in these are that in one a rack rod carries the carbon, while in the other the same function is performed by a chain or cord running over a spring drum. In other essential respects the mechanism of these two kinds of lamps is the same. This controlling mechanism is exceedingly simple. The feed is by gravity. The downward movement of the carbon is arrested by a pawl and ratchet movement and controlled by a single electro-magnet in the shunt circuit. There is no series coil. The normal position of the carbons is not the usual one of contact between the carbon points; upon trimming the lamp and bringing the two carbons together the upper carbon springs back when let go, a distance of about 1/8 of an inch, when the current passes through the shunt circuit, actuating the electro-magnet, the pawl is freed from its ratchet, and the positive carbon descends into contact with the negative, thus establishing the circuit. The moment the flow of current is thus diverted from the shunt, the positive carbon springs back again a distance of about 1/8 of an inch and the arc is established. It is in its former position, and in this position it remains until, by the burning away of the carbons, the resistance of the arcs becomes sufficient to abnormally increase the flow through the shunt coil again, and thereby again causes the feed mechanism to operate. The positive carbons used in direct current lamps are cored and the negative solid; in this matter of the use of carbons this company has likewise left the beaten path and adopted the practice which has now more advocates—namely, of using a larger positive and smaller negative carbon. The advantages of this practice are easily recognized, the principle being that a longer life can be obtained for a given combined length of carbon, a larger upper carbon gives more crater surface, and consequently more reflecting power, while a smaller carbon furnishes less obstruction to the downward passage of the rays of light, and consequently much less shadow. As the comparative length of carbons used affects the life, the requirements of which are various for different purposes, the company has adopted as a standard three sizes, which are called, respectively, "Short," "Standard" and "Long," which give 8, 11 and 14 hours of life. The total length over all of the Rack Feed type of lamps varies from 35 inches for the first to 51 inches for the last, and in Chain lamps or Bijou lamps, from 27 to 31 inches over all. The standard voltage adopted by the company at which the arc for the various amperages are adjusted, is as follows: Four amperes, 41 volts; 6 amperes, 42 volts; 8 amperes, 43 volts; 10 amperes, 44 volts, these being the principal standard capacities that are used. The lamps are connected two in series on 110-volt circuit, or five in series on 250 volts, 10 in series on 500 volts, and so on. The excess of current is taken care of by a resistance, which, as constructed by this company, is made of German silver wire, mounted on a metal frame and insulated entirely with porcelain, the construction of which is very simple and at the same time thorough and effective. The resistances are either placed on the lamp or furnished separate, as may be desired. In the standard form of lamps the mechanism is thoroughly insulated from the frame at every point so that it is impossible when handling the lamp to come in contact with the current.

In addition to its exhibit above described, this company had 170 lamps of the "Railway" type, which were used by the Exposition management to illuminate the Annex to Machinery Hall.

These lamps operated ten in series across a 500-volt power circuit, and were provided with an extra resistance, thrown into circuit by an automatic cut-out when any individual lamp was cut out, so that the continuous burning of the remainder of the lamps was not interfered with. Referring generally to the operation of the lamps, it may be said that the light which they give is very steady and reliable, entitling them to the name which the company has given its lamps—namely, the "Incandescent Arc Lamps,"—and in practice it has been found that its steadiness is not interfered with even by very abnormal variation in the voltage of the circuit on which the lamps are operated. They claim the range of 30 volts as a margin within which the perfect operation of the lamps is not interfered with. Of course such a variation in voltage affects the candle power, but it does not, as might be expected, cause the arc to break or become unsteady.

The Alternating-Current lamps are a more recent development, but these were also shown and seemed to be as reliable and as steady in operation as the latter. They regulate perfectly on a 30-volt circuit, with no waste of current in resistance whatever. For currents of about 50 to 100 volts, economy coils are supplied, which transform them to the proper pressure without appreciable loss.

The operating mechanism in its general design does not vary in important details from that of the direct current lamps. Of course the construction of the magnet and its core are different, and in this particular, several new and ingenious improvements have been introduced. No more metal in the spool has been used than necessary to make it rigid, and the iron of the core is carefully laminated, so that the induction of the alternating current cannot cause too much heat.

THE NUTTING ARC LAMP.

The peculiarity of this lamp consists of a ring of wax-like material, two and one-half inches in diameter, mounted between the upper carbon rod. A metal pin is mounted on a stationary holder in such a manner as to have one end embedded in the surface of the wax one-eighth or three-sixteenths of an inch, the cross section of the ring being about one-half inch square. The remaining surface of the pin has wound on it fine German silver wire of 150-ohms resistance, which, with an additional resistance of about 600 ohms, forms the shunt or feeding current of the lamp. It is evident that with the pin imbedded in the wax the carbon rod is held rigidly in place; but on the separation of the carbon points the shunt circuit will at once take its share of the current, the proportion depending on the voltage at the terminals of the lamp, *i. e.*, the carbon points. The metal pin becomes heated by the passage of the current through the resistance wire wound around it, the heating effect varying with the square of the current passing through the shunt circuit. As the pin heats, it melts the wax-like material in which its end is embedded, and allows the ring to slowly revolve on its shaft, thus feeding the upper carbon and maintaining the arc at a uniform length as the carbon is consumed. The melted material of the ring at once solidifies behind the pin, leaving a perfectly smooth and even surface for the next revolution. In actual service it is found that the movement of the slowly revolving ring is so uniform that it is practically impossible to detect any movement whatever, and that the carbon is being constantly fed forward at the exact rate at which it is being consumed. The feeding mechanism being entirely devoid of magnets or other inductive mechanism, and depending on a principle extremely more sensitive to current fluctuations than any combination of magnets, makes it possible to attain in an arc lamp a simple and perfectly constant feed. The alteration of adjustment necessary for extreme change of temperature has been thoroughly worked out, and is accomplished in an extremely simple and positive manner. The cut-outs are all mechanical and cover all possible cases in which they may be called into use. Cut-out magnets are not used, thus doing away with what is often a never-ending source of repair.

THE EXCELSIOR ELECTRIC COMPANY.

The rows of arc lamps suspended above the railing that surrounded the exhibit immediately attracted the attention of passing visitors. In the exhibit, were various styles of arc lamps the company manufactures, lamps adapted to the different places in which the arc lamp is used, weatherproof lamps for the street, lamps for interiors with high ceilings, short lamps for low ceilings with the arc but fifteen inches from the top. In this lamp both the upper and lower carbons move. Lamps for use on steamboats with lenses, focusing lamps for photographic purposes, etc., were also shown. The latest duplex street lamp was shown, which is so arranged that while the first set of carbons is burning the other set is held up. The top of the rod of the first carbon is provided with a button which trips a lever when that carbon is burned out and throws the second set of carbons into use. Special attention has been given in the construction of these lamps to the facilities for cleaning them easily. The switch on the lamp when turned throws the lamp entirely out of circuit, thus the safety of the trimmer is assured.

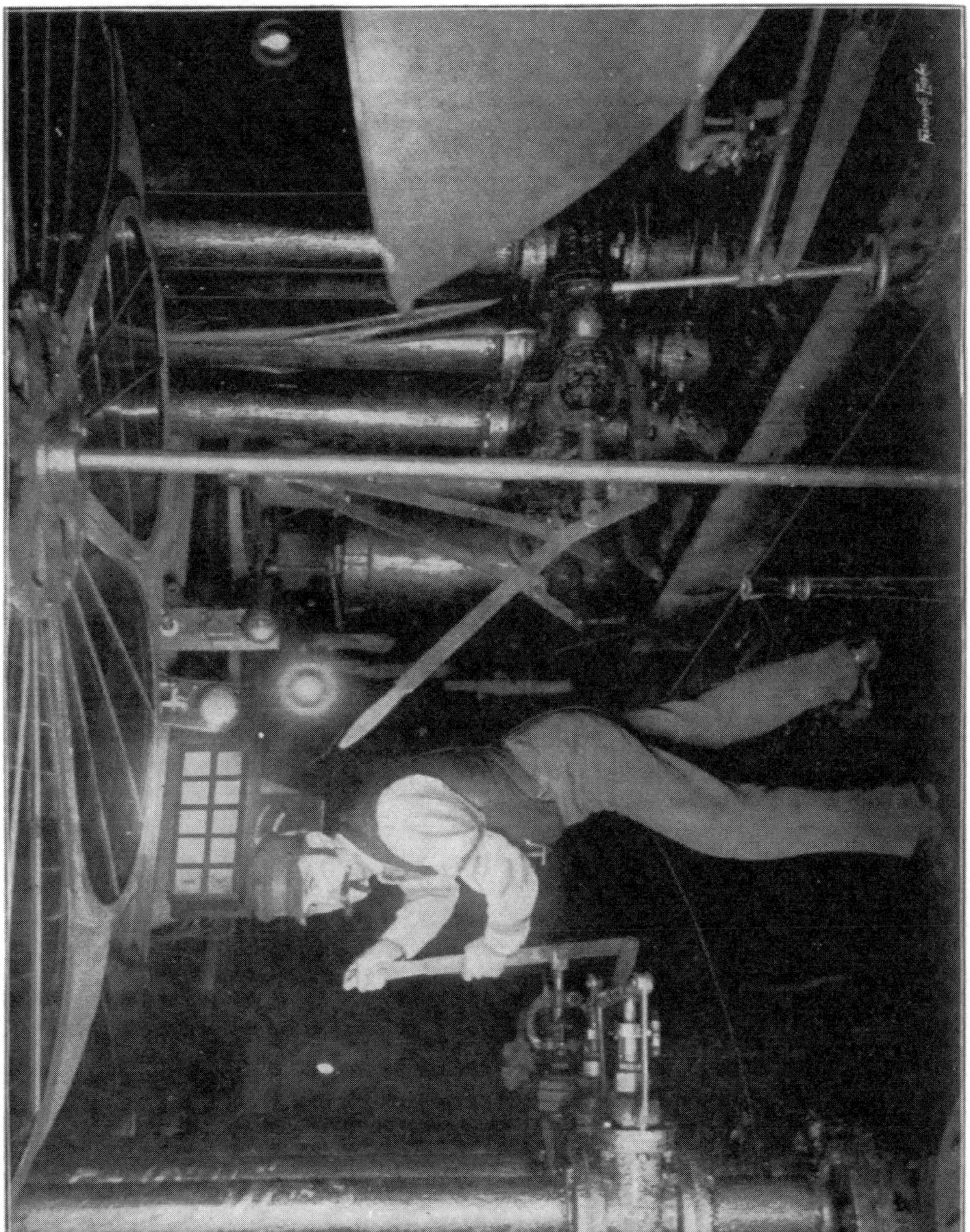

Interior of Fountain.—Working the Pump Valves.

THE MATHER ELECTRIC COMPANY.

The lighting of the Libbey Glass Works in the Midway was arc and incandescent lamps, operated on the same circuits by the Mather Company. The arc lighting was furnished by forty Ward arc lamps, distributed as follows: twelve byzantine arc lamps, located in the dome of the furnaceroom; six nickel-plated arc lamps in dynamo and engine room; eight regular arc lamps lighting up different parts of the building; fourteen regular arc lamps distributed on the outside of the building and for lighting the grounds; two twin-carbon arc lamps located in the tower. All the arc lamps were provided with switches placed in convenient locations in the building, by means of which any set of lamps might be cut in and out of circuit.

GENERAL ELECTRIC COMPANY, LIMITED, LONDON.

On exhibition there were several Byng arc focusing lamps, adapted for series or parallel circuits, continuous or alternating currents for high or low candle power, and much used in England and on the continent. The lamp belongs to the class whose operation depend on differential magnet action, acting directly on the carbon holders, without any clock work, springs or other mechanism. All the working parts are carried on one plate suspended over a pulley and passing along rollers. There is an electro-magnet between the tubes, and an iron core inside the tubes adapted to short circuit the magnetic field. The core of the lower carbon holder tube is wound with fine wire and is in series with the shunt coil, while the other core is wound with a coarse wire and is in series with the series lamp coil of the electro-magnet. When the lamp is out of circuit, the two carbons and their holders balance each other. As soon as the current passes through the lamp it causes the carbons to separate and to cut out the series wire coil, while the fine wire magnet is acting alone for an instant, attracting the bottom carbon holder until the carbons touch again. At this instant the main current passes again and the shunt ceases to act. Now the series coil or magnet is acting alone and raises the top carbon and the arc is formed, causing a difference of potential at the two carbons and causes a current to flow again through the shunt magnet The two windings of the electro-magnet are in opposite directions, and tend to neutralize each other, and the magnetic action only takes place when the one or the other is predominant. If they are equal in strength the carbon holders remain balanced in the same position. On the slightest change of the length of the arc or current in the series winding, the balance no longer exists, and a different action takes place, endeavoring to reëstablish equilibrium. In order to prevent violent jumping of the light at the formation of the arc, there is attached to the pulley a brake, to which the above named tubes are suspended, which allows the tubes to travel up and down within a very limited space until the lamp has regulated itself and has arrived at normal condition. The lamp, which may be a single or a double carbon lamp, is provided with a waterproof cover and the internal parts are nickeled or lacquered to prevent oxidation. The object of the brake is to prevent the carbon points from separating too much when the arc is first established, and by tightening or loosening the screw, the proper length of the arc can be adjusted when the current is on; the screw must not by any means touch the carbon holder. For indoor lighting, and especially where the customary heavy shadows are objectionable, this company has adopted the system of indirect lighting, or lighting by diffusion. This they accomplish by using the lamp above described in the reverse way, and supplying a large reflector which deflects all the light to the ceiling.

ALLGEMEINE ELEKTRICITATS GESELLSCHAFT OF BERLIN.

These lamps are provided with shunt as well as with different regulation for any strength of current. They work safely and continuously for direct and alternating currents, and burn steadily without flickering. Exterior damages are rendered almost impossible by the simplicity of their construction and by the careful closing up of the case.

SPECIAL FIXTURES FOR ARC LIGHTING.

WESTERN ELECTRIC COMPANY'S COLUMBIAN ARC POSTS.

The Columbian lamp posts are designed for use in street lighting, the arc being suspended at a distance of twenty feet from the ground. The construction of these posts deserves especial mention. The base of the post, which is six feet high, is an ornamental iron casting, through the center of which is placed a heavy wooden turned post, which serves as a support for the surmounting hood in which the lamp is placed. This hood is made with but a single arm running by the lamp, which is, in in actual use, put towards the building, thus preventing objectionable shadows being cast in that portion of the area which is desired to be illuminated. The wooden post also serves as an insulator from ground, so that an operator can trim the lamps when in service without being subjected to shocks. Near the base of the iron casting is a door opening to a double-poled switch which entirely disconnects the lamp from circuit without opening the dynamo circuit.

THE STAR IRON TOWER COMPANY.

Tower System of Arc Lighting.—The system of electric lighting from towers is of very recent date. Their construction has been subject to gradual change and development. At the outset a square pyramidal tower was most used, made with inclined posts at the corners, united by horizontal struts and braced in the plane of the sides and horizontally. These horizontal braces prevented the ready arrangement and operation of an interior elevator, so resort was had to a triangular tower similarly braced on its exterior faces, but requiring no interior bracing. This left free space for an elevator, but in order to insure strength, the tower was tapered from the top to the base, so that with an altitude of (say) 150 feet, the spread at the base was about 28 feet. This formed no great objection where the tower could be located in a park or unoccupied public square, but to light a city or town these towers had to be located at street corners. It was found necessary, therefore, to have the tower span the street or a sidewalk, so that one corner of the tower might rest near the corner of the street and the two other corners of the tower rest adjacent to the buildings. This was unsightly and inconvenient, as frequently in cities the space between the sidewalk and the building is occupied by vaults or openings, and tower foundations placed there would often be very much in the way, necessitating the shifting of a tower location from a point where it ought to be to some other locality where it was possible to erect it. Again, with a pyramidal tower not guyed, the wind pressure caused great strains and required heavy members, and necessitated frequent adjustment and repairs. Assuming the same wind pressure per foot of height on the pyramidal tower as is permissible for the prismatic one, and that the former is not guyed and that the latter is perfectly guyed, then for equal dimensions the unguyed tower would be strained at least four times as much as the other for the same wind pressure.

The development and improvement of pyramidal and other forms of towers gradually led to the one exhibited in the Electrical building. It was essentially a triangular prism, six feet on a side, built in seventeen vertical sections, each eight and one-half feet long, supported on a single base column and guyed at two points. The center lengths of all sections are equal, and members, to a considerable degree, interchangeable. The foundation is generally located just inside the angle of the curb at the street corner, and consists of a double platform of two-inch oak plank laid crosswise, carrying a pier three feet square of hard-burned brick set in a mortar of hydraulic cement and coarse sand; the pier is capped by a stone three feet by three feet by six inches that receives the base casting, the latter secured by six one and one-eighth anchor bolts through the masonry and bearing on an iron ring underneath the platform. A tubular column fourteen feet high screws into the base casting, and receives struts at the top connecting it with the prismatic tower. A light elevator platform runs in the interior of the tower and is carried by a single endless cable passing over an upper fixed sheave and a lower sheave attached to an adjustment lever, pivoted at the short end and counterweighted on the free long end; the outer part of the cable carries a counterpoise nearly balancing loaded elevator. Two stationary guide-cables pass through the car and are engaged by a clutch which holds the elevator stationary and only permits ascent or descent when released by pressure of the occupant's foot. The towers have suitable lower and upper platforms with railings, and mast and arms for the lamps. All the compression members are of the best lap-welded tubing, screwed into connection blocks of first quality malleable iron, that have ribs to which are bolted the forked ends screwed to each end of all the wrought iron diagonal rods. There are two sets of four guys each, one leading from a point seventeen feet below the upper platform and the other from a point about thirty-four feet lower down on the tower. The guys are each seven-sixteenths inch galvanized wire cable, connected by a turn-buckle to an eyebolt in an oak post fourteen inches square by fifteen feet long, set vertically six feet in the ground at about 150 feet from the base of the tower and receives two guy ropes in the same vertical plane. The iron work is painted one coat of asphaltum and the posts two coats of lead. The towers are erected by first putting together the top section, then hoisting up and building on the next beneath it, and so on until completed. The weight of a complete tower, elevator, etc., including guy-ropes, is about 7200 pounds. The entire wind surface, including lamp-hoods and mechanism, is calculated at eighty-three square feet, but this should probably be increased fifty per cent. for oblique and indirect exposure. This form of tower has been subjected to terrific storms. At Evansville, Indiana, twelve of them were subjected to three cyclones, one of which destroyed large numbers of houses, barns and shops, and, in fact, demolished some of the heavy business blocks in the vicinity of the towers. The towers were not injured in the slightest degree. The other storms were likewise very destructive of other surrounding property. They have withstood severe storms at Detroit, Michigan, where 122 of them have been in service for nine years. None of them has ever sustained the slightest injury, nor have they required any repairs; not even a bolt or nut has required tightening or adjusting since the towers were erected. Exhaustive experiments have shown that 150 feet is the most satisfactory height for these towers. Increasing this height impairs the illumination near the foot, and does not perceptibly increase the total lighted area, while diminishing this height diminishes the illuminating area and affords unnecessary brilliancy at the base. By reason of very lofty buildings in the immediate surroundings a height of 175 feet might be, in some instances, desirable. The towers should, so far at practicable, be arranged in a triangular system. The distance apart in business sections may be 1200 to 1500 feet; in the best residence sections, such, for instance, as may be found in Detroit at a distance of half to three-quarters of a mile from the business center, the towers may be 2000 feet apart, and in the less densely populated sections and suburbs they may be 2500 to 3000 feet apart. Thus at Detroit, Michigan, twenty-one miles of territory having about 4000 street intersections, is thoroughly and brilliantly lighted by 122 towers, and there are no gas or naphtha lights within the corporation limits. In a general way towns of three to six thousand inhabitants, occupying, say, a square mile of space, may be thoroughly lighted in every quarter by seven towers, one at the center and six at the angles of a hexagon, the towers being 2000 feet apart. So, also, an excellent illumination may be had by five towers, there being one at the center and four at the angles of a square, the towers being 2000 feet from the middle tower; and where greater economy is desirable there might be four towers, one at the center and one at each of the angles of a triangle, and about 2000 feet from the center tower. In all cases it is desirable that the towers should have but four lights of 2000 c.p. each. The central tower might have six lights of 2000 c.p. More lights will not materially improve the effect at a distance from the tower, while a less number of lights will scarcely afford illumination sufficient in the vicinity of the tower. All spaces are lighted by the tower system—front yards, back yards, alleys, and streets; and in street lighting alone the four lights on a tower will do the service which would require sixteen to twenty lights on poles, while the pole lights will illuminate only the streets and not the yards and alleys. A number of towers will serve to produce an atmosphere of light so blended in every direction as to neutralize the shadows, so that dense shadows exist in no quarter, giving the effect of moonlight over the entire area. It is stated that these towers, with 11,000-c.p. lamps give an illumination at 200 feet and 1500 feet equal to that from a gas-light at 7½ feet and 57 feet respectively, assuming the latter at the rather low value of 16-c.p. As these results are proportionate to the inverse squares of the distances, they are probably nearly correct, except as influenced by special local conditions.

Interior Electric Fountain.—Color Slides and Projector.

SEARCH LIGHTS.

SCHUCKERTT & COMPANY, NUREMBURG, GERMANY.

Heretofore, when parabolic mirrors for search lights have been considered, it has been tacitly understood that such mirrors were made of metal brought to the desired form on a suitable block. Such mirrors, however, have faults of different kinds: they cannot be constructed accurately enough in the form of a paraboloid of revolution; the reflection from the surface of the metal reflector is much less than from the surface of silvered glass; the surface of the metal reflector soon deteriorates through oxidation, and from the frequent cleanings disadvantageous changes occur. The difficulties of constructing glass parabolic reflectors, which appeared until quite recently insuperable, led Mangin to invent his well known reflector, and induced Tschikolew in St. Petersburg, and Siemens, in Berlin, to construct their reflectors as an approximation to the parabolic form. Schuckertt & Company exhibited a system of search lights calculated to overcome the difficulties, and by the construction of a special machine, make parabolic mirrors with sufficient accuracy for all practical purposes. The parabolic mirror surpasses all the other systems hitherto in use, and seems destined to supplant them. Quite setting aside the optical superiority of this type, the parabolic mirror—because with it the choice of focal distance is not limited—allows the use of lamps with carbons arranged horizontally, from which the radiation of light takes place much more satisfactorily than with inclined or vertical carbons. Schuckertt & Company are makers and were exhibitors of projectors with parabolic reflectors, and projectors with reflectors 40 cm., 45 cm., 60 cm., 75 cm., 90 cm., 110 cm. and 150 cm. in diameter. These are made in the following types: Projectors for the army, in which, on account of transportation, the smallest possible weight is secured; projectors for use in the navy, which are especially made to secure easy working on board of ships; projectors for the merchant service, which are confined more to ordinary and smaller types, and serve more particularly for the passage of canals, narrow waterways, and for the purpose of sighting buoys, etc., at night. In these classes there are also different types, especially in class I; as, for instance, the type used for coast defense, for use in fortresses or against fortifications, for field service, for turrets or for balloons. It would lead us too far to explain the special construction of all this apparatus; it must, therefore, suffice to describe some characteristic types. The apparatus built for a current of 120 to 150 amperes has a parabolic mirror of 90 cm. diameter, and is intended for the navy. The frame rests with trunnions in bearings in the side frames, which are fixed on a turntable; the table turns on rollers upon a cast iron base. There are, therefore, two rotary motions possible, one in a vertical, the other in a horizontal plane, so that it may be placed in any desired position. For both rotary motions fine adjustments and clamps are arranged so that either fast or slow motion can be given to any point on the horizon. The slow motion of the projector is, for both movements, also controlled from a distant point, such as the bridge of a ship; this is done by using worms which gear into the worm wheels of the slow gear placed on the spindles of small electro-motors in the projector frame, and those may be operated from the bridge. A switch is provided with contacts so arranged that the current can pass the armatures of the motors in either direction, so as to obtain any movement the observer may desire. The projector is thus under the direct control of the officer on watch, so that mistakes in carrying out orders are entirely avoided; the current needed for the movement is obtained from the main current, used in the projector itself. The motor used for the motion in a vertical plane is fixed to the turntable inside of a box, while the motor used for the motion in a horizontal plane is placed upon the fixed bedplate. The system of moving projectors from a distance by means of electric motors has been used for several years. The parabolic glass mirror is fixed to the projector casing by means of screws and nuts; the casing is not completely closed underneath, but allows a space for the horizontal lamp, permitting it to be moved along grooves parallel to the axis of the projector, so that the crater of the positive carbon may easily be brought to the focus of the mirror. The carbon holders reach up to the center of the casing, the carbons themselves being in its axis, in order to make it easier for the attendant to observe the position of the carbons and shape of the crater while the apparatus is in use, and to enable him to put in new carbons when required; small optical projectors are arranged at the side end on the top of the casing, which throw an image of the arc as seen from above and from the side upon the same plate of the ground glass. The image of the crater surface should be covered by a vertical image if the crater is exactly at the focus of the reflector; if it should not be so, then this position will be obtained by moving the whole lamp. By means of side doors arranged in the casting the screws on the carbon holders are reached so that the carbons can be put into the proper position. At the front end, opposite the reflector, the casing is closed by a system of dispersers which consist of single glasses of plano-convex cylinder lenses. The lenses of the inner disperser are completely covered by those of the outer disperser. The inner one can be moved parallel to itself and along the axis of the projector toward the outer one along the slide; the outer one rests on rollers on the same slides, but is kept in position by a number of springs placed at equal distances around the circumference of the casing. If the inner disperser be moved so that the two dispersers are distant from each other a space equal to the sum of their focal length, then the rays of light reflected from the mirror are so refracted that they meet at the focus and diverge thence. The focuses of the two outer dispersers have a shorter focal length ; when the diverging rays pass through these outer lenses they become parallel again. As the outer lenses have a shorter focal distance than the inner ones, they are not struck in their whole width by the diverging beams of light, and thus there remains between the rays which have been rendered parallel by passing through the two lenses a space which receives no light and allows room for the shutter of a kind of darkening apparatus. On the upper half of the figure the lenses of both dispersers are close to each other. The rays coming from the parabolic mirror through the first system of lenses will be refracted as in the former case, but before they reach the focus they strike the second lens and are caused to converge still more, so that the two lenses act as one with a shorter focal distance. As the rays of light leave the lenses, converging toward the focus and there cross and diverge, there remains with this arrangement a dark space for the shutter of a darkening apparatus. Each position of the movable dispersers between their two extreme positions corresponds to a certain angle of spread, so that, therefore, by a simple movement of one disperser by means of the hand wheel, all degrees of spread can be obtained which are between the concentrated beam with about six degrees spread in a horizontal direction and the maximum spread of about 45 to 48 degrees. The darkening or signaling apparatus is mounted in front of the outer disperser, and, as it does not in any way interfere with the light, can always remain in position, so that the projector is always ready for flash signaling, which is of great moment on board of ships.

The outer disperser is not fixed to the casing, but kept in place by springs. This arrangement has been used to allow of elasticity in the direction of the axis of the projector, as if it presented a solid surface it would be in danger of being broken by the atmospheric pressure produced by firing heavy guns. The arrangement just described is known as the "double disperser," and is in use in many vessels of war. One of Schuckertt's combined dynamo and steam engines specially made for service on board of ships was also shown ; the particular one exhibited was made for use on a torpedo boat. A smaller projector had a mirror of forty-five centimeters diameter, and was intended for field service in the army, as well as the merchant marine. The casing is similar to that of the large apparatus, and is movable in two directions around a vertical and horizontal axis, and for both movements it is fitted with fine adjustment.

With a view of making the smaller projector, as well as the larger type, suitable for use on board of ships, it is provided with a "double disperser" and an electromotor. If it is sufficient that a change can be made from a concentrated beam to the greatest spread, without passing through all the various degrees of spread and power illumination, then the disperser may be much simplified by mounting one glass close in front of the other, each composed of alternate strips of plano-convex, and plano-concave lenses, which could be made to slide past each other sideways, the exact breadth of one strip. In the one case two plano-convex or two plano-concave lenses stand opposite each other, and the projector is arranged for the greatest spread; while in the other case a concave glass is in front of a convex, and vice versa, so that the curved surfaces are parallel to one another, and the light leaves the lenses in a concentrated beam without material change. The gear for moving the outer disperser in front of the inner one consists of a screwed spindle placed on top of the casing, and provided with a hand wheel. The lamps of the projectors are generally arranged for automatic regulation, but hand regulating apparatus is always provided as well. The regulating mechanism of Piette and Krizik is used for lamps which have to be moved around frequently, and therefore need to be as light as possible; but for ship use, and for projectors which are subjected to shocks, a new type of lamps is used. One of the pieces exhibited represented a search light apparatus of sixty centimeters diameter as in use for the fortifications of the St. Gothard Mountains in Switzerland. Posted always at a fixed point, it is provided with a sight and the horizontal and vertical movements divided in degrees, so that by night time any desired destined point may be lit up without searching for it, for it has been adjusted by day time and registered in record.

The largest search light apparatus shown was the most powerful ever made. Its mirror diameter measures 150 centimeters and its lamp is built for 150 amperes, whereby 60 volts are needed at the terminals of the lamp. The positive carbon is cored and has a diameter of one inch and a half and the negative carbon of one inch. For the parabolic glass mirror in connection with the carbons being in the geometrical axis of the parabolic, a very simple relation to find out the intensifying power and luminosity. A search light of that size was posted in the World's Fair grounds at the northwest corner of the middle roof of the Manufactures Building at a height of 230 feet. The beam of light was seen in Milwaukee, Wis., 85 miles distant, on July 15, 1893, and in Antioch, Ill., 65 miles distant, on August 5, 1893. By this light a person standing in the beam of light at a distance of ten miles is able to read a newspaper. Standing by the side of that large apparatus one is able to detect at a distance of twenty miles moving or approaching troops or war vessels by aid of a good field glass. Such large search lights are especially used for coast defenses in order to throw their light such an immense distance that it is impossible that they may be reached by gun balls. The trainage and the vertical motion of the beam of light—*i.e.*, the apparatus—is electrically controlled, so it may be operated by the commander himself from any distance desired. The average intensity of the light is 194,000,000 c.p. On the middle roof of the Manufactures Building were three other search lights of Schuckertt & Company in operation, of different sizes and outfits.

In the exhibit in Electricity Building, German section, were search light projectors exhibited with the most complete outfit for war vessels; they were partly in operation. Besides, there were navy dynamos and signaling apparatus—all for use for navy and war purposes. The projectors made by this firm are used in the German army and navy, and also in Russia, Austria, Switzerland, Sweden, Belgium, Netherlands, Denmark, Italy, England, China and Japan, for fortifications as well as for the merchant service and navy; they were used by the Italian expedition to Massowa, mounted on the special wagons made by the same firm, and did good service.

THE GENERAL ELECTRIC COMPANY.

The large light exhibited by this company in operation on the middle roof of the Colonnade at the Exposition stands about ten feet six inches high to the upper side of the ventilator on the top of the drum, and the total weight is about 6000 pounds, but so perfectly is it mounted and balanced that a child can move it in any direction. The reflecting lens mirror used in this projector is 150 centimeters, or 60 inches, in diameter. It has a concave spherical mirror of the Mangin type, free from spherical aberration, reflecting a sensibly parallel beam of light. It was manufactured especially for this projector in Paris, France, and is a most perfect specimen of optical work, three and one-fourth inches thick at the edges and one-sixteenth of an inch thick at the center, and weighs about 800 pounds. The metal ring in which it is mounted weighs about 750 pounds, and the total lens, ring, and cover weigh about 1600 pounds. This great mirror is mounted at one end of the big drum, the outer end of which is furnished with a door, consisting of a metal rim, in which are fixed a number of plate-glass strips five-sixteenths of an inch thick by six inches wide. Inside this drum and sliding, upon ways arranged on the bottom, is placed the electric lamp, the source of the light which is reflected by the mirror. It is entirely automatic in its action, is six feet high and weighs about 400 pounds. The carbons used are also made especially for it. The upper or positive carbon is one and one-half inches in diameter and twenty-two and one-half inches long, with a five-

Schuckertt Search Light.—Northwest Corner Manufactures Building.

sixteenths of an inch core of soft carbon running from end to end through its center. The lower or negative carbon is one and one-fourth inches in diameter, is fifteen inches long and also has a core of soft carbon running through its center. In addition its outer surface is heavily coated with copper. The positive carbon is set a little in front of the negative, and thus almost all the intense light of the incandescent crater is cast upon the reflector. The maximum current at which this lamp operates is 200 amperes, and at this current the lamp has a luminous intensity of about 90,000 to 100,000 candles, the reflected beam a total luminous intensity of about 375,000,000 candles, an intensity which the eye cannot appreciate. In looking at the side of the beam the spectator only distinguishes a stream of light of comparatively low intensity, but in looking at the beam directly its brilliancy is fully seen and the effect is absolutely blinding. Ventilators at the top and sides allow a constant current of air to pass through the drum and dissipate the heat generated by the arc lamp; and they are so arranged that no light can escape through them. All the connections for adjusting the positions of the carbons and the lamps are brought through the drum to the outside, and are arranged in close proximity to one another at one side, so that all may be manipulated by the operator without moving from his position. Through openings in the drum covered by densely-colored glass the operation of the lamp may be watched and its adjustment verified. The drum is supported by trunnions in bearings at the top of a Y-shaped fork, set in a base plate, and the whole supported on a system of friction wheels, forming a turntable resting upon the top of a massive pedestal supporting the whole structure. The drum, fork and base plate may be rotated horizontally on the turntable either by hand or by gearing provided for this purpose. The drum may also be elevated or depressed in a vertical plane by similar gearing.

ELECTRICAL SERVICE OF THE SEARCH LIGHTS.

Among the last pieces of apparatus installed in power work were the search lights located on each of the four corners of the Manufactures building, on the west side of the roof of the Casino and Music Hall, and on the roof of the Colonnade, a total of seven large search lights, six of which played upon the MacMonnies Fountain, and were so arranged that they could throw any of the prismatic colors. For carrying the current over to the Manufactures building for the four search lights on its roof, two No. 0000 circuits were run in the subway to the Manufactures building, and then up to its roof. This required 15,000 feet of No. 0000 cable and 200 feet of No. 0. The current for these lights was taken from the Siemens Brothers generator. This dynamo was a 120-

volt machine, with an output of 1500 amperes, and was directly connected to a Willans engine. This plant was installed in the British section in Machinery Hall as an exhibit, and was loaned to the Exposition for this work without cost.

The circuit to the two search lights on the Casino and Music Hall was carried out on the Intramural from the south side of Machinery Hall as far as the southwest corner of the Agricultural Annex, where it was carried under the Agricultural building over to the center of the Peristyle and from this point to the roof. It consisted of two No. 0000 cables to this point, where it fed into two No. 0 circuits running to the search lights on the Casino and Music Hall respectively. It required 75,000 feet of No. 0000 cable and 2300 feet of No. 0 wire to complete this installation. The current for these two lights was furnished by the 45-k.w. Edison generator without cost to the Exposition. This machine was furnished and operated by the General Electric Company for the operation of the search lights furnished by them. This dynamo was 110-volt, direct-connected machine coupled to a Willans engine. For the large search light on the Colonnade a No. 0000 circuit was run from the Siemens Brothers machine in Machinery Hall, comprising about 2200 feet of No. 0000 cable.

Below is given a table showing the current, potential, diameter, weight and size of carbon, candle power, and make of the seven search lights in use upon the grounds of the Exposition :

SCHUCKERTT'S SEARCH LIGHTS.

Location	N. W. Corner Manufactures.	S. W. Corner Manufactures.	S. E. Corner Manufactures.	N. E. Corner Manufactures.
Diameter	60 in.	44 in.	36 in.	24 in.
Weight	3000 lbs.	2000 lbs.	1500 lbs.	1000 lbs.
Current	150 a.	140 a.	90 a.	50 a.
Potential	60 v.	58 v.	52 v.	46 v.
Size of Positive Carbon	1½ in.	1½ in.	1 in.	$\frac{15}{16}$ in.
Size of Negative Carbon	1 in.	1 in.	¾ in.	⅝ in.
Candle Power	194,000,000	104,300,000	66,000,000	30,000,000

GENERAL ELECTRIC COMPANY'S SEARCH LIGHTS.

Location	Music Hall.	Casino.	Colonnade.
Diameter	30 in.	30 in.	60 in.
Weight	1300 lbs.	1300 lbs.	6000 lbs.
Current	90 a.	90 a.	200 a.
Potential	54 v.	54 v.	96 v.
Size Pos. Carbon	1¼ in.	1¼ in.	$1\frac{9}{16}$ in.
Size Neg. Carbon	1 in.	1 in.	1¼ in.
Candle Power	130,000	130,000	100,000,000

F. BARBIER & COMPANY, PARIS.

This firm made an exhibit of lens lighthouse reflectors. The principal light shown contained the largest lens made up to the time of the Paris Exposition of 1889. The firm also showed flash light apparatus for use either with arc light or mineral oil; also what is termed a permanent lamp for lighthouses, which is capable of equipment for fifteen days constant operation without renewal..

SAUTIER, HARLE & COMPANY, PARIS.

This firm exhibited several styles of the Mangin mirror lights, for lighthouse and naval operation. They also showed several (4) lens flash lights for coast-guard service.

HENRY LEPANTE & SONS, PARIS.

The light for lighthouse operation shown by this firm was a bivalve, two-lens light, built on the principle of the new flash lights adopted by the Lighthouse Board of the French Government.

THE STIERINGER ELECTRIC FOUNTAINS.

At a very early date in the planning for the World's Fair, it was universally agreed that one or more electrically illuminated fountains should be built and operated as part of the evening entertainment. The great Paris fountain of 1889 had been one of the most interesting features at that Exposition, and Chicago people had come to know the drawing power of such a feature because of the popularity of the Lincoln Park fountain. Both of these fountains, however, were the very acme of the art of fountain building, and there was a serious question whether the World's Fair could improve on them—and of course mediocrity was not to be thought of. Lincoln Park fountain looked grand, looming up as it did above all its surroundings in the beauty of its bright colors, yet the fact was appreciated that those surroundings were a dead level—grass, shrubs, and a very few trees at some distance away, alone served to belittle the proportions of the fountain; whereas, at the Exposition grounds, the immensity of the buildings, rendered all the larger to the view because of the monotonous whiteness of their exterior, would serve to dwarf any fountain not truly geyser-like in its power. Almost every conceivable location on the grounds was considered in turn, the architects and landscape gardeners naturally being averse to allow any such unholy mechanism mar the simple grandeur of their handiwork. After all had been said, however, it was decided to make the fountains two in number, and to locate them one on either side of the Columbian fountain at the west end of the Court of Honor. Then it was discovered that the cornice lines of the Electricity building on the one side and the Machinery Hall on the other, were seventy feet above the ground. The Lincoln Park fountain, amid such surroundings, would assume the relative prominence of a whale spouting in a stormy sea. Only one thing was to be done, namely, build the fountains to comport with their environment, and plans to this end were made. Then tenders of a contract were made, but only one response came in, that being from an English firm, and the figures were not only exorbitant but only contemplated the actual lighting effects, the whole engineering feat incident to the performance being burdened upon the Exposition.

The consulting electrical engineer of the Exposition, Mr. L. Stieringer, was given the task of making the plans in detail, and the General Electric Company was given a contract for $65,000 for their execution, all material to revert to that company at the close of the Exposition. Not the least of the difficulties of preparation was the fact that the caissons necessary would fall on a line far below the lake level, and it would be expensive if not impracticable to wall out the water. Moreover, the illumination planned for the grounds of the Court, and for the interior as well as the exterior of the buildings, would necessitate the employment of light of great intensity in order that color lenses should show to advantage. All these difficulties were fully appreciated, however, and in being appreciated were the more readily overcome. In the plans the designer developed many new features, the entire installation being larger and more comprehensive in every respect than had ever before been attempted, with the additional feature of a marked economy in construction and operation over all previous fountains of a similar character. In the construction work many natural obstacles were encountered. Excavations had to be made in the sandy ground, and a water-tight chamber built eight or nine feet below the lake water-level. The projected use of an iron caisson was abandoned in the face of the many difficulties which presented themselves, and a special construction was adopted. Piles were sunk into the ground, and upon the shorter ones a floor of interlacing beams was laid, upon which concrete was run to form the floor proper of the chamber. Longer piles served as columns for the support of the wooden water-deck, and also for the framing against which the Wakefield sheet-piling, in three layers, three inches thick by twelve in width, twenty feet in length and lap-jointed was allowed to rest, the same being sand jetted in, thereby securing a water-tight chamber, to which entrance was gained by a water-tight passage. Within each of these subaqueous chambers were located the projector lamps, the color screens, the switchboards and the signal indicators and ventilating apparatus.

In each operating chamber there were nineteen 80-ampere arc lamps, estimated to give a beam of light of 250,000 c.p. each. The lamps were run two in series and were automatic in their action, being provided with electrical apparatus for striking their arc, and automatically feeding their carbons together as they were consumed, so as to keep the arc central to the reflector. The beam was either vertical or at a slight angle to the vertical. The silvered reflector was twenty-eight inches in diameter.

In connection with these arc lights, each of which acted as a holophote at the center of and under each of the nineteen composite groups of jets, was the system of color screens operating in a horizontal plane. In previous fountains the colored screens were laboriously slid in and out like lantern slides. In this instance the screens were so geared together that they could be operated easily by a single attendant and at a high rate of speed. The screens were mounted at the top of standards. As the screens revolved, each of the colored blades or paddles came in the range of the beam of light from the arc lamp reflector. There were six blades—sea-green, green, yellow, blue, ruby and plain white; and the groups were arranged so that they could be run in any kind of combination. The glass in each screen was divided into sections, to obviate the trouble that might have been caused if any plane of these glass windmill sails remained long subjected to the heat of the arc. The vanes were supported by pipe shafts, and the arc lights were carried on standards or bases placed on the floor of the caisson.

Near the entrance to each chamber was placed the switchboard, to which was brought current on the Edison three-wire system at a potential of 240 volts. It was provided with voltmeter and ammeter, so arranged that the voltage on either side of the system and the current flowing in any lamp circuit could be measured by them. The switchboard also carried at the bottom a rheostat for each lamp circuit, by means of which the current flowing in the lamp circuit could be regulated. There were also switches for throwing the lamp circuits on and off, and at the back of the board were grouped fuse boxes for safety.

The signaling apparatus was both efficient and complete. It embraced the color, valve and other signals in each fountain and the signal boxes and switchboard in the northwest tower of Machinery Hall, from which point the operation of the fountains was directed. The color signal box in each fountain was divided into twenty-four compartments or windows, arranged in three horizontal rows of light windows each. In the top row the panes were of ground glass, and numbered to correspond with the color control apparatus to which that vertical row was the signal. The succeeding horizontal rows from the top to the bottom were filled with stained glass to correspond with the colors in the screens; the seventh was fillled with plain ground glass to give white, and the eighth

Great Corona.—Centre Manufactures Building.

with ground glass with a central spot of black to indicate "no change." Behind these windows were located twenty-four incandescent lamps—one for each window. Beneath the color signal box was a simple signal testing switch, by which any of the lamps in the signal box could be thrown into circuit to test the integrity of the circuits. Over the signal box was located a small electric bell, operated by a push button in the controlling station, by which signals were given to "stand ready" to change the colors. The operation of these signals was as follows: When a change of color was desired the operator in the tower, who overlooked both fountains, pressed the signal "Stand by to change." If he desired to have the central geyser red, the six encircling composites blue, and the outside circles yellow, whatever lights had previously been illuminated would be extinguished and the red switched in the center row, the blue in the one marked 2, and the yellow in the one marked 3. This would at once illumine the lamps in the signal box under the fountain which were covered by that colored glass in the vertical rows, and the operator there would move the color screens accordingly. The valve signal box was similar in principle to that containing the color signals. It consisted of a box with two horizontal rows of compartments, each containing an incandescent lamp and each covered with a ground glass pane marked in black with the number of the valve and the words "Project" and "Pulsate." To fill out the box one compartment was left blank. The signal was given by illuminating the lamp in the appropriate compartment, and when the compartment marked "Pulsate" was illuminated, it indicated that the operator was to move the particular valve back and forth so as to give the jets under control a pulsating motion. When the word "Project" appeared, it indicated that the valve was to be opened suddenly so as to give an explosive or rocket effect. Beneath the signal box was another switch for the purpose of testing the signal circuits. An electric push button connecting with the pumping engines enabled the valve operators to communicate their demands on the pumps. There was also telephone communication with the controlling station. The water-tight chambers containing all the apparatus just described were of sufficient height to enable the operators to move easily in an upright position. Ventilation was achieved by means of suitably arranged electric motors and fans. The electric generators furnishing current for the fountains were located in the power plant in Machinery Hall, and consisted of four 175-k. w. 140-volt bi-polar Edison generators, furnishing current through Edison underground tubes on the three-wire system. These generators were driven by two large high-speed engines of the Ball and Arlington-Sims type.

Water for the fountains was supplied from the Worthington Pump station south of Machinery Hall through a 24-inch main at 100 pounds pressure. When operated to full capacity each fountain required 22,000 gallons of water per minute. The basin of each fountain was 60 feet in diameter and contained 19 composite groups of jets having 1,710 orifices. These jets were designed to be of the following numbers and dimensions:

12 Wheat-sheaf rings, containing 1,560 ⅛-inch holes.
7 Little Geyser rings, containing 47 ⅝-inch holes.
1 Curtain ring, containing 90 ¼-inch holes.
18 Nozzles, 1¼ inches, for small geysers and parabolas.
12 Nozzles 1½-inches, called "pulverizers," on the outer edge.
1 Nozzle, 2 inches, called "Great Geyser."

The dimensions of the nozzles finally adopted were in some cases smaller than the above, however, to overcome the failure of the pumps to give a sufficiency of water at all times. At the center of each fountain basin was the great geyser, encircled by a ring of jets called the curtain, and flanked at six points by the little geysers. On the next and outermost circle of all were the parabolas and "pulverizers" combined with the central wheat sheaves. The jet holders were all concealed from view by blast furnace slag, in imitation of rock work.

These fountains had each a greater number of water orifices, a greater number of electric lights and a greater volume of water than anything heretofore produced of a similar type. Each of the fountains was designed to utilize 22,000 gallons per minute when operated to its full capacity, whereas in the Lincoln Park fountain but 4,000 gallons per minute were available.

The Paris fountain of 1889 had only 17 orifices, while each of the Chicago fountains had 19, or a total of 38. The Chicago projector lamps were one-third more powerful than those in use at Paris, and the volume of water utilized in the two Chicago fountains was five-fold that used at Paris. When the lights on the Exposition grounds and the exterior of the adjacent buildings were extinguished, and high winds, which operated to disturb the sprays, were not prevalent, the results attained were highly satisfactory.

Plans were prepared to illuminate the MacMonnies fountain by electricity, by light from eight orifices similar to those in the electrical fountains, but with a different arrangement of streams. These orifices were to be located in a circle between the outer and inner figures. The scheme was abandoned; although if it had been carried out the results would, undoubtedly, have been superior to that attained with the ornamental fountains at the last Exposition at Paris.

While this fountain had no electricity directly connected with it, its beauty was enhanced by the aid of the powerful search lights.

THE ORIGIN OF FOUNTAINS.—The origin of the illuminated fountain is veiled in obscurity, but the principle was without doubt crudely applied by the ancients long before the Christian era. The sources of light were in all probability originally fagots, candles or oil, supplemented by reflectors. Later, gas and the calcium light were used, and color was added to further embellish the display sought for.

The principles governing the luminous fountain were first stated by Prof. Colladon in a communication to the Academy of Arts and Sciences in 1842. M. de Bovet mentions having seen a small luminous fountain in 1879 or 1880 at Ouro-Preto, the capital of one of the provinces of Brazil. This fountain was exhibited at the City Theater by a passing showman, and embraced a reservoir placed on the roof, a few tin pipes and a Drummond light.

Electricity was first employed in the luminous fountain by Sir Francis Bolton, of England, in 1884, and the perfected result of his labors was exhibited at the Health Exposition in London the same year. In 1885, the first electrical luminous fountain ever seen in America, was constructed and put in operation at Staten Island. It was subsequently removed to Chicago, where it now stands in Lincoln Park. Similar fountains have since been exhibited in Manchester, Liverpool and Glasgow.

The electrical fountain at the Paris Exposition of 1889 was the largest and most elaborate which had been constructed up to that date. It consisted of two parts, one of French, the other of English design, embracing similar plans to those embodied in the previous fountains shown at Manchester, Glasgow and elsewhere.

CHAPTER IV.

THE POWER PLANT.

MACHINERY that supplied the Exposition with arc and incandescent light and power for all purposes, occupied about one-third of the entire floor area of Machinery Hall. The power plant space was about 1000 feet long by 100 feet wide, making a total floor area of nearly seven acres.

The boilers were placed in a fire-proof building made mostly of iron, with galvanized iron roof, and extended for 827 feet along the east side of Machinery Hall and was 86 feet wide. All the principal makers of boilers were represented, and there were in use the standard forms of feed water heaters, condensers, feed pumps and the auxiliary appliances. The floor of the boiler house, of granatoid, was nine feet lower than that of Machinery Hall proper, with a gallery running the entire length, on a level with the Machinery Hall floor, for a walk-way for visitors. The water was converted into steam and passed from the boilers at a pressure of 125 pounds through overhead pipes into a main feeder 800 feet long and 36 inches in diameter, supported under the gallery and divided into seven sections, each adjoining section being connected by a ten-inch steam equalizing lop arranged to take up the expansion. From the top of this main feeder the steam passed through pipes laid under the aisles in Machinery Hall to the engines in the respective blocks. From the steam chest of the engines, the exhaust passed through pipes supported beneath the flooring to the feed-water heaters on the visitors' gallery. The equivalent of about 25,000 horse power in steam was generted by batteries of almost all the standard boilers.

In the distribution of the engines, several considerations had to be taken into account—ease of access to the steam pipes, convenience of arrangement of the electrical generators and accessories, and the arrangement of engines so that exhibitors could get the best possible advantage out of their exhibition, by giving visitors a good opportunity to get at them. The space of 1000 feet by 10c was divided into fourteen blocks, each 95 feet in width, and each separated from the others by spacious aisles. In these spaces were the engines, dynamos, switchboards and all auxiliary apparatus, all easily within view for examination by exhibitors. In treating of the power plant, it will be interesting to begin at the beginning and take up first the temporary plant built for the service of the Exposition in its construction period.

THE TEMPORARY PLANT.

Early in the construction period it was deemed advisable to establish a temporary station for the purpose of furnishing light and power through the grounds of the Exposition, and in execution of this plan, on June 6, 1891, letters were written to all of the light and power supply companies doing business in the neighborhood of the Exposition grounds, asking for bids, the basis of the propositions to be the supply of 1200 and 2000-c.p. arc lamps, the prices to be figured on the cost of each lamp for all night service; the Exposition Company to erect and maintain all construction, bidders to supply lamps, generators and service. The first installation was to be for 50 lamps, the ultimate capacity being fixed at 200 lamps. Later the estimate was increased to 600-h.p. capacity. The Grounds and Buildings Committee authorized a temporary power house, the cost of building not to exceed $6,000. The estimated cost of equipping this plant complete for operation, including all steam and electric machinery, motors and all outside lines and connections, was not to exceed $50,000. In July, 1891, specifications were issued, for the furnishing of the electrical apparatus for the plant to all the principal electric companies, the terms to be the rental to the Exposition Company of the apparatus specified. Three companies responded, and the bids were as follows: Thomson-Houston Co., $8600, f.o.b.; Western Electric Co., $4750, f.o.b.; General Electric Co., $2932, f.o.b. The contract was given to the General Electric Company, and accepted on August 31, 1891, for all the apparatus, and provision was made for future increase on a pro rata basis. The estimate for this service included 7 miles No. 6 insulated wire, 5,000 feet No. 12 wire, 3,500 No. 1 wire, and smaller lengths of other wire, all to be of Brown & Sharp gauge, waterproof insulation, grade of insulation to be at least equal to what is known as K. K. The first work done was for incandescent lighting in the service building, barns, engine house and hospital. Material for work did not commence to arrive until after the first of September, coming in slowly owing to poor shipping facilities. About the 10th of September, 1891, wiring was started for incandescent lights in the barn located near the service building, with a force of three men. In a few

Western Electric Company's Power Plant. — Machinery Hall.

days the force was increased and work started on the service building and hospital. In the evening of the 8th day of October, 1891, the power plant was first started, one small engine being put in service operating two 300-light Edison incandescent dynamos, supplying service to 160 lights in service building, hospital, engine house and barns. Between this date and October 17 one 50-light arc generator was installed in the power house and wired for, twenty-six arc lamps were put on poles for patrol service, and the lights were in operation. Forty arc lamps were installed, part on poles and part in new buildings, Mines, Manufactures and staff works. Work on power lines for motor service throughout the grounds was begun and a few additional incandescent lamps installed; also temporary arc lighting supplied for use of pile driving and for water supply department. For these temporary lines poles were set, wires strung, and lamps and construction moved from one place to another as occasion required. During the month of November, 1891, the power lines to a number of the buildings were completed and eight motors were put in service, as follows: One 15-h.p. for saw-mill, Electricity building; one 15-h.p., Manufactures building; one 25-h.p., Machinery Hall; one 15-h.p., Mines building; one 20-h.p., Agricultural building; one 15-h.p., Transportation building; one 9-h.p., Horticultural building; one 25-h.p. hoist, Horticultural building.

From this time forward, motors of varying horse-power capacity were installed, temporary overhead construction being used, for the operation of saw-mills, hoists, crushers, dredges, tool sharpeners, and for almost an infinite variety of other purposes. In fact, at the beginning of the great Columbian year, the Exposition grounds were a network of power and light lines, machinery of every variety was in full operation, and already an exhibition of the applications of electricity was in full operation, the like and variety of which had never before been equaled. It is worthy of a brief pause to call attention to the fact, that the newest of all the sciences, that one which people are accustomed to regard as in its infancy, served more purposes, saved more money, and gave better results, than any other one science, art or industry. Imagine, for instance, that every contractor and operator at the Exposition grounds, during the construction period, was obliged to maintain an independent steam plant for each saw-mill, hoist, dredge and crusher; the cost in machinery, fuel and attendance would have been enormous. Thus it will be seen that from the very outset of the Exposition, electricity became a prime factor in all operations, which it maintained to the close. During the construction period extending up to May 1, 1893, the maximum of motors used in the work, reached fifty in number, varying in capacity from 1-horse power to 50-horse power. A maximum of 600 arc lamps of 2000-c.p. were used about the grounds for patrol purposes and at the buildings as their construction progressed, a majority of the arc lights being used to allow contractors to work their men through three shifts of eight hours each. Incandescent lamps were used to the number of 200 in the offices at the service building, in the stables of the company, and at various other points. These lamps were of the Edison make, and were run off the regular Edison temporary plant. The Westinghouse service replaced these everywhere in the grounds about May 1, 1893, at the time of the opening of the Exposition.

THE PERMANENT POWER PLANT.

WESTINGHOUSE ELECTRIC & MANUFACTURING COMPANY.

The Alternating Current Generating Station.—The great central station plant installed by the Westinghouse Company in Machinery Hall, adapted, as it was, to the operation of incandescent lamps, arc lamps and motors, and capable of delivering more than 12,000 horse power, of electrical energy, was a complete exhibit of the most modern type of electrical machinery adapted to various kinds of service. In order to carry out the contract with the World's Columbian Exposition, the company installed in Machinery Hall a complete alternating current central station equipment, by far the largest plant ever built. The plant was located adjacent to the southern wall of Machinery Hall, about the center of the building and occupied a space of some 31,000 square feet. The installation comprised the following apparatus:

Two 750 kilowatt, two-phase alternators, tandem belt driven by a 3000-h.p. quadruple expansion engine, built by the E. P. Allis Company, Milwaukee, Wisconsin.

One 750 kilowatt, two-phase alternator, belt driven by a 1000-h.p. Frazer & Chalmers engine of the Corliss type.

One 750 kilowatt, two-phase alternator, belt driven by a 1000-h.p. McIntosh & Seymour engine.

One 750 kilowatt, two-phase alternator, belt driven by a 1000-h.p. Buckeye engine.

One 750 kilowatt, two-phase alternator, belt driven by a 1000-h.p. Atlas engine.

Six 750 kilowatt, two-phase alternators, each driven direct by a 1000-h.p. Westinghouse Columbina steeple engine.

Two 240 kilowatt, single-phase alternators, each belt driven by a 350-h.p. Westinghouse compound engine.

Three 75 kilowatt exciters, each direct driven by a Westinghouse 100-h.p. Compound engine.

A marble switchboard for twelve two-phase alternators, two single-phase alternators, four exciters, and forty circuits completed the installation.

The total output of the plant was 9480 kilowatts. The records of the readings of current and potential which were taken at the switchboard every half-hour during the time the Exposition was open, show that about 55 watts at the generator terminals was required for each 16-c.p. lamp connected with the circuit. The station, therefore, was capable of simultaneously supplying 172,000 16-c.p. incandescent lamps. The two alternators belted to the large Allis engine—the engine started by President Cleveland when he opened the Exposition—were tandem-driven by belts 72 inches in width. The distance from the center of the engine shaft to the center of the shaft of its nearest dynamo was 47 feet; that to the center of the more distant dynamo 65 feet. All the engines and dynamos were erected upon massive foundations, installed under the supervision of the mechanical engineers of the Exposition. In the case of the direct connected alternators, with their engines, a single concrete foundation was used. The plant illustrated on a very large scale the methods of driving by direct coupling of engine shaft to dynamo shaft and by the use of belting. The advantage of the direct-connected over belt-driven machines, as regards the saving in floor space, will be seen by comparison of the space occupied by one of the 1000-h.p.—"engine dynamos," as they are called, consisting of a 1000-h.p. Westinghouse Columbina steeple engine and a 750 kilowatt alternator—with that occupied by belted engines and alternators. The outside measurement of the direct-connected unit was 31 feet 8½ inches by 13 feet. The space occupied by the smallest belt-connected engine and alternator was greatly in excess of this, being about 65 feeet by 27 feet. Electrically considered, each alternator was a double machine which delivered to its circuits two-phase currents, the difference between the phases being almost exactly 90 degrees. The frequency employed was 60 periods per second, or 7200 alternations per minute, and the potentials at the terminals of the machines varied from 2000 volts to about 2300 volts, depending upon the load, drop in primary circuits, transformers, secondary wire, and upon the potentials required by the lamps. The speed of the alternators, both belt-driven and direct-connected, was 200 revolutions per minute. Each machine of this type had two armatures, mounted upon the same shaft and magnetized by fields electrically and mechanically independent of each other. The fields were of the circular or ring type, having thirty-six inwardly projecting pole pieces, built up of mild steel laminated and cast into the yoke. In these machines lathe-wound coils are slipped upon the poles and securely fastened in position by bolts. The armatures are of the toothed type generally employed by the Westinghouse Company for their alternating current work. The armature coils are lathe-wound, and any one of them may be readily removed and replaced without disturbing the other coils upon the same armature. Each armature has thirty-six teeth, to correspond with the same number of poles on each field. The armature shafts are of forged steel and supported in self-aligning ball-bearings equipped with automatic oilers. The armature shaft carries at each end, between the armature and bearings, two ring collectors, which serve to collect the alternating current from each armature. The shaft also carries two communicators, from which a rectified current is conveyed to the auxiliary winding of the field coils. The excitation of the field is chiefly derived from separate exciters, hereinafter described, but this auxiliary excitation is employed to secure automatic increase of the magnetic field in proportion to the reaction due to the current flowing in the armature. The current for the auxiliary excitation is derived from small transformers in the spider of the armature. The main current of the armature traverses the primary of the transformer, while the secondary current of the transformer is rectified by the commutator. In this manner the automatic compounding of the field is attained without the introduction of high potential current into the field coils. In the belt-driven machines provision is made for adjustment of the belt by sliding the machine as a whole upon foundation rails. The connection between the Westinghouse engines and their alternators is made by rigid coupling. The machines measured 15 feet 1 inch in height. The weight of a belt-driven type of machine was a little over 140,000 pounds, while that of a direct-connected alternator was 106,000 pounds. The following claims are made for these two-phase alternators: Simplicity and solidity of construction, few parts, moderate speed. The standard speed for these machines is 200 revolutions per minute. Use of laminated pole pieces in cast iron yoke; the field poles are built up of sheets of mild steel fifteen-thousandths of an inch in thickness and are cast into an iron ring. The losses in field iron due to reaction of the armature are thus minimized, while the greatest rigidity of frame is secured by the use of a relatively large section in the cast iron yoke or ring. Bearings are self-aligning and self-oiling. Ability of the machines to deliver one-phase or two-phase currents; each machine is a double machine, the two parts being electrically and magnetically independent. By keying the armatures to a shaft in a certain position, the currents delivered will coincide in phase, while by keying one armature 90 degrees in advance of the other, two-phase current, suitable for the operation of Tesla motors, is obtained. As here used, the machines deliver currents differing in phase. Construction of armature—magnetized part laminated steel; center wheel or spider, cast iron; coils protected against mechanical injury; the construction secures the advantages of the toothed type of armature without necessitating the use of the special and expensive steel in the entire armature. The cast iron wheel is surrounded by a ring of laminated steel, this ring being connected to and mechanically mounted upon the wheel in such a way as to avoid magnetization of the cast iron. Sectional winding of armature; the coils used are lathe-wound, and before being placed in position upon the core are thoroughly insulated by mica and by two layers of insulating tape. This tape is applied to the coils by winding in a direction at right angles with the axis of the coil. This method of separate insulation on lathe-wound coils secures far more perfect insulation of the armature conductors than is possible in ordinary forms of hand-winding. The sectional winding makes it easy to replace a burned out coil. In case of accident, the damage is usually limited to one coil, and with this type of armature, a spare coil at the station is as much an insurance as a spare armature in the case of other types of winding commonly employed. In case one coil is burned out or otherwise injured, it may be cut out of circuit and the armature operated without it. This is, of course, true only of machines having a comparatively large number of poles.

Sectional Winding of Fields.—The advantages of this type of field winding are practically the same as those above claimed for a similar construction of armature. It is to be noted also that the use of inwardly projecting poles, each carrying a field coil, secures a very large radiating surface to dissipate the heat due to the field current. As compared with those alternators in which but one or two field coils are used, this sectional construction makes it possible to work the field copper at much higher density.

Efficiency.—In the design of this machine, mechanical excellence, general reliability, and ease of repair have been considered of first importance; at the same time, efficiency has not been sacrificed, and the iron and copper losses have been carefully proportioned to secure good results.

Regulation.—Two currents are used to excite the field. One is derived from a separate exciter, and the other is a low potential current from the armature, which is commutated by a commutator upon the armature shaft. Each field of the double machine is thus independently compounded. It should be noted that the method of auxiliary excitation employed introduces no high potential current in the field coils. Arrangement of tension screws and sliding base for belt-connected type. Shape of phase; the form of the current wave from an alternating current armature of the toothed type is greatly affected by the relative form of the field poles and the armature teeth. The form of field pole and armature tooth selected secured a symmetrical phase and a proper relation of maximum e.m.f. during a cycle.

Exciter.—Current for exciting the fields of the alternator was obtained from 375-kilowatt continuous current machines of the multipolar

Belt Driven Block.—Westinghouse Generators, Atlas, Buckeye and McIntosh & Seymour Engines.

type, which were driven by Westinghouse automatic engines. These exciters made 300 revolutions per minute, and delivered current at 200-volts. Two of them furnished current to excite ten alternators and a third was held in reserve. At the close of the Exposition one of these machines had been running continuously for about three months, and in that time had made over 37,000,000 revolutions.

DISTRIBUTING SYSTEM.—The contract awarded the Westinghouse Company included not only the machines for generating the electricity but also the wiring of the buildings and the laying of the supply circuits. In order to carry out this contract the company sublet the wiring to the New York Insulated Wire Company. From the switchboard the circuits were run under ground to the various points at which the current was utilized. A part of the circuits was run in a large subway used in common for lighting and power circuits and extending from Machinery Hall to the more important buildings in the grounds. The subway lines were connected with distributing circuits wherever it was necessary to supply current for lighting. These distributing circuits consisted of duplex lead-covered cables, placed by the Standard Underground Cable Company. The supply wires were run to converter pits placed near the buildings or objects to be lighted. The converter capacity in each pit varied according to the number of lights that were to be installed, but any single pit was never allowed to supply over 800 lights. In case more than that number of lights were required in any building, pits were placed along the side of the building near the portion that was to be directly supplied with current. The converters used in the pits were in 100 or 200 light units. The secondary circuits extended from the pits to main switches located in the building. They were placed in a special cupboard, together with the safety fuses and distributing switches. The switchboard from which this plant was operated is described under "Switchboards," found elsewhere.

WESTERN ELECTRIC COMPANY.

The plant consisted of 10 Western Electric arc dynamos, each having an easy capacity for 50 nominal 2000-c.p. arc lamps. These machines were built with the latest improvements, being entirely automatic in their action, permitting of change in load from nothing to full capacity of machine without injurious heating or excess of sparking. The machines were of constant-current type, operating the lamps in series. In building these machines it was the design to use a construction which would require a minimum amount of attention from the operator. In order to do this, it was necessary to discard oil cups and provide a means for lubricating the commutator. To accomplish this end, self-oiling ball and socket bearings were designed, which not only lubricate their own bearings but also take perfect alignment with the shaft, thus preventing the bearings from becoming hot. The old style copper brushes are done away with, and carbon brushes substituted in their place, which in themselves lubricate the commutator sufficiently to permit it to run twenty-four hours without attention. The automatic regulator which is at all times connected with the machine will operate to detect the variation of ¼-ampere of the normal output of the machine and correct the same at once, the result being a machine which requires little or no attention from the operator. By means of a switch located on the machine the normal output of the dynamo can be varied from 10 amperes to 8 or 7, as desired, so that with the same dynamo and the same lamps any candle power desired can be produced. The machines are also constructed with an end to make each part accessible, and, with this in view, the bearing seats are so constructed that one bolt is sufficient to hold a bearing rigidly in place. It is possible, with this construction, to stop a machine which is doing full duty and place the armature upon the floor in not to exceed three minutes' time. The regulation of this machine is accomplished by shifting the brushes around the commutator as the load changes, thereby producing a constant-current output from the machine. The regulator is strong, simple, durable and positive in its action. It consists of but few parts, which are at all times accessible to the operator. Any variation in the current output is immediately remedied by a small electro-magnet, through which passes the full current of the machine. The dynamo is mounted upon a wooden sliding adjustable belt tightener, the wooden frame serving to insulate the machine from the foundation. The commercial efficiency of these machines at full load is not less than 90 per cent., and the heating, when operated for any length of time, will not exceed 90 deg. Fahr. above the surrounding atmosphere.

Arc Lamps.—The dynamos were connected to the lamps through Western Electric arc-lighting switchboard, made of Tennessee marble. Upon this marble were mounted springjacks, to which were connected the various circuits. Half the jacks in this board were devoted to the service of the Exposition, and the other half to that of the exhibit of the Western Electric Company in Electricity building. These springjacks were constructed in such a manner that any circuit could be short circuited at the switchboard by means of what is called a disconnecting plug. By means of this device any circuit can be cut into or out from service while the dynamo is in operation without interfering with the lamps not intended to be affected by this operation. In connection with this board was a testing device, by means of which the insulation resistance to ground of any circuit could be measured while the lamps were burning. The following diagram of parts and connections fully explains the method employed in measuring the insulation of a live circuit, and is as follows: The circuit is short circuited at the dynamo through a pair of condensers connected to a drop which rings a bell when the shutter falls. When the insulation of the circuit drops to a point low enough to charge the condensers, the shutter falls, which rings a bell, thus notifying the operator of a grounded circuit. The switches are then opened and the plugs inserted. The resistance consists of various resistances of 100 ohms each. The plug connection with the switch is then inserted through the resistance of 100,000 ohms, which places the Wheatstone bridge in connection with the circuit in exactly the same manner as though the circuits were dead. When the switch is closed the current has two courses of travel—one through the plug, the resistance of the Wheatstone bridge to ground, back through the ground on the arc light circuit to the dynamo. This causes two currents to pass through the galvanometer in opposite directions at the same time. Consequently, when the plug is inserted in resistance at a point where these currents neutralize each other, the galvanometer will not be affected, and the Wheatstone bridge can then be used as though measuring a dead circuit. After measuring the insulation resistance, the fixed resistance of 100,000 ohms is deducted from the reading, and the result is the true insulation resistance of circuit to ground.

SWITCHBOARD.—This switchboard consisted of brown Tennessee marble fourteen feet long and six feet high, surmounted with Japan lacquered cap and a beautiful eight-day clock in grille work located in the center of the board. The front of the board was finished in the most perfect style of art, no detail being overlooked to make the board complete. The back of the board was not less beautiful than the front. All connections for the arc circuits ran up through the floor and were fastened to small porcelain insulator held to the board by means of small brass brackets, and thence ran to their respective springjacks. All the connections for the incandescent apparatus were made of heavy brass castings finished in Japan lacquer, to which all the circuits were connected beneath the floor. This board stood near the center of the space and was held rigid by means of large grilled bronze brackets, making on the whole a very perfect piece of work.

The apparatus in Machinery Hall devoted to the use of the Western Electric Company's exhibit in Electricity building, consisted of two dynamos, each operating 50 nominal 2000-c.p. arc lamps. It was agreed between the World's Columbian Exposition and the Western Electric Company, prior to the opening of the Fair, that the Exposition should furnish foundations for engines, dynamos, and free steam during the period of the Exposition, and that the Western Electric Company should provide engines, dynamos and operators. As a result of this agreement the Russell Engine Company, Massillon, Ohio, installed two 200-h.p. engines and one 100-h.p. engine for this service, and Graton & Knight Manufacturing Company, Worcester, Mass., supplied the belting. These generators and dynamos were directly connected to the switchboard.

STANDARD ELECTRIC COMPANY.

The main service plant of this company occupied Section "S" in Machinery Hall, and consisted of 20 5-light 2000-c.p. Standard Arc Dynamos. Sixteen of these dynamos were operated from a countershaft beneath the floor of the building and driven by two Russell engines. The remaining four dynamos were direct-belted to an Erie City compound engine running 275 r.p.m. The dynamos were arranged in two long rows extending the whole length of the Section. Each dynamo was provided with an ampere meter placed immediately opposite the dynamo and mounted upon a nickel-plated stand. In the center of this plant was a general switchboard, built of marble, of the combination plug type, having a capacity of 21 circuits and 21 dynamos. This board was of an improved design, occupying very small space and being easily understood and operated. On the back of the cabinet surrounding this switchboard was mounted lightning arresters for each circuit. All wires and leads running from dynamos to switchboards were placed beneath the floor, care being taken to properly insulate.

In Section "Q" were two Standard Arc Dynamos connected with one Ideal engine by means of an Ideal transmitter. One of these dynamos had a capacity of 50 2000-c.p. lights, the other being a 4000-volt machine. In this section was installed by this company a twelve-circuit, twelve dynamo marble plug switchboard of the same general type and character as the large one in the main service plant. This board was used by the operating department as an exchange board, connecting circuits and dynamos in different plants so as to provide facilities for furnishing current from one plant to circuits being operated from any other. The total number of Standard machines were operated on circuits containing Standard single and double service lamps, the principal circuits being Machinery Hall, Agricultural building, Forestry, Shoe & Leather buildings, the Grand Basin and other points. The record of the operating department shows a total output by the Standard service plant of twenty-five and a half million arc lamp hours, being the largest output made by any service arc lighting plant in the Exposition. The complete service run of this plant from May 1st to November 1st was made without the loss of an armature coil or the removal of any of the armatures from the machines. The service was maintained regular hours every day during the Exposition.

GENERAL ELECTRIC COMPANY.

The Power Plant of the General Electric Company, from which current was received to operate their numerous exhibits, the "Model Arc Plant" and Tower of Light in Electricity building, the electric fountains in the Court of Honor, and the various search lights, was located in Machinery Hall. This plant showed in active operation three types of generators manufactured by the company. There were four standard bi-polar Edison type machines, two of which were belt-driven from a high speed cross compound condensing engine built by the Ball Engine Company of Erie, Pa., and two from a simple engine of the Armington & Sims type. These generators were shunt wound and had a capacity of 150 kw. each at 140 volts and 450 r.p.m. They were beautifully finished in white with polished brass trimmings. The headboards were of polished white marble with polished brass and copper switches and connections. On the headboard of each was mounted a "three-wire" changing switch, for throwing the machine on either side of the three-wire system, on which they were operated in multiple. The plant ran twenty-four hours a day, and these generators were frequently overloaded as much as 25 per cent., which they carried with safety. In this space was also a 45-kw. generator of the General Electric type, with iron-clad armature direct connected to a Willans vertical engine, built by the M. C. Bullock Manufacturing Company. This dynamo was a six-pole machine, running at 460 r.p.m., overcompounded for 5 per cent.line loss, and gave 116 volts at full load. Its frame was of soft cast steel with a very high magnetic permeability, giving a very sensitive regulation and large output per pound of weight. It was on a circuit by itself, and was used for lighting the Power Plant Space and for operating the search lights of the General Electric Company on the Casino, Music Hall and Colonnade. The main feature of the Plant was the large direct-connected set, composed of a vertical triple-expansion condensing engine of the marine type, on each end of the shaft of which was mounted the armature of a General Electric Company's multi-polar generator. The set operated at 100 r.p.m. and the engine was particularly noteworthy from the fact that it was the first engine of this type ever constructed with a Corliss valve gear. The combination of Corliss valve gear, triple-expansion, and condensation made it possible for this engine to show an economy never before attained in an engine of its indicated horse power. The generators were twelve-pole machines, shunt wound, and had a capacity of 400 kw., each at 150 volts. The armatures were wound with flat, bare copper bars, carefully insulated from each other with mica. There was no commutator on these armatures, the radial part of the winding (of very heavy copper bars) serving that

Brush Electric Company's Power Plant.—Machinery Hall.

purpose, the brushes bearing directly on it. Standing beside this immense set was one of the smallest direct-connected generators made by the General Electric Company, a four-pole machine, with iron-clad armature of 4 kw. capacity, and connected direct to a Case engine running at 800 r.p.m. This generator had a small but beautiful polished marble switchboard mounted on a floor-stand near by.

The main switchboard of the Plant, and to which all the other generators were connected, was a good example of modern practice. It was of white marble with polished brass trimmings, and the regulating and controlling apparatus were of polished brass and copper. The board proper was made up of five panels. On the two end panels were mounted the ammeters, feeder switches, and comparative pressure indicators for measuring and controlling the service on the feeders and mains. On the other three panels were mounted the ammeters, voltmeters, main switches and field rheostats and switches, for measuring and controlling the output of each individual machine. The center panel had also a Weston Standard Voltmeter, and switch for throwing it in circuit with any dynamo, for comparative purposes. Above this center panel, on top of the board, was a clock in the form of a small multi-polar generator of the General Electric type, the head of the armature forming the dial. All connections to instruments were made from the rear, where all busbars and wires were carried, so that only the instruments proper showed on the face of the board. In front of the board, and forming a part of it, was the operating stand, of white marble, and surrounded on two sides by a polished brass railing. This stand was raised about two feet from the floor, its front and sides being also of white marble with polished brass trimmings. There were two switches on this board of especial interest, being double-pole, double-break, and having a capacity of 4000 amperes each. The blades were very heavy straight copper bars, mounted in a frame sliding between two guides and thrown in or out of contact with the terminals by a screw feed.

The Arc Lighting plant installed by the General Electric Company in Machinery Hall consisted of 27 dynamos of the Thomson-Houston type, each having a capacity of 50 2000-c.p. lamps. Of these seven were driven by a cross-compound Corliss, rated at 350 h.p.; six by a tandem-compound Corliss, rated at 350 h.p., and four by a simple Corliss, rated at 200 h.p., all manufactured by Lane & Bodley. The other ten were driven by an Atlas Machine Works tandem compound engine. These dynamos were of the standard T. H. type of the General Electric Company, having Gramme ring armatures, and finished in white and gold. The armature coils were separately removable, being wound in a form and afterwards slipped over the armature core, making the time required for replacing a coil very short should same be necessary for any reason. The switchboard to which these machines were connected, and from which the lamp circuits were run, was a safety switchboard of polished white marble with connections for forty machines and forty circuits, so arranged that any circuit might be thrown on any machine. The machine leads were connected to horizontal brass bars with metal bushings set about three inches back of horizontal bars. As there were no circuit connections on the face of the board it was impossible for the operator to accidentally put himself "in circuit." Above this switchboard proper were the panels on which were mounted the forty ammeters for the forty machines and the forty pairs of lightning arresters for the forty circuits. On separate marble panel boards, one on either side of the switchboard, were mounted the wall controllers which brought into action the regulators attached to the dynamos, by which the current was kept constant regardless of the number of lamps in operation at any time. By this arrangement the lamps were kept noiseless and of uniform candle power. The arrangement of the whole plant was such as to make it most easily handled, and the switchboard was a model of its kind.

FORT WAYNE ELECTRIC COMPANY.

All machinery in motion at the Exposition was driven by electric motors, which derived their power from generators situated in Machinery Hall. In the Fort Wayne block in Machinery Hall, among their arc machines, was situated a 160-h.p. generator, wound for 500-volts potential, driven by a tandem compound Buckeye engine of 175 h.p. Wires from this machine were led to a marbleized slate switchboard, on which were arranged a full complement of safety and regulating devices. From this switchboard the current was carried over to Electricity building, on a pair of 000 b. and s. wires, and was led to the distribution switchboard, also made of marbelized slate, in the Electricity building. On this switchboard were properly arranged controlling devices and safety appliances, including switches, reading instruments, etc., for the control of five dynamos, three of which were direct-current generators, the remaining two being the main motors which operated the exhibit. The wires, after they left the switchboard, were led to a 125-h.p. 500-volt "Wood" motor, belted in turn to a substantial line of shafting, manufactured expressly for this exhibit by the Falls River & Machine Company of Cuyahoga Falls, Ohio. On this shaft were mounted six other pulleys, driving a 1500-light "Wood" slow-speed iron-clad alternator, a Slattery 1200-light alternator, two 200-light "Wood" 110-volt incandescent dynamos, one 400-light "Wood" 110-volt incandescent dynamo, one 40-light 2000-c.p. "Wood" arc dynamo, and one of the new "Wood" 3000-light slow-speed iron-clad alternators, and, lastly, in case of lack of power in the larger motor, an

80-h.p. "Wood" motor was always on hand, being belted to the other extreme end of the line shaft. Among the stationary dynamos grouped around the aisles was exhibited a 120-light "Wood" arc dynamo, its voltage being, when loaded, 6550-volts; an 80-light and a 60-light dynamo followed. Spare armatures for each type of dynamo and motor employed in the exhibit were arranged in order on x-shaped yellow pine stands. A small direct connected set comprising a 3-h.p. motor, and a one-light arc dynamo, was in operation, furnishing current for a marine projector or a locomotive headlight, which latter also formed part of the exhibit. Small direct-current generators and motors, together with small arc machine and a spare alternator, were on another aisle. For service lighting and power in the building the company had an aggregate converter capacity of 2000 lights. These were arranged in batteries of 500 lights each, connected to separate bus-bars on the distribution switchboard. This switchboard was of the skeleton type, and had combinations for twelve machines and twelve circuits. The transformers were mounted on fireproof platform, thoroughly insulated and isolated. On a broad shelf-like construction around this transformer platform, were different instruments and apparatus manufactured by the company.

Lightning Arresters.—Two forms of lightning arresters were exhibited in connection with the operating plant, one for alternating and one for arc or any form of direct current. The alternating arrester was composed of laminated iron core, around which several turns of the main line circuit and current were carried; one connection from the main line was led inside between two slate discs, and about an inch away the ground connection was similarly disposed. Two gun-metal rollers were placed equally distant between these two points, the lightning thus having only a small distance to jump, the rollers presenting a new and clean surface after each discharge, and the magnetic field in the coils effectually blew any arc out. The arc lightning arrester lately designed by Mr. Wood is very simple and efficacious, the general principle being two binding posts, one on the top and one on the bottom, carrying discharge combs, and pivoted between them a substantial lever about six inches long, at the upper end of which, near the pivots, is secured an iron armature. Both ends of this lever are also fitted with discharge combs, placed in such a position that gaps of but one thirty-second of an inch are left between them. Across one of these gaps is shunted a pair of heavy spools of wire, fitted with projecting pole-pieces. The action of the arrester is as follows: provided that during a lightning discharge the dynamo current should follow over the gaps, a difference of potential, and consequently a magnetic field, due to such arcing, would naturally energize the said coils and attract the iron armature. The effect would be to raise the lever, and so interrupt the circuit, the field also in the magnets helping to blow out the arc.

THE MATHER ELECTRIC COMPANY.

This installation consisted of the following apparatus: Two compound wound electric power generators, capacity of 225 kw. each, and working at a potential of 550 volts, 450 revolutions per minute; two compound wound electric power generators, capacity of 120 kw. each, and working at a potential of 550 volts, 525 revolutions per minute. The 225 kw. generators were six-pole machines, with field magnets and armature supports cast in one piece without joints, except between the upper and lower half of the field. The winding of the armature is such that there are only two paths for the current through the armature, and yet opposite bars of the commutator are connected and so maintained always at the same potential. The commutators are cross-connected so that the opposite bars are maintained always at the same potential, so that it is possible to use either all the sets of brushes, or any two sets, as may be desired. The generators were mounted on sliding iron bases fitted with four screws, connected together in pairs by means of a chain, and operated by ratchets so that the generator could be kept in perfect alignment and the tension of the belt regulated. The base frames were firmly bolted to the foundation, which in this case were made of heavy timber and hard-burned brick laid in Portland cement. The four generators were connected to one set of bus-bars, so that all the machines could be operated in multiple when desired.

The switchboard to which these generators were connected were made of slate slabs set in a steel frame, making the switchboard self-contained. On the switchboard were mounted four triple-pole switches, through which the generators were connected to the bus-bars; four ammeters which recorded the quantity of current generated by each machine; one voltmeter, so connected by means of voltmeter switch that readings could be taken from either generator, or from the bus-bars by moving the switch-bar on the respective points; one differential indicator, which might be connected by means of indicator switch to any one of the generators when it was desired to throw one or another of the generators into circuit while one or more were in operation; and this instrument indicated correctly when the machine to be thrown in circuit attained the proper potential. There were four double-pole single-throw switches through which the circuits were connected to the bus-bars; four ammeters, which showed the quantity of current used on each circuit; and the circuits were protected by means of double-pole safety cut-outs, which carried a fusible strip of suitable capacity, and were further protected by means of the automatic adjustable circuit-breakers mounted on separate frames behind the switchboard. The switchboard was encased in a massive framework of pure Doric design, and finished in white enamel and gold. The bus-bars and all the connections between the switches and instruments were of the best quality of copper, of ample capacity, all polished and lacquered. These generators supplied the current for operating the motors in the Mines and Mining building, Agricultural building, the movable sidewalk on the pier, the Mather Electric Company's exhibit plant in the Electricity building, and other motors. The generators were all finished in white and gold, the whole plant presenting a handsome and imposing appearance. These generators were particularly adapted for street railway work, mining work and to operate motors for factory purposes; they regulate very closely and their efficiency is of the highest.

BRUSH ELECTRIC COMPANY.

This plant consisting of sixteen No. 8 65-light 2000-c.p. arc light dynamos and five Ball & Wood engines, occupied the space immediately east of the large Allis engine. The plant was the first arc plant installed and operated for sewer lighting on the floor of Machinery Hall, and the only arc plant in service twenty-four hours a day during the Exposition. In the months of March, April, and May, 1893 ,the entire arc lighting was done from this plant. In the rush of April and May, when nearly all the lighting was put in temporarily, this plant carried the load; when circuits were necessarily in very bad condition from the time of starting to the end of the Fair, not one moment was lost, continuous service being rendered throughout that time. The plant as erected showed a model high-speed plant, the dynamos being driven by means of belts, direct from engines, and two 150-h.p. tandem driving three 65-light dynamos each, the remaining four being driven by 1-200-h.p. cross-compound engine. Eleven hundred and twenty 2000-c.p. lamps were run from this plant at one time. The excellent service given the Exposition Company by this plant spoke well for its durability.

C. & C. ELECTRIC MOTOR COMPANY.

For supplying current to the power mains of the Exposition the C. & C. Electric Motor Company furnished to the authorities a plant consisting of four of their standard 80-kw. compound wound generators with switchboard for the same. These dynamos had each an output of 160 amperes at 500 volts, the total capacity of the plant being 640 amperes, or about 430 h.p. The compounding was designed for an increase of 10 per cent. in voltage at full load when the machines were driven up to their maximum speed. In addition to the plant just

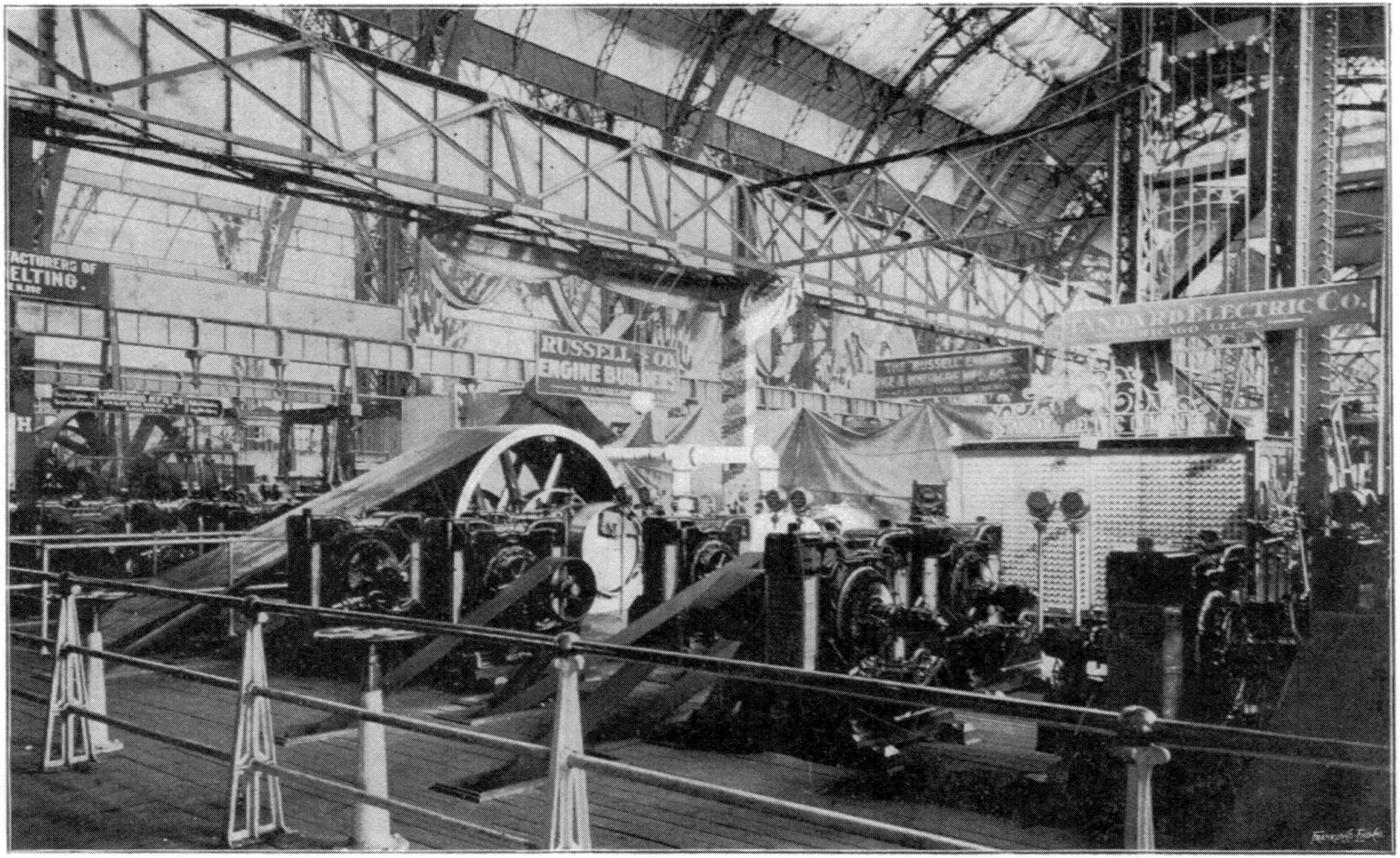

Standard Electric Company's Power Plant.—Machinery Hall.

described the C. & C. Electric Motor Company supplied the Exposition two of their standard 80-kw. dynamos, compound wound, for 250-volts. These machines were similar in every respect, except voltage, to those just described, and were also in operation in Machinery building supplying power circuits. In both of these installations the dynamos, switchboards, and apparatus pertaining to them, were furnished by the C. & C. Company, the setting up and wiring of the same being done by the Exposition authorities. It may be added that the four-dynamo plant has a record during the Exposition of 760 hours continuous run. The switchboard operated in connection with this plant was back of the dynamos and between the two engines driving the plant. There were three heavy copper bus-bars running across the back of the board, to which were connected the mains from the four generators through three-pole switches. The three bars were for the positive, negative and equalizer connections. It was thus possible to run any combination of the four machines which might be desired, the general practice being to run one machine until the load became too great, when a second was started, and so on. Each dynamo was connected to the main circuit through an ampere meter. There was also a voltmeter and multiple switch arranged to measure the pressure on any of the circuits or dynamos. On the bottom of the switchboard were the rheostats for changing the voltage of the generators. These rheostats were made entirely of slate, metal, and asbestos, being absolutely fire-proof. The switchboard had four external circuits carried through the buildings and grounds to supply power for various exhibits.

EDDY ELECTRIC MANUFACTURING COMPANY.

The company had a service plant in Machinery Hall consisting of four machines of 200-kw. capacity each. These generators were six-pole machines, of the standard Eddy type. A new slow-speed motor of 15-h.p. capacity was also exhibited. It is known as type "H." The magnet of this machine is of the Kennedy type, with two salient and two resultant poles. The material of the frame is soft cast steel. The armature has a toothed drum core, in the slots of which the coils, previously wound on forms, are placed. The core is thoroughly ventilated by perforations and air ducts. Pillow blocks support universal sleeve bearings at either side to give a self-alignment to the shaft, and the lubrication is performed by oil rings. End contact brush-holders are supplied for this machine, but two sets being necessary. This machine ran at 600 r.p.m. Driven by this motor was one of the dynamos which the Eddy Company manufactures for the deposition of metals. Three machines for this purpose were shown, designated as Nos. 0, 1, and 2. These machines have the Mather ring bi-polar magnet, hand wound, and a Siemens drum armature, wound and supported similarly to type "C" already described. The field winding consists of a number of wires in parallel, connected to a plug switch on the base of the machine in such a way that more or less of the parallel conductors may be cut into circuit as the resistance of the external circuit diminishes or increases, thus preventing the "cutting out" of the fields within the capacity of the machine. The driven machine of this type had a heavy copper wire short circuiting its terminals, which was maintained at a white heat by the heavy current traversing it.

SIEMENS & HALSKE COMPANY, BERLIN.

This exhibit furnished electric light and power in the following places, namely: German section, Electrical building, with circuits for power, arc and incandescent lights; Wooded Island, circuits for 125 incandescent lamps; Choral building (or Festival Hall) 3000 incandescent lamps; arc lamps in train yards of Illinois Central R. R.; arc and incandescent lamps in terminal station; the charging of storage batteries in the German section, Electrical building, and the operating of electric motors in German section, Machinery Hall. All this work was done on the Siemens & Halske 5-wire, direct, system—an entirely new feature in America. The dynamo which furnished the initial energy for this system was built for an output of 1400 amperes at 500 volts. It has an outside armature, revolving at the exceedingly low rate of speed of 100 revolutions per minute around stationary fields fastened to the bed plate of the engine, and by this arrangement the armature is mounted directly on engine shaft, thereby obtaining the desired peripheral speed with a small number of revolutions. The machine requires very little space compared with its output, and each part can be reached without any difficulty, and can be taken apart and remounted quickly. The machine is provided with an outboard bearing, which arrangement gives great stability to the shaft and avoids trouble with the bearings. From the arrangements of the field magnets and relative position of the armature, nearly all lines of force are cut; therefore, very little magnetism is wasted or dissipated. The magnet core is very short and thick; therefore, the magnetic resistance becomes almost nominal. The armature is made of solid copper bars, which render the apparatus strong and massive, and by the style of winding employed the potential difference between each bar on the commutator is reduced to a very small factor, thereby making it impossible to burn out. By turning off the outside winding of the armature and using it as a commutator, a special commutator, with all connections thereto, is not necessary, and, moreover, the resistance of the armature is reduced to the lowest degree; consequently the machine requires but very little regulation on varying loads, and gives a greater output than other machines that have special commutators. From an electrical and mechanical standpoint the construction and workmanship on this machine represents what might be called an ideal dynamo, and through its elegant appearance and novel construction it attracted a very considerable attention. The dynamo was coupled direct to a 1200-h.p. triple-expansion, jet-condensing engine, manufactured by F. Schichau, of Elbing, Prussia. The lamps upon Wooded Island were operated in series of four, and incandescent lamps in the buildings upon the 5-wire system, with equalizing motors at the centers of distribution. The feeders were of armored cable, laid direct in the ground. This plant was interesting as illustrating the difference between European and American practice, and was the only plant of its kind in practical operation in this country.

CHAPTER V.

DYNAMOS.

POWER, electrically transmitted, is largely responsible for the fact that the Columbian Exposition was essentially an electrical exhibition. By its use, energy was generated to supply the motive power for the machinery about the grounds and buildings, to light the Exposition, and to render possible the magnificent night illuminations that proved so attractive. Dynamos of all conceivable kinds were used and shown in the Electricity building, and in connection with transformers, currents were generated, ranging from 2 volts up to 1,000,000 volts, and these currents were used for almost every conceivable purpose.

Dynamos were broadly divided into two classes : those producing direct or continuous currents and those generating alternating currents. Late advances made in electrical science permit the use of either kind for the same purposes, but for the utilization of electric energy at any considerable distance from the source of power, the alternating system possesses advantages of flexibility that make its use imperative.

In the line of direct-current dynamos, various kinds were shown, their capacity ranging from ⅛ of 1 h.p. up to 2000 h.p., and of a voltage ranging from 2 to over 6250 volts. They were used for arc and incandescent lighting, for railway and other power purposes, for supplying current to telegraph and telephone lines, and for electro-plating. As a rule the machines used for arc lighting were of the direct-current type and of high voltage, the same depending on the number of lamps in circuit. Of these machines the smallest was made for 1 arc light and the largest for 125, the voltage ranging from 45 volts to 6250. Standard arc light machines were series wound and the regulation controlled by an adjustment of the brushes, dependent on the number of lights being used, from one up to the capacity of the machine. For incandescent lighting, compound wound dynamos were used; that is, the current generated in the armature is carried directly around the field coils when the resistance encountered in the circuit is slight, as is the case when many incandescent lights are being burned in multiple. When, however, the resistance is high a smaller shunt wire connected to the positive and negative terminals of the machine and carried around the field coils, equalizes the loss of current in the series winding. Such machines give a current of constant potential but of varying quantity and the voltage is usually 110.

For power work, other than running small machines such as fan motors, motors for dental and sewing machines, etc., compound wound dynamos of the direct-current type, of a voltage of 500, were in general use. As the work to be done in such cases requires a large amount of current, it is found advisable to use this higher voltage, as it makes a large saving in copper conductors. The dynamos of this type on exhibition ranged from 50 to 2000 h.p., and supplied power for various purposes throughout the grounds. The largest of this type of dynamo was used to supply power to the Intramural Railway, and many other large ones were installed in Machinery Hall in the electric station.

THE FIVE-WIRE SYSTEM was shown by the Siemens-Halske firm, and consisted of a generator delivering current of a constant potential at 440 volts. The five wires were connected in multiple series by four equalizing dynamos with one of the wires from each terminal of same. Any drop of potential on the circuit of any two contiguous wires was met by its corresponding equalizing dynamo and the potential brought up to standard; any increase of potential was met by the equalizer which in turn became a motor, and thus the potential on the entire system was constantly equalized.

The alternating machines on exhibition were of one, two, three or multi-phase current. The discovery by Henry of the principle of electrodynamic induction made it possible to generate currents in impulses or waves and by means of induction coils to transform them into currents of high or low tension. By means of alternating-current machines, currents of low tension and great quantity may be generated, transformed to high tension and small quantity, and sent long distances over very small wires, re-transformed into large quantity and low tension, and used for all the light and power purposes.

With single-phase dynamos and motors made to act synchronously with them, these currents were in some instances utilized for power purposes. With dynamos generating two-phase or multi-phase currents, that is, two or more alternating currents differing from each other by 180 or less degrees, motors with stationary armatures and rotating fields were

Thomson-Houston and Excelsior Plant.— Machinery Hall.

used which do not have to maintain a synchronism with the generating machines and which are self-excited.

The Tesla alternating machine shown does away with the revolution of the armature and uses instead a reciprocating motion, by means of which the armature is vibrated back and forth in the field coil by means of compressed air, and alternating currents are produced.

All the methods of driving dynamos were shown; the steam engine, direct connected, and belt driven, and water power direct connected and belt driven. Among the interesting exhibits in that line was the first direct-connected dynamo ever made, and another the largest ever made in this country, of 3000 h.p.

The history of the dynamo was well illustrated by historical apparatus. When Sturgeon in 1825 discovered the principle of the electro-magnet, and Henry and Faraday in 1831 simultaneously discovered the principles of electro-dynamic induction and magneto-electric induction, the world had in its possession all the principles necessary for modern dynamo practice. Magneto-electric machines, with permanent magnets, were early constructed, but were used mainly as experimental apparatus until in 1866 Siemens constructed the first machine with self-exciting electro-magnets and demonstrated that enough residual magnetism remained in the field coils to excite feeble currents in the armature, which still further excited the field coils, and thus reciprocally built up a current to the full capacity of the machine. Siemens predicted then that this action was reversible, or, in other words, he foreshadowed the modern electric motor. Gramme, at Vienna, in 1873 proved that one dynamo could be made to run another similar machine, and actually did furnish power enough to run a small force pump. Of the alternating-current machines the original apparatus of Galileo Ferraris showing the method of application of multi-phase current, and a rotating field was also shown.

The special features of the different dynamos exhibited, were in the direction of mechanical improvement, the principle of the machine being practically the same as the Siemens dynamo of 1866.

FORT WAYNE ELECTRIC COMPANY.

The Fort Wayne Electric Company, of Fort Wayne, Indiana, made the following exhibit of generators, Wood System.

Arc Dynamos.—These machines are made in capacities ranging from 1 to 125 lights, and constructed for any candle power. The machine has an armature of the Gramme type, which is best adapted for machines of very high potential, on account of the fact that adjacent sections have no extreme electro-motive force between them. The armature may also be repaired very easily, and should an accident happen to any individual coil, the same may be cut out and the commutator short circuited, so that its loss is not felt in the regular running of the machine. Ventilation of this armature is also perfect. The machine runs in gun metal bearings, with thread-like oil-ways cut in the same; the tendency of these spiral oil-ways being to continuously return the oil again to the center of the bearing. This insures perfect lubrication. The machine has four field magnets, with but two pole pieces; the whole machine being rigidly bolted together, renders it impossible for the shaft or bearings to get out of line, when once put together. The regulating device which automatically shifts the brushes to the proper point of commutation, is secured to the machine itself, making the wiring exceedingly simple, there being but two binding posts to attach the main line wires to. This regulator enables the machine to be run from a dead short circuit to its full load, without the slightest tension to the brushes, the power being in exact proportion to the number of lights burned. It is constructed in such a manner as to allow of compensation in every direction, for wear. This regulator is designed to have practically no inertia, and therefore instantly comes to rest under any and all changes of load. There is also an attachment by which the regulator can be thrown out of action, but is automatically thrown into action, should the external circuit in any way be changed, or the speed of the machine increased. The fact that the regulator derives its action from a spur-wheel on the end of the shaft, connected with two friction rollers, and connected with a long lever, which is influenced by a series coil on the top, and a permanent steel spring on the bottom, renders the regulator absolutely reliable, instantaneous, and above all things, automatic.

INCANDESCENT DIRECT CURRENT DYNAMOS.—These machines are furnished in sizes from 10 lights to 1600 lights, and are compound-wound, rendering them absolutely automatic, and obviating the necessity of touching the brushes under any change of load. Power generators and motors for any voltage and horse power, are made in this form. A description under one heading will suffice.

The following features are worthy of special merit: The magnetic circuit is exceedingly short, having but two joints in its entire length, and consequently very low magnetic resistance, requiring a minimum amount of copper, as well as current, for its excitation. The current necessary to maintain the full field being between one and two per cent of the out-put of the machine. Mechanical construction of this machine leaves nothing to be desired, as it possesses uniqueness and simplicity of design, with the greatest attainable mechanical strength. It is so constructed that the armature may be easily and safely removed, in case of necessity. The boxes are fitted with expansion bronze bushings, which make it impossible for the bearings to stick. They are also self-oiling, and self-aligning, holding sufficient oil to run for several months, without any addition to the original supply. These machines are mounted on an insulating sliding base, which moves on anti-friction rollers, making it possible to move the heaviest dynamo against the tension of the belt with the least possible exertion. The binding posts, which are on the lower front portion of the dynamo, are amply large for the currents to be delivered. They are fitted with multiple fuses for each end of the circuit, as well as equalizing connections, intended for use when more than one dynamo is connected in multiple.

They are wound for 110, 220 and 500 volts, the 220 and 500-volt generators being fitted with carbon brush holders, a very desirable feature when generators are used for the transmission of power, especially where the load is subject to large and constant variations.

Alternating Current Incandescent Dynamos.—These machines are made in sizes ranging as follows: 750, 1500, 3000 and 6000 lights respectively, voltage being as desired either 1040 or 2080 volts. This machine is a most interesting one, on account of the radical departure made in its construction and design. In constructing this dynamo, the chief aim has been to reduce the amount of material employed to a minimum, keeping the speed, of course, within reasonable limits. The wiring of the armature is as nearly perfect as possible, all copper being encased in solid mica conduits one-sixteenth of an inch in thickness. These ducts are set well into the iron, and are thus protected from injury from any cause. In building up this armature a wide departure has been made from the beaten track. In all other armatures solid stampings are used, and in large machines the expense thus entailed may easily be seen. In the armature in question, however, small c-shaped stampings are used, giving great ventilating properties, and at the same time allowing scrap-iron to be employed to advantage. By using this type of armature, three points are gained: economy in construction, enormous surface, copper necessary to generate the requisite current reduced to such an extent that one volt is produced for from three and a quarter to three and six-tenths inches of actual copper conductor a result never attained in previous dynamos. These machines may be over-compounded as high as 25 per cent, to compensate for losses in long feeders. The general construction of the machine is as follows: The bed-plate is cast in one solid piece; pedestals carrying the bearings are massive, and well able to carry the strain imposed upon them. The bearings proper, which are of bronze, are mounted in flexible iron sockets, giving elasticity, and rendering the boxes free to expand in case of heating. A 1500-light and a 3000-light, machine of this type, formed part of the service lighting in the Electricity building. A few figures given will suffice to show the economy in the new machine, over machines built previous to this. The new Wood Alternator runs 350 revolutions slower, and has 1000 pounds less copper wire on it and complete weighs 650 pounds less. The insulation on each coil is carefully tested to 5000 volts between the coil and the core, before the armature is placed in the machine for final test. The iron-clad principle of the armature does away with all bands and other external fastenings, which always have a tendency to become loose or burst, besides being a constant source of danger from sparking through. The style of laminating the armature also, gives enormous radiating surface, which in this machine is equal to about forty times the actual surface of the armature, and as the armature revolves, powerful draughts of air are created, and forced through the sleeve-like periphery of the armature, rendering it almost impossible for the wire to become overheated, even though it be run on an overload for a considerable time. The connections of the armature are seen to be exceedingly simple. In this machine the coils are connected in multiple series, one end of the coils being connected to one end of the commutator segments the other commutator segment and the other ring being connected together. The machine is compound wound, the series excitation being due to current furnished from a small separate exciter, which passes around all field magnets in series. The main line current is then rectified, and passed around two halves of the field magnet, which are connected in series multiple. By this method the voltage of the machine is automatically raised as the drop in the line increases. Each field magnet is wound with four layers of wire, the first two layers being supplied by current from the exciter, and the two outside layers by the commuted alternating current. The small exciter above referred to is also of the slow-speed type, with iron-clad armature, which gives practically constant potential, without the necessity of compounding on all changes of load. It also has a maximum capacity of double the current required to excite the alternator, which it accompanies, thus giving considerable working margin. The construction of the brush yoke on this dynamo is also a new departure, the yoke being made somewhat less than half circles, and placed below the center of the shaft, and adjusted from an independent pedestal. The advantage of this arrangement, is that it permits the removal of the armature without disconnecting the yokes, brush holders or connections. The field magnets are constructed with slotted pole pieces; in order to prevent Eddy or Faucault currents, these are well ventilated by the currents of air blown across their surfaces by the armature. The field shunt, or non-inductive resistance, is used to regulate the over-compounding, and also to prevent sparking. This piece of apparatus is hung inside the pedestal where it is removed from all danger of destruction of alteration, and is ventilated by a force draught from the armature, which is carried in ducts cast in the main frame of the machine. This pedestal has a perforated removable door, from which the shunt may be inspected from time to time. A sliding base also accompanies these alternators, which not only enables the attendant to removed the dynamo, so as to properly adjust the belt, but at the same time, move the exciter, so that the tension of the exciter belt is not altered by changes of position of the alternator. This is a very convenient improvement, as it enables this part of the work to be done quickly, and by one man. On the commutator two sets of brushes are used, each side in multiple, so that one set can be removed for trimming while the machine is in operation. On the collecting rings of the large size alternators two sets of brushes are used in tandem, so as to give increased carrying capacity, and also allow of the removal of a brush which may be necessary from any cause.

THE BRUSH ELECTRIC COMPANY.

A full display of standard sizes of Brush arc dynamos was made from the smallest one-light to the latest arc dynamo made by the company. The new No. 9 arc dynamo is the largest direct current arc machine ever made, its capacity at 525 revolutions being 6250 volts at 9.6 amperes. The regular Brush armature with its good ventilation, its separate bobbins, good mechanical construction and improved commutator, makes this four-pole arc machine an admirable piece of mechanism. The company showed this dynamo direct coupled to a Willans engine, the output of the set at 460 revolutions being 5200 volts at 9.6 amperes. The small space taken up (4x11 feet) and the combined efficiency from boiler to the lamps about 75 per cent. make this outfit most desirable. The incandescent direct current dynamos made by the Brush Company, show careful designing and the best form of construction, the larger machines having but one turn of wire for each bobbin. A recent improvement in the armature of these machines gives a perfect running commutator, there being practically no lead to the brushes under all varying loads. The alternating current dynamos made by the company are distinctly "Brush." The armature is stationary, the fields revolve. Both armature and fields are built up in sections, giving flexibility. A 3000-light 2000-volt machine was shown coupled direct to a motor; the armature of the alternator was connected for 110 volts, and the machine operated incandescent lamps direct without the use of a transformer. Five hundred and one thousand-volt direct current power generators formed an interesting part of the exhibit. The construction of Brush generators is such that the higher

Ft. Wayne Central Station Exhibit.—Electricity Building.

voltages can be reached easily, the commutation of the machine being perfect. Slow speed (100 to 200 revolutions) motors for crane work, 250-h.p. 220-volt motors and steel-clad running motors, 15x18 inches, 9-h.p., were shown in operation.

THE GENERAL ELECTRIC COMPANY.

This company exhibited a full line of generators for all purposes, both in the Electricity building and at the power plant. In the Electricity building, prominently located, was the model arc light plant and exhibit designed to show in actual operation all the standard types of arc generators and lamps. Near the main isle was a pit about 16 feet square and 6 feet deep, in which stood a 220-volt 200-h.p. Edison type motor. In conjunction with the motor was a starting switch-board of slate, mounted on a floor stand and contained two rheostats, one of which was for starting the motor and the other for controlling its speed under load. A little above these were the main switches, with ammeter between them. Two pilot lamps of 110 volts each, mounted the whole. Two bare copper rods, arching gracefully, carried the current to the motor, whose armature shaft was direct connected to the line shafting extending beneath the floor across the entire space. To this shaft were connected, by means of Webster Steel Plate Friction Clutches, eight Thomson-Houston type arc dynamos. Close to each machine was a shifter, mechanically connected to each clutch, by which the dynamo tender could throw on or off any machine desired. This plant served not only as an exhibit, but also furnished current to 105 lamps of 2000 c.p. each, wired up for service lighting in the building, in addition to those operated for other exhibitors. The isolated lighting exhibit was a most notable one. Its most noticeable feature was the directly connected sets of engines and generators, of which three types were shown. One set consisted of a 45-k.w. 110-volt generator directly connected to a Williams Vertical Compound Engine, run at 460 revolutions per minute. There was also a 75-k.w. generator giving 110 volts at 275 revolutions per minute, and directly connected to a vertical compound engine built by the Lake Erie Engineering Works. Perhaps the most interesting of the directly connected sets was the Edison Vertical Triple Expansion Engine, driving two 25-k.w., 125-volt generators, an armature being on each end of the shaft. All these directly connected generators were of the multipolar type, with iron-clad ring armatures, drum wound. Generators of the Edison bi-polar type were shown in capacities ranging from ½ k.w. to 45 k.w. capacity, and near by were the Thomson-Houston Bi-polar type, ranging in capacity from 2 k.w. to 30 k.w. There were also some generators of the Thomson-Houston Spherical Armature type shown. On one of the spaces was a system of four dynamos, all driven by one motor and countershaft, which supplied current for the World's Fair Quadruplex Telegraphic service of the Western Union and Postal Telegraph Companies. The switchboard arrangement enabled the companies to take off either positive or negative potentials with a range from zero up to 350 volts. The special machines for various purposes, mining and railway dynamos, etc., are described under their special headings. For alternating incandescent lighting were shown belt-driven generators of 30 k.w., 60 k.w., 120 k.w. and 300-k.w. capacity, and a direct connected generator of 225-k.w. capacity at 180 revolutions per minute. These machines are separately excited, and except in the case of the direct connected generator, composite wound. The 225-k.w. dynamo was connected direct to a Mackintosh, Seymour & Co. Tandem Compound Condensing Engine. Instead of a composite field winding, a potential regulator on the switchboard was used to automatically keep the voltage at the right point under varying loads.

THE EXCELSIOR ELECTRIC COMPANY.

This company showed a good line of generators. One of the machines was a thirty arc light dynamo with case opened so as to show the armature and bearings.

THE NEW HOCHHAUSEN ARC LIGHT MACHINE.—Beginning with the magnetic frame of this machine, it consists of two field cores connected by a cast-iron yoke at one end, and provided with pole pieces at the other, which encircle the armature on three sides. The armature is so placed that its shaft rests in bearings which are on the interior of the machine, so that the armature as exhibited can be entirely exposed. This is facilitated by the pole pieces, which are hinged and which can be turned out so as to expose the armature completely and allow it to be drawn out a sufficient distance for thorough repair, if necessary, without removing the shaft from the journal bearings.

The armature core is built up of iron wire, insulated by paper, the wire being wound on a cast-iron frame or skeleton, having a T-section which divides the core into two parts. The arms of the spider, which hold the armature, are insulated from the core and frame, and fit into notches which are cut tapering from both sides towards the center, so as to keep the whole concentric.

The coils of the armature, which are rectangular in section, are wound automatically upon the core by an ingenious machine devised by Mr. Hochhausen for that purpose. When placed upon the armature core the coils are separated on the outer circumference by wooden wedges, which are secured to the cast-iron skeleton of the armature, and they are held in position by sections of fiber band which are screwed into the wooden wedges just referred to.

The regulation of the machine consists both in the shifting of the brushes in conjunction with a regulating resistance, both of which are simultaneously operated by an auxiliary motor. This motor, which is entirely hidden from view, is situated in the hub-like projection bearing on the arms which span the two pole pieces beside the end of the armature shaft, and which might be mistaken, except on closer scrutiny, for the bearing of the shaft. This small motor is fixed in a magnetic field, which is produced by attaching the cast-iron arms shown, to each pole piece and bringing them together so as to surround the armature of the small motor. It might be thought for an instant that such a bridging of the magnetic circuit would take a large number of the magnetic lines of force away from the armature, but this has been provided for by mounting the cast-iron arms, not directly on the pole pieces but by separating the two iron surfaces by a good thickness of hard rubber, so as to make a considerable break in the magnetic circuit. The magnetism passing through the arms, therefore, is very weak, but nevertheless sufficiently strong to produce a field for the small regulating motor, which acts with the greatest promptness. The regulation of the machine is effected in a very ingenious manner by means of a wall regulator which serves as a controller for the regulating motor. The second size of this machine constructed is designed for 50 lights of 2000-c.p. each, and has 90 pounds of wire on the armature and 970 on the field. The third size, designed for 30 lights, has 55 pounds of wire on the armature, and 400 pounds on the field. In general appearance the machine is exceedingly compact, and the open construction adopted allows of ready access to every part for examination and repair, if necessary. The machines have already been placed in several stations where their operation has been attended with marked success. There were also in the exhibit plating dynamos of various designs and sizes; also machines for mining work, pumping, etc.

THE WESTINGHOUSE ELECTRIC & MANUFACTURING COMPANY.

The company exhibited a full line of generators for general and special work, both in the power plant (see heading "Power Plant") and in the Electricity building. There were two-phase alternators of several styles and various sizes. One class of these machines was exhibited in power plant in Machinery Hall and was used for supplying the lighting circuits of the Exposition. Another type of this class of generators, together with Tesla motors and revolving transformers, was exhibited in Electricity building. The two-phase alternating generator was a six pole machine making 600 revolutions per minute. The frequency employed was 30 periods per second, or 3600 alternations per minute. The machine was of the multipolar type and similar in general appearance to the railway generators manufactured by this company, except that in addition to the commutator there were four collector rings provided, carried upon the opposite ends of the shaft. Pole pieces were built up of mild steel .015 inches thick and cast into an iron yoke. Losses in the field due to reaction of the armature are thus minimized and great rigidity is secured by a relatively large section in the cast iron yoke or ring. The bearings were self-aligning and self-oiling. The coils were thoroughly insulated and protected against mechanical injury—the wires, enclosed in slots in the armature core, were insulated by the use of paper tubes. No band wires were used on the armature, but the winding was maintained in place by the form of the slot in which the wires were placed. Sectional winding of the armature, allowing a burned out coil to be removed and a new one inserted without rewinding the armature. In the design of this machine, mechanical excellence, general reliability and ease of repair, have been considered of first importance, at the same time efficiency was not sacrificed, and the iron and copper losses have been made such as give the best results. Two methods could be used to excite the field, one by means of a separate exciter; and the other by the rectified current from the armature. This two-phase generator was directly connected to a Pelton water wheel, which assisted the non-synchronous motor to drive the machine. The two-phase current, after being conducted from the generator to the receiving station, as above described, was conveyed to the ring collectors of a 500-h.p. rotary transformer. This machine was identical with the one just described as a two-phase alternating current generator. The two-phase current was also used to operate a 60-h.p. two-phase motor of the synchronous type. The advantages of both these motors, which operated synchronously after speed was attained, are as follows: Simplicity and solidity of construction—few parts—the bearings and lower half of the field constituting a single casting, which insures great rigidity. The circular form of field secures maximum strength for given weight of material. Mechanical, magnetic and electrical symmetry: bearings are self-aligning and self-oiling; use of laminated steel fields in cast iron yoke or ring. This permits use of a very powerful magnetic field, and minimizes losses in field poles due to reaction of armature. The high induction also means large output for a machine of given dimensions, reduction of sparking. Mechanical strength and simplicity of armature; wide separation of conductors differing greatly in potential; mechanical excellence of armature winding.

Conductors are rigidly held and each conductor transmits strains due to magnetic drag directly to the iron core. The conductors are mechanically protected by the iron core. Conductors are separately and very highly insulated. This machine may be used as a generator of two-phase alternating currents, as a direct current generator, as a commutating machine, receiving, alternating and delivering direct currents, or as a commutating machine, receiving continuous and delivering alternating currents. It should be noted that the capacity of the machines when used for commutating currents is about 60 per cent. greater than when used as a generator or a motor. The 500-h.p. rotary transformer, as above stated, was belted to an arc alternator which supplied twenty-five lamps located at the south end of the space. Here the wonderful regulating powers of the arc machine were illustrated, as the full load was thrown instantly on and off by the closing of a switch and the current was maintained constantly at ten amperes, no matter what the load, as indicated by an ammeter. The Worthington pump was driven by the large rotary transformer or motor, and served to supply the Pelton water-wheel directly connected to the two-phase generator on the north side of the space. Thus it is seen that a part of the electrical energy generated in this machine was sent through transformers over a transmission circuit, through transformers again and into the large rotary transformer. Here its form was changed and by means of the belt it operated the Worthington pump and through pipes was conducted back to the other side of the exhibit where it operated a Pelton water-wheel, thus being returned to the generator from which it originated. One of the unique exhibits made by the Westinghouse Company was a model of dynamo and steam engine of almost microscopic size. The dynamo stood 5-16 inch in height, and yet was so perfect in every detail as to be capable of operation. The engine matched the dynamo and was run by compressed air.

THE MATHER ELECTRIC COMPANY.

The apparatus exhibited consisted of six Mather Patent Ring Type constant potential, direct current, compound wound dynamos, each of capacity as follows: 55-kw., 30-kw., 17-kw., 1c.5-kw., 7-kw. and 3.5-kw., all wound for a potential of 125 volts, and complete with bases and rheostats. The 55-kw. dynamo, and the 10-kw. dynamo were in operation The 55-kw. dynamo furnished the current for all the incandescent lights used for lighting or decorative purposes in the exhibits at a potential of 110 volts; speed, 750 revolutions; the 7-kw. dynamo was connected with the lighting circuits in the exhibits, so as to supply a portion of the incandescent lamps used for lighting purposes when the large dynamo was not required; one multipolar slow-speed compound wound, constant poten-

Westinghouse 15000 Light Alternator.—Belt Driven.

tial dynamo, capacity 17-kw., and wound for a potential of 125 volts, complete with base and rheostat; one multi-polar compound wound, constant potential, electric power generator, capacity 125-kw., and wound for a potential of 550 volts, at a speed of 525 revolutions per minute; these generators had a commercial efficiency of more than 90 per cent. and regulated absolutely from full load to no load; the generator was set up on an iron base, fitted with four screws, connected together in pairs, by means of a chain and worked by ratchets, so as to control the alignment of the machine and the tension of the belt. These machines were provided with a rheostat of suitable capacity, and had also a commutator-turning device, adjustable on the base of the machine; one 225-kw. ventilated armature for electric power generators, of latest improved style was shown. This armature has a slotted core and is built of copper bars, and the winding is such that there are only two paths for the current through the armature, and yet opposite bars of the commutator are connected and so maintained, always exactly at the same potential. The commutator is cross-connected, so that the opposite bars are maintained always at the same potential, so that it is possible to use either all the sets of brushes or any two sets as may be desired. The generators for this armature are built with six poles and carry six sets of brushes; one 120-kw. armature for electric power generators. This armature has a slotted core, and is wound with wire on the same principle and lines as the armature above described. The commutator is connected as described above; one 25-kw. armature for Mather ring-type lighting dynamo; one set of field coils for 225-kw. electric power generator, 550 volts; one set of field coils for 120-kw. electric power generator, 550 volts.

THE MATHER ELECTRIC COMPANY'S PLANT AT THE LIBBEY GLASS WORKS, MIDWAY PLAISANCE.—This plant consisted of the following apparatus and material: Three Mather patent ring-type constant potential, direct-current, compound wound dynamos with a capacity of 55-kw. each, and wound for a potential of 125 volts. Each dynamo was set up on a cast iron base, fitted with a belt-tightener attachment operated by a ratchet. The three dynamos were connected in multiple, so that all the dynamos or either one or two might be operated together. The bus bars and instruments to which the dynamo were connected were mounted on a switchboard made of slate slabs, set up in a steel frame, making the switchboard self-contained On the switchboard were mounted three rheostats, connected with the fields of the dynamos; three double-pole Mather switches, through which the dynamos were connected to the bus bars; three magnetic Vane Mather ammeters, which indicated the quantity of current generated by each dynamo; one voltmeter, which was so connected by means of voltmeter switch that readings could be taken from either dynamo, from the bus-bars or from a pressure wire by placing the switch-bar on the respective points; one differential indicator, which might be connected by means of indicator switch to any one of the dynamos when it was desired to put one or another of the machines into circuit while one or more are in operation. This instrument indicated correctly when the machine to be cut in circuit had attained the required potential; four double-pole S. T. Mather switches through which the circuits were connected to the bus-bars. The circuits were protected by means of double-pole safety cut-outs, mounted in four sets of three circuits each. All the connections between the switches and instruments and the bus-bars were made of the best grade of copper of ample capacity, and all mounted on the bank of the board. These dynamos supplied current for all the incandescent and arc lamps, and also for the electric motors used in the works.

THE EDDY ELECTRIC MANUFACTURING COMPANY.

This company manufactures constant potential, direct current motors, generators for lighting and power, and dynamos for the electrical deposition of metals. Machines of various types for all purposes constituted the exhibit, and several were shown in active operation. Type "C," a style of machine used by this company, consists of a two-pole cast-iron magnet of the "horseshoe" type, with a Siemens wound drum armature. The yoke and limbs of the magnet are separable and have machine-wound exciting coils slipped on the latter. The armature revolves in yokes which are supported by studs screwed into the poles of the magnet. Oil rings revolving with the shaft supply the lubrication from a reservoir in the bottom of the yoke. There were two of these generators shown as follows: One 5-h.p. 110-volt compound generator; one 25-h.p. 110-volt compound generator. The 50-h.p. motor was bolted to a 40-kw. 110-volt multi-polar generator, which operated incandescent lamps throughout the exhibit. This generator is of the company's type "E." The frame is of cast-iron, octagonal in shape, with four salient poles. The exciting coils are machine-wound on spools and have separate compartments for the series and shunt windings. The armature is of the four-pole drum type, wire wound on a smooth core. The bearings rest on pillow blocks, which are bolted to projections of the frame casting. The lubrication is obtained by oil rings as with type "C" above described. Four sets of tangent brushes collect the current from the generator. Their type "C" compound-wound dynamos are especially adapted to isolated lighting plants, having a very powerful field and few turns on the armature; this type of dynamo gives the highest possible efficiency. The dynamos manufactured by the company for the deposition of metal are capable of depositing forty ounces of silver per hour.

THE C. & C. ELECTRIC MOTOR COMPANY.

The transmission of power by an electric current at a relatively high pressure and its reduction for use to a lower pressure was the demonstration in this exhibit. Beginning with the dynamo, as the source of electricity, the following is a brief description of the machinery of the company contained in the exhibit: The dynamo is bi-polar, having circular consequent pole field magnets and drum armature. The commutator has heavy, cast copper bars with lugs, to which armature connectors are screwed. The bearings are self-oiling and self-aligning, and are provided with sight oil-gauges and means for preventing oil from being carried along the shaft. The bed is of cast-iron, being of turtle-back form with stiffening ribs underneath, a form of construction giving great strength and rigidity with minimum weight. The bed rests on the usual slide rails with belt-tightening bolts and the rails are bolted to a solid wood base framework. Each dynamo is provided with a terminal board bolted to the top pole piece. On the face of the terminal board are the armature, series coil, shunt coil, equalizer, and main circuit terminal connections in plain view, where they can be easily inspected and changes made for special cases in installation. For example, when compound dynamos are run in the reverse direction it is necessary to reverse the series coil connections; this can be done in a few minutes with these generators. The terminal board is also provided with a main circuit switch, which opens the external circuit at the dynamo. Each dynamo is provided with a fire-proof hand regulator for adjusting the voltage to the required pressure. In this exhibit the dynamo was set on the face of the main switchboard and formed an integral part of the same. On all machines where carbon brushes can be used—viz., all except the very heavy current dynamos of low voltage—the new reaction brush holder of the C. & C. Electric Motor Company is used. In this holder the brush is bevelled at both ends and rests on an inclined plate so that it bears on the commutator at an angle with the commutation diameter. A curved arm, actuated by an adjustable spring, bears on the free end of the brush. The action is such that which ever way the armature runs, viz., to the right or left hand, the brush is forced into contact with the commutator by both the spring and the tangential pressure of the commutator on the brush. With this holder various widths and thicknesses of brush may be used in the same holder, and the removal and replacing of a brush takes but a moment. The pressure on the commutator is also very slight and the consequent heating due to friction greatly reduced.

STANDARD ELECTRIC COMPANY.

There were in this exhibit lighting dynamos, lamps, switchboards and accessories. There were on exhibition five series arc dynamos varying from 20 to 50 arc lamps capacity each, and of 1200, 1600 and 2000-cp. The dynamo exhibit was operated by means of a 40-h.p. 500-volt motor, which received current from a generator located in Machinery Hall. The exhibit was so arranged that all of the dynamos could be operated at one time by the motor. A comprehensive showing was made of the methods of constructing and winding the armatures used in these dynamos, one armature of 30-light capacity being shown in an uncompleted condition, and a 50-light armature of modern type, complete, illustrated the revolving part of the dynamo. In explaining in a cursory way the method of the construction of the Standard dynamos and the advantages of the Standard system in its entirety as claimed, it may be stated that the dynamo in design presented some special features in which the form of the well known Manchester type of field has been carried out. The pole pieces which form the main portion of the frame are of special cast iron, and the field magnet cores as employed are of wrought iron. These pole pieces with their cores, which contain the magnetic circuits of the apparatus, are insulated from the base by a piece of brass, which is utilized to reduce the magnetic leakage to a minimum and give uniformity of lap to the brushes on either side of the commutator. The heat developed in these machines never exceeds a rise of more than 60 deg Fahr. above the surrounding atmosphere. This type of field magnet permits a very intense magnetic field with the use of a limited amount of large size field wire. The ohmic resistance, therefore, is very low. The armature employed is of the ring or Gramme type, and has a large open interior which affords better ventilation and radiation of heat units developed than is possible in any other type of machine. In the Standard armature and dynamo a large amount of iron is used, thereby reducing the quantity of wire required to produce the given output. The Standard armature is of the closed coil type, and contains a large number of sections or segments, thus reducing the p.d. between the neighboring coils or wires. Only one set of brushes is necessary, the electrical proportions of the machine being such as to maintain a constant current through all changes of load without the necessity of adding differential brushes or other devices. The neutral field of the dynamo is quite extensive and permits a wide range of the brushes on the commutator without perceptible sparking. The regulator of this dynamo is electro-mechanical in its nature and arrangement. It permits the entire load being thrown on or off the machine without subjecting it to any undue strain, the action of the regulator being quick and positive, thus insuring the safety of the armature even in event of an absolute short circuit. The switching device on the terminal board of the dynamo is of such construction that the full load of the machine may be switched on or off without any sparking or kindred trouble at the brushes. This switch was devised particularly for use on closed coil machines of high voltage. The mechanical construction of the dynamo is such as to permit the removal of the armature in a very few minutes. By removing one of the journal supports at either end of the shaft, the armature may be removed or replaced safely and quickly. The journals are made self-oiling, self-centering and self-aligning.

THE JENNEY ELECTRIC MOTOR COMPANY.

This company made a very comprehensive exhibit covering a complete transmission and service station. They used current from a 40-kw. 500-volt compound-wound generator in Machinery Hall, for the 35-h.p. and 25-h.p. motors in the exhibit, which were coupled direct to a 30-kw. and 20 kw. 110-volt, compound-wound generator, from which current was supplied to two sets of bus wires on the switchboard, and there, by means of double-throw switches, was supplied to arc and incandescent lights, and motors; 5-h.p. also being furnished a neighboring exhibitor. Current was also taken for the French Beacon Light on Electricity building. Their dynamo exhibit inclnded a 40-kw. 500-volt generator, 30-kw., 20-kw., and 3-kw. 110-volt generators and a 300-ampere 5-volt electric plater. In addition to the 35-h.p. and 25-h.p. 500-volt motors, their machines run from the 110-volt generator, one each of 12-h.p., 6-h.p., 1½h.p. and ½ h.p. motors. The 6-h.p. motors operated a 24-ampere dynamo, and the ½-h.p. motor coupled direct to a flexible shaft was used in buffing, drilling, and general shop work.

WESTERN ELECTRIC COMPANY.

The exhibit of dynamos manufactured by the Western Electric Company was in the extreme southern end of the exhibit, and was as follows: One 60-light arc machine, one 50-light, one 40-light, one 30-light, and one 20-light. The incandescent dynamos consisted of the following sizes: 900 amperes, 650, 550, 400, 300, 230, 165, 100, 50 and 30. All of these were wound to 110 volts. For full description of dynamos see "Power Plant."

THE SIEMENS-HALSKE COMPANY.

During the past two or three years there has been a strong tendency towards the direct combination of dynamos and engine on one shaft and foundation. This has been noticeable especially in the heavier machines of the railway power house, and it has grown in favor with central stations

SIEMENS BROTHERS DYNAMO AND WILLANS VERTICAL ENGINE.

and isolated plants. As early as 1887 the Siemens & Halske Company commenced the equipment of the Berlin central station with direct connected dynamos and compound engines, and this station has grown until now there are twenty of these direct connected machines of the same size and type as the one shown in this Exposition. On the five wire systems, as shown here, stations have been constructed or are under contract in a number of cities. It will be seen that up to the present date European engineers have given more attention to this class of apparatus than those of our own country. The exhibit of Siemens & Halske Electric Company was located almost in the center of Machinery Hall and consisted of a large vertical engine, a dynamo connected directly to the end of the crank shaft, and the equalizer for distributing the current to the five wires used in the system of conductors. The engine built by F. Schichau, Elbing, Prussia, is of the vertical triple-expansion type, running 100 revolutions per minute and rated at 1000 to 1200 h.p. The high-pressure cylinder is 22.86 inches diameter, the intermediate 37.4 inches, and the low pressure 57.08 inches; all having a stroke of 27.56 inches. The valves are balanced pistons with the cut-off valve surrounding the inner piston, which is called the expansion valve, and being controlled by the governor regulates the speed of the engine. The control of speed is very accurate, and it is said to vary but one to two per cent. from no load to full load. The governor is one of the ordinary centrifugal type, which through a simple system of levers turns the inner piston valve on its axis, the stem of the valve extending through the upper end of the steam chest for that purpose. The ports in both valves are diagonal slots, somewhat after the style of the old Ryder cut-off, and the turning of the inner valve changes the relative position of one opening to the other very rapidly, and this cuts off the steam earlier or later in the stroke as the case may require, in order to lower or raise the rate of revolutions. This valve is said to admit steam for seven-eighths of the full stroke when necessary. The framing of the engine is light and graceful, but is so braced as to make the most of every pound of metal contained in it. For the capacity of the engine it takes but little space, about 9 by 18 feet, and runs under full load without the slightest noise or jar. The lubrication, as in all machines of this type, is accomplished by means of small tubes leading to each and every joint from oil tanks on the rear of the cylinders. All stuffing boxes, bolts, nuts, and in fact every part that could in any way work loose, have been secured in some special manner, advantage having been taken of the firm's very extensive marine practice and experience on these points. The dynamo known as type "I" is shunt wound and is of the regular external armature type made by Siemens & Halske, having ten inside pole pieces, ten brush holders, about 1700 windings on the armature, which also serves as a commutator, the external surface of the drum having been turned true for that purpose. The method of taking the current directly fròm the armature in this manner is said to increase the output. At 100 revolutions the total output is 1400 amperes at 500 volts, and as but 22 amperes are used in the shunt at full load and the resistance of the armature is extremely low, the electrical efficiency of the machine is rated at 98½ per cent. and the mechanical efficiency approaches 96 per cent. The construction of the dynamo is especially interesting. A heavy ring of soft iron forms a continuous yoke and support for the ten field magnets which radiate from the outer surface. The ends of the pole pieces spread out so that they form almost a continuous circle, thus presenting a large field surface to the armature revolving about them. This frame, with the radiating fields, is secured to the engine foundation. The engine shaft passes through the center of it and supports the arms to which the outside armature is attached. The armature core is constructed of thin soft iron plates that overlap each other, thus breaking the joints in the construction of the ring secured on the transverse pins which extend out from one side of the radial arms. This frame is supported on a large bearing, on which it may be moved away from the fields, thus giving plenty of room for inspection or repair. The windings of the armature are thin copper bars shaped to fit the core and securely soldered in place. By having no lead wires or connection, many of the small attending evils are avoided and great simplicity of construction secured, allowing the armature resistance to be so low as not to require compounding. The same bearing that supports the armature spider also supports two other frames, one having radial arms, supporting from its ends the brush holder rods, which are of steel and extend across the face of the armature. This frame rotates a short distance for adjusting the brushes to the non-sparking point. The other frame supports a gear and short levers attached to the brush holder rods in such a way that when the frame and gear are rotated a short distance all the brushes are lifted from the armature surface at once. The insulation for all exposed parts is thick hard rubber, and the connections from the brushes to the connecting rings and from thence to the terminals are layers of thin sheet copper. The connecting of all the brushes of like polarity is accomplished through a very substantial ring supported on the armature bearing and to which the flexible leads from the brushes are connected, heavy flexible connections leading from these rings to the regular machine terminals. Where it is necessary to make connections for very heavy currents, the surfaces of the contact blocks are grooved to give better contact and permit of greater pressure. Three brushes, about 1½ inches wide and of small, straight copper wire, are used on each brush rod. It is said that the commutator surface of the armature of the machine first installed in the Berlin station in 1887 is as good to-day as when installed and shows but the slightest signs of wear.

From the dynamo heavy conductors are led to the small marbleized switchboard on which are placed two large single-pole switches, one for each side, an ammeter showing the total output of the machine, a voltmeter recording to 600 volts, and a rheostat for the shunt, the contact wheel being on the front of the board and the resistance back of it. There is also a special carbon contact switch for breaking the shunt field circuit gradually, to avoid trouble from extra currents when stopping the machine. When the machine is running the switch is closed, but when it is desired to stop the switch is opened, throwing the current through the carbon pencils, which are then gradually separated. The current is carried from this board by two feed wires to the centers of distribution, where are placed the equalizers. These equalizers are specially constructed shunt motors with very low resistance armatures, all connected on one shaft, the four separate machines being exactly alike. The fields are connected in series from one terminal of the 500-volt circuit to the other, the armature circuits being also connected in series between the same terminals. A resistance is placed in series with the armature circuit and used only for starting, in exactly the same manner as that in use with a shunt wound motor, and is cut out after the machines are up to speed. By this method the action of the equalizer is briefly as follows: The fields, being in series across the terminals, are practically independent of the machines themselves, and exert the same influence all the time; in case one of the four circuits is loaded heavier than the others the particular machine on that circuit acts as a generator, and being driven by the others adds its current and makes up the balance; if the circuit is underloaded as compared with the others the machine is carried as a motor, and thus balances for the overflow. Of course if the circuits are all equally balanced no action takes place in the machines, the counter electro-motive force simply balancing the consumption of the current excepting that needed to overcome the friction and machine losses, the machines need only be of sufficient capacity to take care of the largest difference in load between any two of the circuits, care being taken in mapping out the system to as nearly as possible balance the different circuits one with the other. In this special case the equalizers are wound for 125 volts pressure, 50 amperes of current, and run 1650 revolutions per minute. At full load they are said to have 90 per cent. efficiency, the armatures being of very low resistance, and the brushes taking current directly from the end face of the windings without special commutator, the machines are practically automatic, the variation in pressure being not over 1½ per cent. from nothing to full load. It may be said that where the load is very uneven storage batteries are advocated by this company in place of the equalizers, as having a somewhat higher efficiency. They are placed across the terminals and connected up in precisely the same manner as the motor equalizers and have the same effect. At the center of distribution the equalizer is placed between the terminals, and five wires are run from it as distributing mains, one from each of the outside terminals and one from each junction of the armature circuit. As the potential difference at the dynamo is generally 500 volts, and the lamps used are 110 volts, a drop of 60 volts may be allowed between the primary dynamo and the lamp socket. It is not necessary to run five wires in every case, as side streets and buildings requiring but few lights need but two or three wires, the system being balanced at some other point. The comparative sizes of the five wires are about as follows: 1, ¼, ½, ¼, 1; the first and last being the outside terminals, the second and fourth the intermediates, and the third the central wire which is used as one of the outsides in cases where only three wires are run. The system permits of a great degree of flexibility and the fact of needing but two feed wires tends to reduce the apparent complexity due to the great number of wires. On the continent these circuits are laid directly in the ground about two and a half feet below the street surface, and without other protection than the iron or steel armor of the cables themselves. The insulation consists of a layer of jute saturated with some ozokerite compound next to the wire, then lead, this being covered with another layer of jute and wound over all with two layers of iron or steel ribbon, both in the same direction, in order to cover the entire surface and give the proper amount of flexibility. All connections for customers are made through junction boxes buried in the street, and are quite similar to those made by some manufacturers in this country. Large feeder boxes are placed at street junctions for connecting up and testing the system. The plant exhibited was used for lighting the terminal building, Festival Hall, Wooded Island, German sections in Machinery Hall and in the Electricity building. Equalizers were placed in Festival Hall, Terminal building, Electricity building, the one for Machinery Hall forming part of the exhibit. On this system were carried 450 arc lamps, 3000 incandescent and 250 horse power in stationary motors. It was first intended to erect a complete generating plant for tri-phase work and show a street-car truck equipped with a tri-phase motor running on a piece of track, but the proper space and concessions not being obtainable, the generator, transformers and motor truck were located in the German section of the Electricity building. Outside the plant just described was also shown a 600-ampere 110-volt bi-polar direct-current dynamo. The fields of this machine are very heavy, and are flattened on the insides where they face each other, otherwise the machine is of much the same type and design as bi-polar machines of American make. The speed is 650 revolutions. This exhibit, although not as extensive in its variety as some others, showed the great care and skill of the company in the design, construction and erection of its machinery.

M. E. DESROZIERS.

Direct-Current Dynamos.—One of the most interesting exhibits in the French section in the Electricity building, was that of M. E. Desroziers. Among the principal applications of the Desroziers machine may be cited the lighting of large trans-Atlantic steamers, warships and torpedo boats of the French navy; their employment as generators in the central light stations of Paris, and a large number of private isolated plants. This would seem to indicate a large measure of success when it is considered that the figures relate only to France and her colonies, where the activity in electrical work is not very great. The Desroziers dynamo is a multi-polar machine with a disc armature, or a circular flat and thin armature on which are placed the appropriate windings and which pass between the parallel and closely juxtaposed faces of the pole pieces. The lines of force going from one magnetic circuit to the other will be nearly straight lines. If the wire, for instance, turning about the axis, passes between these two poles, it will cut the lines of force at right angles and will be the seat of an e.m.f. Every other wire situated in an analogous position, will, likewise, be the seat of an e.m.f., the direction of which will depend on the direction of the movement with respect to the poles, and on the polarity of the latter. Each of the radial wires will be thus the seat of an e.m.f., the intensity and direction of which at each instant will depend on their positions, respectively, with relation to the two magnetic fields. Evidently one may multiply the number of poles about the circumference of the circle which are passed over by the wires, and multiply the radial wires themselves to any extent. To collect the current thus produced, the different wires are connected together in such a manner as to obtain a continuous circuit in which the partial e.m.fs. add themselves to make up the required effect, and the armature sections are then properly joined to a commutator upon which the brushes rub. The radial wires are arranged on an interior crown, on two sides of the armature, on hard pasteboard, which holds them in position. They are joined one to the other interiorly and exteriorly by means of curved wire connections. The different sections of the armature are joined to the segments of the commutator, where two brushes take off the current by means of a special connector. This device is rendered necessary by the multi-polar arrangement of the machine and the series winding of the armature wires. It served to facilitate the interconnection of the sections

NEW WOOD ALTERNATOR.—FT. WAYNE COMPANY.

which are symmetrically situated with respect to the poles, with the result that only two brushes are required. The poles on the same side of the disc are of alternate polarity. The number of poles can be varied within the widest limits; thus, there was exhibited a small four pole machine, also a six pole machine, giving 16,000 watts, and in the background a large ten pole machine, of the following capacities:

At 150 revolutions per minute, 350 volts and 275 amperes.
At 215 revolutions per minute, 500 volts and 320 amperes.
At 300 revolutions per minute, 730 volts and 350 amperes.

This great flexibility of operation of the machine and the low speed at which it revolves are some of its great advantages. These slow speeds admit of these machines being connected directly to the prime mover as is now being done, not only in the special cases of ship lighting, but in central station work also. In the case of direct connection, there is generally inserted, between the shaft of the prime mover and the dynamo, a Raffard elastic coupling. There were shown three types of the direct coupled arrangement. The first represented a steam dynamo, adopted in the French navy. These steam dynamos have three uprights for the bearings. A combination of this type is able to furnish 4200 watts at 5,000 revolutions; it weighs only 1320 pounds, and has a total height of about three feet. In the station of Popp Company, in the Boulevard Richard Lenoir, in Paris, four Weyner-Richmond engines, running at 145 revolutions, driving eight dynamos of 100,000 watts each, are used. Notwithstanding their large output, these dynamos operate at a potential difference of 2400 volts, and often at 3200 volts and over. This plant is considered a model one, owing to its elegance and the small space occupied by the generator. More recently, by the aid of Raffard elastic coupling, it has become possible to drive directly with a Crossley two-revolution gas engine, at 250 revolutions a minute. These "gas-dynamos" have been very successful in small private plants, and also in central stations. The light is absolutely constant, and, what will appear more extraordinary, the regulation is obtained automatically and instantaneously at the dynamo and at the engine. In order to collect a continuous current with a series winding and only two brushes, it is necessary to have a winding which satisfies the same conditions that exist in the Gramme ring; that is to say, there must be arranged two series starting from the same wire, for which the e.m.f. developed is zero, and each formed of wires in which the e.m.f. developed goes on increasing without interruption up to a maximum, from which point it decreases again to zero, at the second junction wire of the two series.

All the regularly closed circuit windings can evidently be reduced to two types: (1) The regular polygon winding, and (2) the star polygon winding, in which the sides of the polygon are replaced by wires of any form ending at the same terminal point. But if an attempt is made to connect the different wires among themselves, in order to satisfy the electrical conditions set down above, considerable practical difficulties are met which result, for the most part, in the connecting wires cutting at each instant the radial conductors, and these different conductors are superposed. After a thorough analysis M. Desroziers discerned that they could be separated into successive alternate identical parts. By numbering these elements according to the natural sequence of the numbers, all the even elements—that is to say, beginning from an even top—are identical, and, likewise, the odd elements. It is, therefore, only necessary to separate on one surface, and the even elements on a neighboring parallel surface, but preserving for each its relative position. Thus the corresponding terminals of these elements touch each other, and it suffices to solder them in order to complete the original closed circuit winding. Under these conditions, the successive, separated like elements can be formed of convenient parts and simply placed opposite or side by side without cutting one another, as, for example, by radial parts, or those in the form of a cycloid or parts parallel to each other. The placing of the like elements can be done separately under practical conditions on two plane parallel surfaces. These two surfaces are then placed one against the other so that the wires of the two planes are so situated that the closed circuit winding can be reëstablished by simply joining the ends of the corresponding elements which then adjoin one another. In practice, these plane surfaces are rings or crowns of compressed insulating material, which are fixed on a thin metallic disc in the shape of a star inserted between the crowns before the ends of the corresponding wires are soldered. There is thus formed a symmetrical whole, strong in construction, forming an armature in excellent condition to resist all disturbing influences. In the case of conductors in the form of wire, M. Desroziers employs two thin crowns of compressed card board to hold the odd and even wires. The crowns are divided into a certain number of appropriate sectors and in three concentric crowns. They are perforated in regular order on their outer and inner circumference. In passing the wires through the holes, which constitutes a system of switching, the radial parts are placed on one side of the disk and the curved, or cycloid, part on the other side of the disc. The crowns of card board are fastened on the star; the card board where it projects above the circumferential cycloids and below the central ones, is cut away. In the other cases of winding, the same method of mounting is employed. The method of winding admits of making a number of hanks. It is this system of multiple winding which admits of the obtaining of the desired voltage, however high it may be, because it can be done methodically in several layers.

The above explanations relate to the case of wire conductors; when sheets are employed the card board crowns can be dispensed with, more especially in certain cases where the radial and curved parts can remain in the same plane, as then they do not intersect each other. To recapitulate: With this method of construction there is obtained a winding, methodically classed and joined, thoroughly well insulated and ventilated, besides fixed by small groups passing in the same hole on an excellent insulator, compressed card board. As to the strength of the armature, experience has fully demonstrated that it is able to withstand successfully the most prolonged, heavy strains. From the physical standpoint also the Desroziers machine possesses advantages; thus the absence of iron makes the generation of Foucault currents impossible, and hence the heat usually generated from that cause is absent. The ventilation is such, that the machine can be heavily overloaded without causing destruction of the armature wires, while its stiffness enables it to withstand all lateral disturbing influences. The extreme lightness of the armature also presents an important advantage. Thus suppose a case of a number of dynamos coupled directly to their engines by the Raffard coupling with rubber rings, the machines being joined in parallel. Any accident to, or slowing up of, the engine would cause an inverse current from the other dynamos to flow into the affected machine; the latter would slow up still more, the inverse current would increase enormously, the rubber rings would break, the armature would be set free, as well as the engine. The armature would then immediately increase its speed up to the point where it would generate sufficient e.m.f. to cut down the current of the other machines practically to nil, and there remain without further mishap. Experience has shown that the reaction of the armature on the field magnets is very weak, in consequence of which the machine can be run within wide load limits without regulating the brushes. The self-induction of the armature is also very small, since the variation in the magnetic flux is insignificant even for large variations of the armature load. Should repairs become necessary they can be easily effected owing to the symmetrical winding adopted. The practical results obtained with the machine are very interesting, and especially the low weights attainable with this form of construction. Thus at normal speeds these machines weigh only from 55 to 65 pounds per horse power. The efficiency is also high. At its normal speed of 300 revolutions per minute this machine produces 210,000 watts and at an efficiency close on to 93.5 per cent. On the machines at the Boulogne-sur-Mer central station, running at 350 revolutions, an efficiency of 93.5 per cent. has been obtained.

DYNAMOS.

GENERAL ELECTRIC COMPANY, LIMITED, LONDON.

This company showed a compact and well executed dynamo of the Kapp upright type containing all the latest improvements. The dynamo was two-pole and compound wound. Its speed was 410 revolutions, at which speed it will yield 200 amperes at a pressure of 100 volts. Its efficiency is given as 96 per cent. It was coupled direct to a Chandler high speed engine.

BREGUET COMPANY, PARIS.

Dynamos with Gas Engine.—This exhibit was a gas engine of 250 revolutions, coupled to the dynamo by a rubber joint, type Raffard. For 220 lamps of 10-c.p., they use 12 cubic meters of gas per hour, instead of 24.2 cubic meters which would be used for the gas light of same power.

HISTORICAL DYNAMO EXHIBIT.

In contrast with their new and improved machines, most of the exhibitors had on exhibition one or more types of old-time machines. In almost any other department of the Exposition, the phrase "old-time" would carry one back many decades, or even centuries. In electrical practice, however, "old-time" will run back ten or twelve years, and the crudities of even so recent a past will appear ludicrous almost in the light of to-day. While the dynamo principle has not been disturbed since its first conception, yet in important details the machines of to-day are radical improvements, as the ensuing descriptions of historical machines will discover.

GENERAL ELECTRIC COMPANY.

This company made perhaps the most unique contrast in regard to present and past dynamo practice. They exhibited the "Jumbo," the only survivor of the six dynamos of same size built by the Edison Company in 1882. It is direct connected to an Armington & Sims engine, has a capacity of 1400 amperes at 125 volts, and runs at 208 revolutions per minute. It has been in daily service five hours per day for ten years, has made enough revolutions to roll its armature eighteen times around the earth, and is in excellent condition.

Another dynamo of almost as much interest as "Jumbo" was one of the first dynamos built by Mr. Edison early in 1880. It was used in the first exhibition of incandescent lighting ever given (at Menlo Park), and has been in service constantly since, up to the time it was brought to the Columbian Exposition.

Among the historical dynamos exhibited by the General Electric Company, one of particular interest was the original arc lighting dynamo built in 1879, first used to light the "Philadelphia Bakery." It is the first three-coil armature machine ever built, and was wound by Professor Thomson himself. Another arc machine was one built in 1880, and which was the first spherical armature dynamo built, its form being substantially the same as the present Standard arc machines of the General Electric Company. A number of models were shown, many of which have been steps in the development of modern apparatus, including a number

HISTORICAL DYNAMO, BUILT BY JAMES J. WOOD, MAY 1879.

of early forms of arc lamps. An alternating dynamo was shown having two field coils on a laminated core with poles composed of a number of teeth. Another model was one of the first iron-clad dynamos, somewhat similar to a number of the "Waterproof" motors recently developed, the windings of fields and armature being completely protected by the iron of the field, and the whole being stood on end like a barrel.

FORT WAYNE ELECTRIC COMPANY.

This company exhibited a machine designed and built in May, 1879, by James J. Wood, their electrician, and which was awarded a medal for superiority at the American Institute Fair in 1880. The fields are cast in one piece, and are multi-polar. The speed is exceedingly low for such a small machine, being 1700 revolutions a minute. The efficiency is very high,

New Brush No. 9 Generator.

the internal resistance from binding post to binding post being but one ohm. It is also fitted with an inter-locking commutator, similar to that found in recent dynamos. This commutator, together with the multiple brushes, enables the machine to be cross-connected, making surprising combinations. It may be run as a self-exciting alternator, a tri-phase alternator, and a direct-current machine. Three independent circuits, either alternating or direct, can be derived from this machine without stopping it or making any extensive changes. This machine is seen in operation running an old-style Wood focusing arc lamp, driven by one of Mr. Wood's iron-clad motors, which is in itself quite an advance in dynamo designing, considering the capacity is but 1½ h.p., and the speed 1300 revolutions a minute, taking 110 volts and 10 amperes at full load.

BRUSH ELECTRIC COMPANY.

This company had in its exhibit the first dynamo made by C. F. Brush in 1876; also an old 16-light dynamo, sold in 1879, in complete running order, furnishing current for the arc-testing rack. This dynamo had been in constant service in Fulton Worsted Mills, Fulton, New York, since 1879. Every bobbin and field was intact. Another remarkable exhibit was an old 40-lighter, sold to Berkeley Mills, Providence, Rhode Island, in 1881. Constant service from that time to 1893 had not in any way injured either the dynamo or the lamps, and after twelve years of constant service it was still good as ever, not a bobbin or field being lost in all that time.

CHAPTER VI.

MOTORS.

POWER electrically transmitted was demonstrated by types of motors in great variety and for all purposes, many of these machines possessing originality of design and adaptation for the work for which they were intended. The motor is, as is well known, merely the reverse of the dynamo ; for, in the case of the latter, mechanical motion is transformed into electric energy, and in the former case electricity is changed to mechanical action Any dynamo can be used with more or less success as a motor, and many of the dynamos on exhibition were shown in motion for exhibition purposes by merely utilizing them as motors. As in the case of dynamos, they may be broadly divided into two classes, direct-current and alternating-current machines. Motors were shown for the use of direct currents, constant potential and varying quantity, and for constant quantity and varying potential, to be used, respectively, on incandescent or power circuits, and arc circuits.

For street railway purposes, a variety of motors were exhibited, both for high speed and low speed, to be used with gearing or mounted directly upon the axle. Several motors of over 100 h.p. were shown, and numerous motors, ranging in size from one-eighth of 1 h.p. up to 50 h.p. were used throughout the Exposition grounds wherever power was required.

In the line of alternating-current motors, some were shown for use by the single-phase current. These were synchronizing motors, and were brought to the proper speed by supplementary motors until they were in exact synchronism with the generating machine. Other motors, using two-phase, or multi-phase currents, and involving the principles of a rotating field, were shown for all purposes, including street-railway use.

THE GRAMME MOTOR.—The first ever shown in operation, that which created such a sensation at the Vienna Exposition in 1873, was on exhibition in the French section. The German Government also showed in its historical display a number of motors designed by Siemens, which included the first electric railway motor ever successfully used. For convenience, motors for ordinary power purposes will be described under this heading, and those for railway and other special uses will be described under heading of the machinery for which they were used.

WESTINGHOUSE ELECTRIC & MANUFACTURING COMPANY.

MULTI-POLAR MOTORS.—These machines are similar in general appearance and mechanical construction to the railway generators, and are constructed to run on circuits of 125, 250 and 500 volts. Two of these 500-volt 150-h.p. motors were in continuous use during the Exposition for the operation of the machinery in the Mines and Mining building.

HORIZONTAL TYPE MOTOR.—This machine is like the horizontal type generator in both mechanical and electrical construction. As the points of excellence and the claims for the two machines are practically the same, it will be unnecessary to here repeat them at length. The slow speed of this motor makes it particularly available for certain classes of work.

LETTER TYPE MOTORS.—This machine is similar to the letter type generator both in mechanical and electrical construction. The description of the Letter Type Generator and the claims made for the same apply equally well to this machine.

MANCHESTER MOTOR.—The machine is of the consequent bi-polar type, with two separate field coils wound on metal bobbins. The bobbins are slipped over their field cores before the upper and lower halves of the field are bolted together. The compactness, efficiency, and general adaptability of the motor makes it unsurpassed for most stationary work. It is already favorably known in connection with crane work. The machines are wound for either 110, 220 or 500-volt circuits. The motor may be safely overloaded as much as 50 per cent. for moderate lengths of time.

TESLA MOTORS.—The 500-h.p. non-synchronous Tesla motor was shown in operation in space "J-2." The machine consists of a circular yoke, within which is a laminated ring, pierced with holes through which are threaded heavy copper bars, which are short-circuited upon themselves. This constitutes the armature of the machine, which is stationary. The field, which is the rotating part, consists of a laminated iron-toothed drum, wound with wire, which is connected at one end of the core to four collector rings. These serve to introduce the two-phase alternating current to the revolving part. The small Tesla motor used to drive the exciter for the generator was of the regulation Tesla motor type; that is, the current is supplied to the field winding, which is stationary, and the armature revolves, the windings being short-circuited upon themselves.

The 500-h.p. Tesla motor drove, by means of a 24-inch belt, a 500-h.p. two-phase generator, which was used to produce both alternating and direct currents. The two-phase current from this machine was utilized in driving a two-phase self-starting synchronous 500-h.p. motor and revolving transformer. The power from this machine was used in driving a Worthington pump and several smaller generators designed for various kinds of work. The 500-volt direct current from the commutator end was utilized in running arc lamps and in driving some direct-current apparatus in the same space.

MATHER ELECTRIC COMPANY.

LUNDELL MOTOR.

There were in this exhibit two multipolar self-regulating motors, capacity of 50 h.p. each, wound for a potential of 250 volts. These motors were operated in series on the 500-volt circuit supplied by the Mather Electric Company's generators in their electric power plant in Machinery Hall. They were belted to and run a 35-kw. dynamo. The motors were connected through an automatic starting box, wound for 500 volts. There was also one self-regulating motor, capacity 10 h.p., wound for a potential of 500 volts, complete with base and starting box. This motor operated on the 500-volt circuit supplied by the Mather Electric power plant in Machinery Hall. This motor was belted to and runs the 7-kw. dynamo mentioned above; also one self-regulating motor, capacity 6 h.p., wound for a potential of 220 volts, complete with base and starting box, and one self-regulating motor, capacity 1 h.p., wound for a potential of 125 volts, complete with base and starting box. This motor operated on the 110-volt lighting circuit, and worked the commutator device used for operating the changes in the lighting of the colored lamps in the signs.

At the Libbey Glass Works, in the Midway, the Mather Company had self-regulating electric motors, capacity of 10 h.p. each, wound for a potential of 125 volts, and operated by current at 112 volts. Each motor was set up on an iron base, fitted with adjustable belt-tightener, and connected through an automatic starting box. Three of the motors operated exhaust fans used for the ventilation of the furnace room, and one drove the shafting conveying the power to the glass-cutting tools.

INTERIOR CONDUIT AND INSULATION COMPANY.

Lundell Motors.—The chief characteristics of these motors are embodied in the novel construction of the field magnet, whereby a single energizing coil magnetizes all the pole pieces and a strong protecting shell for the windings of both field and armature forms the magnetic circuit. The two field magnet halves, when bolted together, form a shell which completely protects the armature, the whole of the shell being utilized in the magnetic circuit. These field-magnet halves are so designed that they are readily withdrawn from the mould in casting. The cross-section of the shell equals the cross-section of the base of the pole-pieces, so that no choking of the lines of force can occur at any point. Cast steel is employed for all dynamos and motors. The single energizing coil not only reduces the construction of the machine to the simplest point, but by having only two terminals it is especially adapted to be placed in the hands of inexperienced users, as it avoids any possible mistake in connecting up, or any bother or delay in examination when such becomes necessary. Another feature of the design is that the pole-pieces are magnetized directly, and the resistance of the magnetic circuit reduced to a minimum, whereby a considerable economy in the ampere turns of the energizing coil is effected.

The openings around the armature Pacinotti, at the ends of the field magnets are, in the larger machines, covered with metal screens, to protect the working parts and to secure at the same time perfect ventilation. The commutator portion of the armature alone projects outside the screen, and is well protected by a strong bracket which carries the outboard bearing. The brushes are arranged diametrically around the commutator, and the opposite brushes are in all four-pole machines connected together. They can be raised from the commutator when the machines are not in use. The brushes of the two-pole type have a unique spring arrangement for pressing the carbon brushes against the commutator. The spring has its rear end placed in a curved slot at the end of the brush-holder; while its forward end, in the shape of a T, presses against the carbon which slides in the ways in the brush-holder. To remove the carbon, it is merely necessary to press the spring upwards out of the

New Standard Arc Light Generator.

slot, which releases it so that the T can be withdrawn, and the carbon slides out.

GENERAL ELECTRIC COMPANY.

This company made a most comprehensive exhibit of motors for all purposes, not alone in the Electrical building, but also in service about the Exposition grounds and in the buildings. The stationary motor exhibit showed the standard types and sizes manufactured by the General Electric Company, and also some of their applications to special uses. There were graduated sizes of Thomson-Houston type motors, ranging in capacity from 2 h.p. to 60 h.p. With the smaller sizes of the Thomson-Houston type motor were shown the new automatic switch and starting-box brought out by the General Electric Company, by which the resistance is thrown in and circuit broken in case either of an overload or of stoppage of current from any cause. The point at which it operates from an overload is adjustable, and may be set for any load desired. The Edison type motors are equipped with marble motor boards, mounted on floor stands, and containing incombustible starting rheostat, ammeter and two single-pole main line switches. In the line of special applications of motors referred to above, there was an eight-ton derrick, operated by a 6 h.p. "F-12" motor, series wound, and equipped with controlling rheostat and reversing switch. In addition to the ordinary hook for hoisting there is a magnetic lift, for use when iron is to be handled, the one exhibited having a lifting capacity of 2000 pounds. Near the derrick was a 3 h.p. iron-clad motor, mounted on a hand truck, with a flexible shaft connection to a drill set for working in iron, the whole so arranged as to be easily moved from place to place. On an elevated platform along the south side of the exhibit were placed an electric-launch motor, with its controller and rheostat, a multi-polar, slow speed, 1½ h.p. motor, direct connected to a 10-inch Sturdevant blower, and three motor generators, which were designed for use on existing commercial circuits to supply current for telegraphic service. In the Transportation building the motors of the General Electric Company were to be seen in connection with the transfer table installed by The Industrial Iron Works, by which over 52,000,000 pounds of freight were installed. The General Electric Company's motors were also to be seen in many of the main buildings and State buildings on the World's Fair grounds, where they were driving machinery and passenger elevators of various styles and manufacture, but which were not of sufficient special interest to merit individual description. In the north aisle of Machinery Hall was the electric traveling crane installed by the Morgan Engineering Works, equipped with seven General Electric motors, all of which were under the control of the operator in his cage at one end of the crane, and which could be operated together, separately, or in any desired combination. Near the crane were two 135-h.p. General Electric motors driving the countershafting of that part of the building.

S. S. WHITE DENTAL MANUFACTURING COMPANY.

The motors exhibited by this firm were almost exclusively applied to the operation of electro-therapeutical and dental apparatus; hence their description will naturally fall under those classes of apparatus.

OTTO RAUDA, OF NEW YORK.

The exhibit contained small motors, standard type, for various domestic purposes. They were the "Gem," exhibited first as plain motor; second, mounted on iron stand, with 6-inch transmission wheel, by which the speed of the motor is converted into strength; and, third, as Geissler tube-rotator, capable to rotate tubes from 4 to 10 inches in length.

WESTERN ELECTRIC COMPANY.

This firm exhibited a complete line of motors for all purposes, more properly descriptive under other headings treating of the special service required. One motor shown in the exhibit proper was of the slow-speed type, and consisted of four coils placed in a vertical position, making a very compact machine, the floor space it occupied being extremely small. This machine was mounted on an iron sliding-base with belt-tightener, and provided with self-oiling ball and socket bearings and carbon brushes. The other motors manufactured by this company are 10, 5, and 2 h.p. respectively, and are multi-polar in type and circular in form. These motors are slow speed, and adapted to meet all the requirements which may be made upon an electric motor.

ALLGEMEINE ELEKTRICITATS GESELLSCHAFT, BERLIN, GERMANY.

For the purpose of showing their types of motors, the firm had in operative exhibition a "Type G," 600 x 100-h.p. continuous-current machine, wound for 120 amperes at 500 volts, the motor running at 500 revolutions. The field magnets are of cast iron, cast in one piece with the foundation plate and pedestals. A peculiarity of the magnet system is the so-called pole ring, a strong hollow cylinder of cast iron. This arrangement has this advantage, that the reversing of polarity in the armature of the motor takes place gradually, and so insures perfect freedom from sparking at every alteration of load. The armature winding, being of the drum type, consists of only one layer of solid insulated copper bars of rectangular section. To prevent heating during long runs, the spokes which support the laminated core are screw-shaped, thus producing a strong blast of air through the inner hollow of the armature when revolving. For starting the motor a fluid-resistance is used. This motor drives, by means of a belt, a large three-phase dynamo Type D. M. 600, which produces three alternating currents of a phase of 120 deg. difference. As is generally known, the Allgemeine Elektricitats Gesellschaft has, in conjuction with the royalty holders, the "Maschinenfabrik Oerlikon," undertaken the experiment of transmitting water power of 300 h.p. to a distance of 175 km., from Lauffen on the Neckar to Frankfort-on-the-Main, by means of this kind of current. Notwithstanding all previous doubts and prophecies of evil, the brilliant results of this experiment have not only yielded the final solution of this problem, but have also proved conclusively the advantages of this new current over the continuous and simple alternating currents known hitherto. From this time the Allgemeine Elektricitats Gesellschaft has had continually in view the perfecting of this system, so full of promise, which was first practically applied by this company. The rotary current machines exhibited prove their perfect construction. The rotary current machine, Type D. M. 600, has been built for an output of 72,000 watts at 428 revolutions per minute, the pressure between each main being 120 volts. The current frequency is 50 complete periods or 100 reversals per second. The maximum effective strength of current in each main is about 400 amperes. The number of poles is fourteen, only 700 watts being required for the full-load excitation; that is, less than 1 per cent. of its whole output. The commercial efficiency of the machine is 92 per cent., including excitation. The ends of the armature coils lead to three sliding rings, from which the current is taken by brushes. The current is used for driving a number of rotary current motors. The largest of them, D.T. 500, is drum-wound, with 8 poles, and makes 750 revolutions per minute at the above stated frequency. Its output is 50 h.p., with a consumption in the single main of 280 amperes at 100 volts, the total number of watts being about 40,200. The commercial efficiency amounts to 94 per cent. The next motor of the same system, D.R. 50, of 5 h.p. is coupled direct with a Sulzer pump; the next, D.R. 10, of 1 h.p., has a brake for showing the full torque of the motor when starting. A switch proves the facility with which such rotary-current motors may be instantly reversed. Motor D.R. 5 of ½ h.p. has no special application, and runs without load. These last three motors are, of course, wound for the same pressure of 100 volts, but the number of poles is only 4, thus giving a double speed as compared to the large motor, namely about 1500. The winding of the field differs also from that of the 50-h.p. motor, inasmuch as it is of the ring type. The smallest motor, D.R. 1 of $\frac{1}{8}$ h.p., drives a fan. It has two poles, and makes 2700 revolutions per minute. All rotary-current motors of the Allgemeine Elektricitats Gesellschaft have a special construction of the armature, which is protected by a patent. A series of copper bars is drawn through the iron core of the armature, being mutually connected at both ends by a copper ring. There is no supply of current to the armature at all; it is driven mereiy by the indication of the revolving magnetic field. The small motors can be started without any further means than simply closing the circuit; the larger are started by a special apparatus called a "starter." These rotary-current motors have, unlike all others, neither brushes nor commutators, the ordinary and not always harmless sparking of the brushes being here entirely avoided, and all attendance may be dispensed with. They start with full load without taking excessive current. Their efficiency is superior to that of continuous-current motors of the same power. On account of their simplicity, durability and economy, the rotary-current motors and the complete rotary-current system of the Allgemeine Elektricitats Gesellschaft are a superior form of apparatus. The largest rotary-current motor is coupled direct with a continuous-current dynamo, Type G 300, by a brush coupling. In this mode of coupling the groups of brushes are manufactured of comb-shaped punched steel plates, for better working and facility in manufacturing. The machine G 300 resembles in its construction the electric motor G 600 already described. It is constructed for 300 amperes at 120 volts, and the current supplied by it served for feeding the lamps for the lighting of the company's exhibits, for the stage-lighting apparatus, the electric clocks, and a series of smaller continuous-current motors of the S patttern. In the latter motors the magnetic core is of wrought iron, forming a joint-piece to the back-plate, which is screwed to the foundation-plate with a base-plate of zinc. The commutator being composed of a large number of sections, and all parts being well proportioned, it is able to work with an entire absence of sparking at any load, without any displacement of the brushes, within the limits of the normal output. The two smallest, S and S, deliver $\frac{1}{18}$ and $\frac{1}{18}$ h.p., each of them driving a fan; another a small boring machine. For the purpose of convenient and certain control, all the above described machines and motors were connected with a central switchboard, placed in the middle of the installation. The currents were led through the measuring instruments, by which their strength and pressure might be read and controlled. From this central switchboard wires led to all the parts of the exhibit, to be connected to the local switches and cut-outs. There were also the usual fuses to prevent any excessive current.

HISTORICAL EXHIBIT OF MOTORS.

The Davenport Motor, Exhibited by the "Electrical Engineer." This was a working model of the motor designed by Thomas Davenport, of Brandon, Vermont, applied to a railway. The motor exhibited was one of a considerable number made by Davenport in 1837. The motor consists of a stationary and a revolving electro-magnet and commutator, ele-

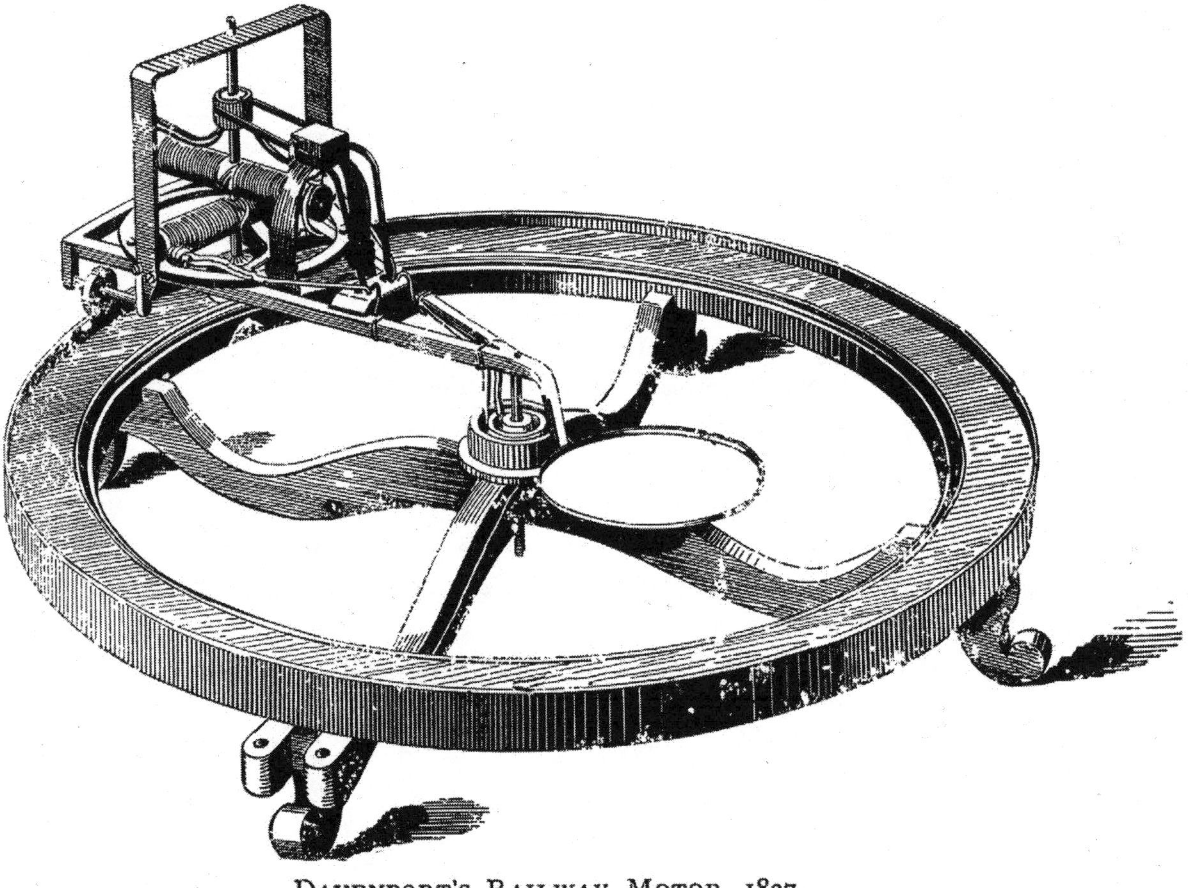

Davenport's Railway Motor, 1837.

ments common to the electric motors that have now become successful for so many purposes. One of the straight electro-magnets constitutes a stationary field, and the other an armature revolving in a horizontal parallel to and above the field.

The armature is connected with the driving wheels by a speed reducing bevel-gear. The field and armature magnets are each 5½ inches long, with cores ¾ inch in diameter. They are each wound with two No. 16 copper wires connected in parallel, there being 24 convolutions of each wire on each core. The commutator is constructed with insulated

Jenney 40 K. W. 500 Volt Direct Current Generator.

springs rubbing upon insulated metallic segments. The connections with the battery are formed by mercury cups. Perhaps the most remarkable circumstance about this motor is that the field and armature are connected in parallel, so that in this model we have a veritable example of a a shunt-wound motor, built in 1836-37. Together with this model *The Electrical Engineer* exhibited a fac-simile reproduction of Thomas Davenport's original draft of his patent specification of January 5, 1835, describing his electric motor; a most interesting document, as setting forth the most essential features of all the electric motors now in use. Although the drawing and specification show and describe a motor employing permanent steel magnets, Davenport writes in a foot-note: "It is thought that there would be an advantage in employing nothing but galvanic (electro) magnets on either the axis or the sides."

THE FARMER HISTORICAL MOTOR.

In connection with the exhibit of the Western Electric Company was the personal exhibit of Prof. Moses G. Farmer, installed under the direction of the venerable scientist, who died just before the Exposition opened. A part of the exhibit consisted of what is supposed to be the first reversible motor ever produced, it having been built and put on exhibition by Professor Farmer in Dover, New Hampshire, in 1847. The apparatus consists of a motor, with trailer attached, upon which are placed two seats. The cars are run upon tracks which were common in those days. The mode of attaching these rails to the sleepers is the same as that employed by Professor Farmer. The method employed in reversing the current was a shifting of the brushes through a small arc around the commutator. One of the seats and two of the wheels are the original ones used by Professor Farmer in 1847.

PROF. GALLILEO FERRARIS, TURIN, ITALY.

Professor Ferraris, in obedience to an urgent invitation from the management of the Exposition, exhibited, among other things of an historical nature, a model of a rotary field motor, with a vertical copper cylinder rotating in the field, produced by means of two coils crossed at right angles. The outer coil, 6 inches high, is wound in a frame having a groove 1¼ inches in width, filled with No. 12 silk insulated wire. The other coil is wound double with two wires of about No. 18 gauge, the two wires being joined in parallel, with their ends connected. The copper cylinder is 3½ inches in diameter. Another valuable piece of apparatus was Professor Ferraris' first rotary field motor, with horizontal axis and copper armature. The two currents with different phases were obtained either by means of a Gaulard & Gibbs' transformer, or by means of two parallel circuits with different inductances and resistances. Thus we find in the apparatus one of the coils composed of two No. 14 wires wound parallel, and the other coil, also consisting of two parts and joined in parallel, of No. 18 wire. The cylinder of this model is 7 inches long and 3½ inches in diameter. Another type of motor with rotary field was a little multi-polar machine, with iron magnetic circuit and having an iron armature. The four field cores of this machine are of wrought iron arranged, as usual, radially around the axis of the armature. To complete the magnetic circuit the outer ends of the magnet cores are joined by having a layer of No. 20 iron wires wound about them to a thickness of ¼ inch. The iron cylinder in this machine is 3 inches long and 1½ inches in diameter, the model itself standing about 8½ inches high. Besides these models there were also shown a number of solid and laminated iron cylinders used by Professor Ferraris. The package containing Professor Ferraris' contribution came across the water in the "Kaiser Wilhelm," from Genoa, and went to the bottom with that vessel in Genoa harbor. After three days' submergence it was rescued, and was again started on its way. As one might well imagine, its three days' submersion by no means tended to improve the looks of the apparatus; but as an historical exhibit its value would have been much decreased, if not entirely destroyed, by undergoing repairs.

MOTORS FOR SPECIAL PURPOSES.

THE ELECTRICAL MACHINE COMPANY, CHICAGO—CLOTH-CUTTING MACHINES.

THE Caldwell Cloth-cutting Machine exhibited, was an interesting adaptation of electricity.

In the construction of the machine the power is applied directly to the work without the intervention of gearing or belting. There are two motors of especially careful construction, mounted on a single shaft, whose wrist-pin carries a pitman connected with an oscillatory cross-head; this cross-head holds a keen blade by the tang, with the edge from the operator. The cross-head is given three motions by an arrangement of adjustable levers, its range being from an absolute perpendicular thrust to a combination approaching a perfect curve section in each reciprocation of the blade. In other words, a curvilinear line of travel so timed that when the blade is at its extreme lowest point the edge is perpendicular to the base-plate, which passes under the goods to be cut. It will be readily seen that with this action the entire length of blade will be brought into actual use, drawing across the goods with a shearing motion, and withdrawing on the upward stroke. In cutting clothing in quantity, the patterns are laid up and chalked on one layer of the required length; other lengths of the material are piled up, and this marked plate placed on top; then, with knives or machines, the entire depth is cut according to the pattern. It has been ascertained that any removal of these goods after being so piled, and before they are cut, results in a tendency to crawl out of place, thus changing the sizes. With a view to overcoming this defect, and that the goods may be cut where laid up, an ingenious trolley system has been devised in connection with this machine. A light iron track, supported from the ceiling of the cutting room, is charged with current; on this track a trolley of peculiar construction (which is in fact a traveling crane) is mounted, and not only affords means of conducting current from the rails, but also, through the medium of an adjustable spring coil, sustains the weight of the machine. At the will of the operator he may allow one pound or more of the machine's weight to rest on the surface of the cutting table, to steady it. This crane allows of the free movement in any direction of the machine, and without any attention whatever to overhead matters; in fact, the operator has but to follow the chalk lines by advancing the knife along them, and puts forth but very little effort in so doing. Owing to the fact that the work is done on the downward stroke of the knife, ample opportunity is given to relieve it from frictional heat, as it travels in a slotted standard without any bearing whatever at the back; a current of cool air, as from a fan, is directly felt by placing the hand at the back of this slot while the machine is in operation. The speed ordinarily required—about

CLOTH CUTTING MACHINE.

2200 strokes per minute—may be increased, without detriment to the machine, to 5000 strokes per minute, if desired. The armatures are wound for 110 volts and 4 amperes; but in order to attain varying speeds, as may be required by the different classes of work, a regulating rheostat is placed in the line at a convenient point. Through this rheostat the speed may be regulated in a range from 1000 revolutions to 5000. There are many manufacturing establishments that are not situated where direct current can be rented; in such cases a dynamo of 640 watts (nearly 1 h.p.) has been designed expressly to run this machine. It is possible to run from six to eight incandescent lamps, in connection with the machine, from this dynamo, which can be placed in any convenient spot where a belt may run to main or counter shaft.

BAY CITY INDUSTRIAL WORKS, BAY CITY, MICHIGAN.

Motor for Operating Railway Transfer Table.—The large transfer table at the south end of Transportation building, used for installing exhibits, was the product of this company. The table, 70 feet long, was built in the usual manner, and carried on seven pairs of wheels, all of the wheels on each side of the table being keyed to the same through shaft. The table was designed for a carrying capacity of 220,000 pounds. It was propelled directly from one of the through shafts by a train of gearing actuated by an electric motor. This motor was of the General Electric Company's manufacture, and known as their W.P. 50, being a waterproof motor of 25 h.p. nominal capacity. By suitable arrangement of the gearing, all of which was cut, and friction clutches, the table could be driven at either of two speeds, passing from one to the other without stopping of the motor. A foot brake of large capacity was supplied, so that the table might be quickly and accurately stopped at any point for the purpose of matching the tracks on the approaches. The pulling on and off drum could also be put in motion by another friction clutch. This device was found extremely useful, as the motor was of such large capacity that the drum had been used for hoisting heavy loads, sliding packages of great bulk into cars, pulling several cars coupled together, and similar heavy service, which use was extended far beyond what was originally intended. The current was obtained for the motor by means of the double trolley system, the wires being laid in a protected conduit in the pit, and well insulated. The controller was of the Thomson-Houston standard railway type, but connected directly to the controlling lever. The current was direct, and of 500 volts. This voltage was deemed better for such long runs as this table has, though the usual practice is to put an electric table on a 220-volt circuit. Either system is entirely satisfactory. All of the electrical parts and the machinery were properly housed in, the operator also being provided with a covering with proper windows, so that his work was always clearly before him. A report of the work of this transfer table shows that it gave the best service, installing all the heaviest locomotives in the Transportation building, and it was never found wanting when required for service.

THE HICKS-TROY ELECTRIC DOOR OPERATOR.

There have been door-checks designed by the score which, though having the advantage of closing the door noiselessly behind one, usually required more than ordinary effort to open it. The Hicks-Troy device, shown in the Electrical building, not only does all that the well known door-checks now accomplish, but it has the additional advantage of opening the door before the person passing through, and closing it silently behind him, without requiring a touch of the finger. The medium adopted to accomplish this is a motor controlled by magnets and switches. The door in one of the arrangements is hung on spring hinges, and the tendency of the springs is always to swing the door open. To the top of the door is attached a cord, which runs over a set of pulleys to a little electric motor, the function of which is to wind up the cord and to pull the door shut and keep it shut, at the same time winding up the spring which opens the door. This, in general terms, is the whole mechanism, but a few details are necessary to understand its workings. The person approaching the door steps on an electric mat, which closes a circuit, setting free the cord connected with the motor and allowing the door to be swung open by the springs. In the pulley-box over the door are two switches. A pin on the door-arm cuts out the motor by snapping the switch in the housing over the door, thus stopping the motor when the door strikes shut. The motor is placed within an iron box in any convenient place, and to its armature shaft there is attached a worm-gearing, with a worm-wheel carrying a set of magnets that attract an iron plate. This plate also carries a grooved cam, around which the cord that closes the door is wound. This cam movement pulls upon the cord with a leverage which is weak at the beginning and grows stronger, being at its maximum when the door is closed; that is, when the spring which tends to open it is also at its maximum tension. As soon as a person steps on the mat, the circuit of the motor is opened at a switch in the motor box. This releases the iron cam plate from the influence of the magnets above referred to, frees the cord, and allows the springs to open the door. As the person steps off the mat, the switch at the motor closes the motor circuit, the armature revolves, the iron brake disc is attracted by the magnets, and the door is wound shut. As soon as it strikes the pin at the

C. & C. Power Plant.—Machinery Hall.

top of the door, however, it breaks the motor circuit, but leaves the current still on the magnets, which attract the iron cam plate, and it thus keeps the cord under tension and holds the door closed. This iron disc acts as a brake-shoe, as it were, and forms a flexible coupling between the motor and the cord. The action of the person stepping on the mat will now be understood to release this brake-shoe, freeing the cord and allowing the springs to open the door. The exhibit included a number of doors fitted with the door opening and closing device. The current required for operation does not exceed ½ ampere at 110 volts.

THE ELEKTRON MANUFACTURING COMPANY.

Elevators Propelled by Electric Motors.—The company had for the most important part of the exhibit the electric elevator mechanism. The motor is a 6-pole 500-volt Perret machine commutator, permitting the use of only two carbon pencil brushes, instead of six, and insuring freedom from noise. The motor is directly connected to a worm shaft by a universal coupling, so arranged that the worm and motor shafts can thrust either way independently. The worm is cut directly on the steel shaft, instead of being sleeved, and the gear is of bronze, hob cut, with a ratio of 1 to 40 with the worm, and bolted to a wide web which, besides being keyed to the drum shaft, is belted directly to the drum as an additional precaution in case of the key working loose. The elevator hand rope throws in the starting switch, and at the same instant releases a brake strap extending four-fifths of the way around a brake wheel on the work shaft, which is re-applied when the switch is opened, and stops the car by the action of a heavy weight at the end of a compound lever. The brake shaft is directly connected to, though insulated from, the switch mechanism, and in neither direction of running is the circuit completed until after the pole-changer has been thrown into contact. The starting rheostat is entirely automatic in its action. A solenoid is employed to move an arm which traverses a row of contacts cutting out the resistance step by step in the usual way, and a dash-pot is employed to prevent the moving of the lever at too rapid a rate. The dash-pot resistance-arm and contacts are not essentially novel, being changed but slightly from ordinary forms in order to give greater durability. The novelty is found in the connection of the solenoid terminals to the brushes of the motor so that the solenoid is not excited until the motor commences to speed up. This is an important feature, for in case the car is very much overloaded the resistance will not be cut out, and only a predetermined current can pass the resistance and enter motor; this prevents blowing fuses. The instant the switch is opened to stop the machine the resistance-arm falls by gravity, re-inserting the resistance ready for another start. Thus in case of an accidental break in the line there will be no blowing of fuses when the current is again turned on. It sometimes occurs that an elevator car sticks in the guides or meets with obstruction in descending; in such case, if the machinery is not stopped, the ropes are apt to become entangled and give trouble. This possibility is also taken into account and provided for by an ingenious and simple device. The hoisting rope, as it leaves the drum, passes beneath a loose sheave on a frame whose ends are fixed to loose collars surrounding the drum shaft. At the end next the motor is a catch, holding back one-half of a clutch turning with the shaft, capable of sliding along it, when released, under the action of a spring-actuated collar. The other half of the clutch is fixed to the motor controlling sheave to which the hand rope is attached. When thus, by reason of the car's sticking or being obstructed on its downward trip, the hoisting ropes become slack, the sheave under which they pass is no longer supported and, with its frame, falls, releasing the movable half of the clutch, which immediately unmeshes with that on the controller sheave and cuts out the motor. To guard against "racking," there are two weights inside the drum, acting like governor balls, and, when the speed becomes too great, releasing a weight that sets a heavy auxiliary brake and brings the drum to rest. As an additional safeguard against accident from allowing the car to go too far in either direction, a traveling nut on the controller sheave is set so as to engage, at the desired instant, a fixed nut at either end of its rim, and automatically cut out the motor. The rheostat is not mechanically connected in any way with any of the other apparatus, and may be placed in any convenient spot.

THE CRANE ELEVATOR COMPANY.

Electric Passenger Elevator.—The Crane Electric passenger elevator is designed to afford direct application of electric power for elevator purposes in the simplest and most effective manner. The steel passenger cage to which all cables and safeties are attached, and which relieves the car from all strain, is the same that is known in connection with steam and hydraulic elevators; the safety governor on the cage, guide rails, and general hoisting work being also identically the same as those of the Crane elevators operated by other powers. The operation of this safety governor is worth more than a passing notice. Excessive velocity of the car, from any cause whatever, causes this governor to liberate the safety grips under the platform, and holds the car and its load safely from the guide rails. The action of the governor is thus seen to be simply that of permitting a force always ready to act and perform its duty. Up to the danger limit of speed the car is perfectly free upon the rails, but the instant such limit is passed the full force of safety grips is liberated by the centrifugal force of the governor in removing a small detent, which under ordinary circumstances prevents the grip from acting. This arrangement, it is readily seen, insures positive action and maximum sensitiveness of operation. Positive and direct connection is made from the armature of the elevator motor to the car, by placing this armature directly upon the double screw shaft of the elevator engine. The armature stops when the car stops, and revolves in the proper direction when motion is desired, raising or lowering the car as may be required. This is accomplished by suitable application of the electric current, controlled by the operator in the car, to the machine. The double screw shaft is provided with right-hand and left-hand worms, each engaging worm gears which mesh together. This arrangement transmits the entire lifting force of both worms to the winding drum, which is keyed solid to the shaft of one of the worm gears, without the loss of power inevitable in ordinary single screw connections with their thrust bearing. It is twice as strong, twice as durable, twice as safe, and twice as economical of power as a single-screw machine. The electric current is controlled in entering the machine by a reversing switch of the most durable construction, which governs the direction of current through the armature and its consequent direction of rotation, and a rheostat which automatically graduates its volume to the needs of the machine. This rheostat is constructed in such mannner as to be practically indestructible, and permits the operator to use the full electric current his machine is capable of utilizing for maximum speed, or a less quantity for a slower speed, if he so desires. This rheostat controls the balance of current automatically through the action of a dash-pot and suitable weights, thus rendering the elevator independent of any particular skill on the part of the operator. The controlling gear may be the hand cable, pilot wheel, or lever in the car, as may be most suitable. The elevator engine is provided with positive terminal stops independent of the operator's controlling gear, bringing the car to a stop at its upper and lower limits of travel, and leaving it free to be started by the operator on the return trips. A slack cable stop is also provided, which stops the machine automatically in case of disarrangement or slacking of the cables. The machine brake is connected with a solenoid in such manner that any interruption of the electric current at once stops the machine and holds the car safely in position in the hoistway. The elevator is thus sure to be amply protected against accident or disarrangement. The design of the machine is characterized by the greatest simplicity consistent with necessary safety provisions, accessibility of parts being provided for in a degree hitherto unattained. Reliability and durability in service, as well as maximum economy of operation, are the ultimate aims of elevator construction.

NORTH AMERICAN PHONOGRAPH COMPANY.

THE EDISON PHONOGRAPH.—To most people the phonograph is known as a wonderful toy—a machine for amusing and entertaining the public. For amusement purposes in the home and in public resorts it is all that is claimed for it. But the phonograph is fitted for other purposes as well, and is rapidly finding its way into business offices, where it is found to be the most expert and economical amanuensis, and into educational institutions, where it is an invaluable aid to the teacher of languages. In the Edison phonograph exhibit the utility of the machine for all these purposes was demonstrated. It was made to play the entire gamut "from grave to gay, from lively to severe." So simple is the mechanism of the instrument that any of the attendants can explain it in five minutes. The phonograph proper may be said to be contained in the portion of it known as the "speaker"—a ring of metal about the size of a fifty-cent piece, in which is fastened a glass diaphragm. Attached to the under side of the diaphragm is a cross-head, to which is hung a pivoted lever, and at the further end of the lever are two small sapphires—one a stylus, in the shape of a gauge, for the purpose of recording on the wax cylinders; the other in the form of a ball, to be adjusted to the track made by the recording stylus and to re-produce the recorded sounds. The "speaker" is fitted into an arm in such a manner that, by depressing a small lever, the recording sapphire touches the wax; and when the machine is in position, and the wax cylinder is made to revolve, all vibrations of the diaphragm are recorded on the wax. To make a "record," the voice or music is directed through a speaking tube, or small horn, attached to the "speaker," in such a manner that the sound waves strike the diaphragm and cause it to vibrate. When these vibrations are recorded on the surface of the wax in a fine spiral thread, the "record" is complete. Some fine particles of wax, cut by the recording stylus in the process, must be removed by means of a fine brush, and the sounds can then be reproduced. This process is equally simple. A touch raises a small lever, which throws the reproducing ball into position to fit into the fine track made by the recording stylus. If it does not drop into the proper place, the adjustment can be made perfect by means of a small thumb screw. Then, as the cylinder revolves and the reproducing ball passes over the indentations in the wax, the vibrations of the diaphraghm which produced them are repeated, and the sounds are thrown out and fall on the ear an exact repetition of the originals. So perfect is the reproduction that the finest distinctions in sound are preserved, and a well-known voice can be detected at once. A singular feature of these "records" is that the microscopic indentations in the wax preserve all the sounds which united to cause them. In reproducing a band "record," for instance, each instrument in the band can be heard in the general harmony. In a quartette, every one of the voices comes out clearly and is quite distinguishable from the others. It is this feature of the phonograph which amazes the public and puzzles the scientific world. It is a mystery which attracts every thoughtful mind, and marks the phonograph as the most wonderful invention of the latter half of the nineteenth century.

The process of recording and reproducing sound, it will be seen, is purely mechanical. It is not necessarily connected with electricity. As a matter of fact, electricity is not the first power that was used to run the phonograph, nor is it the only power that is now employed. In the exhibit there were machines run by treadle power and by water motors, as well as by primary and storage batteries and by the direct incandescent light circuit. Any power which produces a steady movement will suit, and electricity, which is the most controllable of all, is found the most available. Most of the phonographs exhibited were run by electric motors. The steadiness of the movement is maintained by means of governing balls, and the spiral movement of the wax cylinders is produced by a fine drive screw (100 to the inch) into which twin nuts fit when the recording stylus or the reproducing ball touches the surface of the cylinder. There is on each machine an ingenious contrivance for removing "records" and preparing the cylinder for further use—a small sapphire knife, which can readily be brought in contact with the wax, and when in this position pares off a thin film, leaving a smooth surface on which fresh impressions can be made. Each cylinder can, by this means, be used to receive at least forty "records," and each "record" can be reproduced as often as may be desired. The "records" are not imperishable. They may be damaged or defaced by careless handling, or destroyed completely by a chance knock, but with reasonable care a "record" may be preserved for years and reproduced an indefinite number of times.

In the exhibit were phonographs fitted up in various ways to illustrate the different uses of the machine. A phonograph to which a large horn was attached gave out the strains of Gilmore's Band or rendered a comic song to a large audience in the hall ; and so clearly is the reproduction accomplished that every note and every word could be heard above the confused sounds of moving machinery and shifting crowds. An exhibition machine, fitted up with a set of 14-way tubes, was always surrounded by an interested throng. Every pair of ear-bulbs found a willing pair of ears to use them so long as the phonograph was kept running. It is this form of the invention with which the public throughout the country is most familiar. Phonograph exhibitors find it the most convenient and the most profitable form of exhibiting, and they have carried it to every

Mather Power Plant.—Machinery Hall.

corner of the world. There were also coin-in-the-slot machines exhibited, but merely as samples of the phonograph in that particular form. In every building on the grounds slot phonographs were to be found, and all collected their quota of nickels from amusement-loving visitors. The advantage of the slot machine is that it requires no attendant. It can only be worked by dropping a nickel in the slot. In every other form of the phonograph an operator is required to manage it, whether the machine is used for amusement or for commercial purposes.

In the business department of the exhibit four operators were employed to show the use of the phonograph in connection with the typewriter. The two machines stood side by side on a cabinet, and, when not in use, were locked up out of the way of dust and busy-bodies. To use the phonograph for dictation purposes, slip one of the blank cylinders on the mandrel; bring the speaker-arm to your left so that you can begin at the beveled end of the cylinder; attach a speaking-tube to the "speaker," and open the switch to set the machine in motion. Then, when the speed is right, drop the recording stylus on the wax cylinder, and dictate at any speed to suit your own convenience. The phonograph will take every word as fast as it can be uttered. It is not necessary to raise the voice above the tone of ordinary conversation. The machine will record even a whisper, but for clearness in reproduction it is best to speak at a pitch of voice similar to that used when speaking to five or six persons in a room—neither too high nor too low—enunciating the words distinctly. Anyone accustomed to dictating to a typewriter operator will find no difficulty in making a perfect "record," and with less effort than his former experience would lead him to believe is possible. Anyone whose utterances lack clearness would find the phonograph a valuable teacher. It will tell him what no amanuensis is frank or bold enough to say; that his enunciation is imperfect and that the errors for which he is disposed to blame his assistants are frequently attributable to his own faulty dictation. A few experiments with the phonograph will show him where his weakness lies, and enable him to cultivate a clear and distinct enunciation. When the dictation is completed, the operator removes the speaking-tube, brushes off the particles of wax deposited in the process of recording, adjusts a hearing-tube, and throws the reproducing ball into position. Then, with the ear-bulbs depending from her ears, she is ready to transcribe from the dictation of the machine. The speed of the reproduction can be regulated to some extent, but not sufficiently to enable an ordinary typewriter operator to keep up with the dictation of the phonograph. This, however, is not considered an objection of any moment; the operator has complete control of the machine, can drop and renew the dictation at will without stopping the motor, and can have the whole or any portion of the "record" repeated as often as may be desired. While skill in the use of the typewriter is a decided advantage, the slowest operator can transcribe from the dictation of the phonograph. To insure the accuracy of the transcript, the operator can verify it from the dictation of the machine; then if any error is subsequently found, it can be traced directly to the one who made the "record" on the cylinder. When the phonograph was first utilized for business purposes there was an impression amongst operators that the use of the ear-tubes would have the effect of injuring their hearing. This created a prejudice against it which was only removed when the most eminent aurists of the country recommended the use of the phonograph for curing deafness. Since then it has been winning its way into law and business offices in every part of the United States and Canada.

One feature of the phonograph that strikes everyone who has used it, is its extreme simplicity. There is nothing complicated about the mechanism, and it is as durable as it is simple. Anyone of average intelligence can with one hour's instruction and practice manipulate the machine and utilize it in the every-day work of an office. Nevertheless, there are people who would not devote an hour of their time to acquiring a knowledge of any machine, and to meet the demands of this class, and to make the machine available in large business establishments, there have been specially devised alterations. These consist of a number of phonographs connected by means of tubes with the principal offices in the building. There are also electrical signals between the phonograph stand and these offices. When anyone who is provided with one of these hearing-tubes wishes to dictate, he touches a button and signals back when it is ready for dictation. Another signal from the attendant warns the office when the cylinder is full, or from the office notifies the attendant when the dictation is done. In the device exhibited by the company there were three phonographs fitted up with tubes connecting with six different offices; and as can at once be seen, one attendant, an intelligent boy or young girl can attend to the whole of these, thus saving the time of three skilled amanuenses.

The value of the phonograph in law and business offices lies largely in these facts—that it can be utilized at all hours, that it will record and reproduce all sounds with unerring accuracy, and that the work of transcription can be done by anyone who can manipulate a typewriter, whether the operator has a knowledge of shorthand or not. The machine will not take the place of an intelligent, skillful stenographer. Many a business man lacks fluency and clearness in dictation, and to such a man the services of a skillful stenographer who understands him, and can put his ideas in good form on paper, are invaluable. But there is a class of amanuenses whose place the phonograph must surely take, and that class comprises the greater proportion of those employed in business offices. It is no fault of theirs that they are not expert shorthand writers. They lack skill because they have had little practice and experience. An intelligent young lady can become an expert operator on the typewriter with three or four months of steady practice. She could not acquire corresponding skill in shorthand with less than five years' training. Most young ladies find their way out of business offices and into homes of their own before they have had time enough to acquire a fair degree of skill in stenography. With a phonograph such an assistant would be from the beginning as useful as a skilled stenographer, while the employer could, if circumstances rendered it desirable or necessary, dictate his letters after office hours as well as at any time. Most men can do more solid and satisfactory work in an hour when undisturbed, than in half a day when frequently interrupted, and could with the phonograph save themselves much worry and loss of time by attending to their correspondence when alone in their offices. The cylinders, when recorded, could then be left on the typewriter's desk, to be transcribed on her return to the office. This is no mere theoretical mode of conducting correspondence; it has passed the theoretical and experimental stages, and is now the method pursued by a large and ever-increasing number of business and professional men.

There was one apartment in the company's space set apart for illustrating the use of the phonograph in teaching languages. Nothing could be more interesting and instructive than this feature of the exhibit. The space was visited by professors and teachers from every part of the world, and there was a remarkable consensus of opinion as to the value and practicability of the phonograph for educational purposes. To acquire a proper accent in learning a foreign language, instruction from a living teacher is necessary. But a competent teacher is generally a busy man. The value of his services is recognized, and the demands on his time are heavy and numerous. He cannot, even if he had the necessary patience, repeat a lesson over and over again for any one pupil or class. But if he dictates the lesson to the phonograph, it can be reproduced as often as the pupil or the class may desire, and thus all have an equal opportunity to acquire a correct accent or to familiarize themselves with foreign words which sound so strangely to the learner when first heard. Illustrations of this mode of instructing were given by attendants who understood French, Spanish, and German, and everyone who saw the system in operation admits its value and practicability.

One feature of the exhibit which attracted much attention was the original phonograph—the first machine shown to the public. Apart from the fact that it enabled one to realize the progress which has been made with the phonograph in fifteen years it has no special value. It is a crude and clumsy machine, not worth more to-day than its value as old iron, but it excited the wonder and admiration of the world when it was first exhibited, within the memory of the present generation. The most casual observer must be struck with the contrast between this original phonograph and the modern machine, so simple yet so perfect in construction, a machine which is destined ere long to revolutionize office work throughout the world.

NATIONAL ENGRAVING MACHINE COMPANY.

Engraving Machine.—This apparatus is designed for the use of jewelers, plate engravers, and for a large number of mechanical trades where engraved work on metal, stone, pearl, ivory, etc., is desired. This object is accomplished by means of an ordinary engraving tool held at the end of an arm, which is controlled by an electro-magnet, energized by a small number of cells of primary battery. The pattern for the work is generally in the form of ordinary printer's type, which can be set up to suit the occasion, or an electrotype or any relief engraving may be used in place of the type. There is a circuit-closing device, consisting of a lever with contact point, which passes over the face of the type and makes and breaks the contact. This controls the circuit of the magnet which moves the engraving tool. The whole machine is mounted on a solid frame provided with chucking and clamping devices, by which the various kinds of work can be held under the tool. Adjusting arrangements are also provided, so that work of any size or description may be accommodated. The movement of a lever back and forth causes the work to travel under the engraving tool, and at the same time the contact lever passes over the face of the type with the result that whenever the circuit is closed by the type, the engraving tool is brought in contact with the work, and immediately the circuit is broken it is raised from the work. It will be readily seen that as the pattern and the work move exactly together that the pattern will be produced exactly on the object which is being engraved. By the simple movement of a switch from one contact to another, the work will be in relief or sunken, as may be desired, from the same pattern. Means are also provided, so that from a pattern of any size the work can be either enlarged or diminished; for instance, from a pattern covering several inches of space a perfect reproduction may be placed within the size of a sleeve button. The character of the work in this reduction is just as good as where it is cut to the full size of the pattern. Perfect work has been done on this machine which cannot be read without the aid of a microscope.

ELECTRICITY AT THE COLUMBIAN EXPOSITION.

DR. A. H. PHELPS.

ELECTRO-THERMOGRAVURE.—The instrument for doing this work consists of a pen, or stylus, having a point composed of a loop of iridio-platinum wire and heated by a 2 to 10 volt electrical current of from 60 to 80 amperes capacity, and regulated by a coil rheostat placed in shunt. The handle of the instrument is composed of thin brass tubing, and has a water jacket, with supply and outflow tubes connected, to prevent overheating. The instrument, when heated and applied to the glass to be engraved, causes such rapid expansion at the point touched as to result in chipping, and, by properly regulating the temperature, producing a smooth chipped line from one-sixteenth of an inch in width and depth to lines so fine as to be microscopic in character. With proper skill (such as possessed by any draughtsman or engraver) ornamental effects and designs equal to the finest copper or steel plate engraving can be produced with as much ease and rapidity as pen or pencil drawings upon paper. When flashed or enameled glass is operated upon, the effect is strikingly beautiful, as the colored surface is cut through, leaving clear crystal lines of the white glass below in view. There were exhibited three automatic engraving machines employing the above process; these were for the production of designs of various kinds upon glass by mechanical means. A geometrical lathe produced a great variety of lace work or "engine rulings," similar to the Belgian acid etchings; another produced a Greek border around circular glass objects; while the third engraved lettering and pictorial designs upon the same by means of a magnetically controlled point which made the electrical contact with the face of the plate and actuated the controlling magnets. This process was operated at two spaces in the Electrical building, and the product sold as souvenirs to visitors. During the continuance of the Exposition over fifty thousand pieces were engraved with various inscriptions, and sold; one of the interesting features being the engraving of pieces with foreign or oriental inscriptions in native characters. Pieces were engraved in Chinese, Japanese, Singalese, Sanscrit, Arabic, Javanese, and Russian languages and characters, with the same exactness and ease as English, French, or German. Many visitors engraved their own souvenirs, and many had society or religious insignia engraved upon them. The souvenirs engraved were sent to nearly every quarter of the inhabitable globe—even to such remote regions as Soudan, Central Africa, Sepaul, Lapland, and Chili.

Eddy Power Plant.—Machinery Hall.

JAMES H. LANCASTER.

HOISTERS AND GRAPPLES, OPERATED FROM TROLLEY LINES.—The Lancaster Hoisting Machines are novel and extremely efficient. They revolve the circle, are adjustably balanced, raise and lower the steel boom automatically while lifting or lowering the grapple, and dig or discharge anywhere within the entire and varying radial limits of the boom. They are made, when required, self-propelling in either direction, and are worked by one man, either by steam, electric, horse or hand power, and may be mounted on a truck, platform, deck of scow or pontoon, or otherwise. These grapples or dredges dig solid earth, gravel, mud, sand, ore, coal, hardpan, etc., with great ease and rapidity. They have been used for digging some on the seashore at Sandy Hook for the Government; digging nine miles of canals at Savannah, Georgia; raising phosphates in the Carribean Sea; delivering coal in Kentucky; sinking wells at Columbus, Ohio, etc.

THE EDISON MANUFACTURING COMPANY.

FAN AND OTHER MOTORS.—These motors are noiseless, have self-oiling bearings, and are entirely enclosed in iron-clad cases, being provided with glass windows at rear for regulation of brushes. The workmanship is of the very highest grade. The motors are made for operation by the Edison-Lalande primary battery, four cells.

A. PIAT & SONS, PARIS.

HYDRO-ELECTRIC RIVETER.—One of the largest items of expense in erecting bridge works is the field riveting. Field rivets are never really perfect and tight, and their strain is figured by leading engineers at 30 to 50 per cent less than for shop rivets. This means an increase in the number of field rivets, and as their cost is greater, the final cost of field riveted connections is increased by 40 to 60 per cent. As large riveted bridges are becoming every day more in favor among bridge engineers, the field riveting amounts to a large part of the total expense. These considerations have led to the construction of a field riveting machine of small size and great power. To obtain the best results a good riveter must be driven not by hand, but by some outside source of power. The combination of hydraulic pressure and electricity have answered the purpose. The main parts of the machine are: A C-shaped cast steel frame; a plunger-pump, with its water cylinder; the rivet punch, with hand contact motion; the electric motor and its intermediate gearing. By means of the electric motor and its intermediate gearing the plunger is made to act upon the liquid in the large cylinder, thus furnishing the necessary pressure upon the rivet punch. A special feature of the machines is the hand contact motion of the rivet-punch, which secures a better and more progressive application of the water pressure than in other hydraulic machines on the market. Another distinctive feature is the special switch, which allows the machine to run in either direction, and also controls automatically the brake, and another switch arranged to stop the motor when the necessary pressure upon the rivet has been reached. Besides the automatic device above mentioned, a gauge allows the operator to ascertain the pressure required and also operate the switch level. The electric pressure ordinarily used in the motor is the pressure used for incandescent or arc lighting, preferably 110 or 220 volts. It is obvious that greater pressures can also be used in the machines. The riveter exhibited will put in rivets up to ¾ inch in diameter under a pressure of 30 tons on the rivet. The firm also make a more powerful type for rivets up to one inch, requiring a pressure of 45 tons. The advantages of the machine are numerous: The liquid used is within the machine itself; therefore, all flexible pipes and other means of transmission of water under pressure from points at a distance are avoided. The difficulties arising from the passage of flexible pipes and other means of transmission of water under pressure from points at a distance are avoided. The difficulties arising from the passage of flexible pipes, through girders, lattice works, and other obstructions naturally found in bridge work, are completely done away with. The well-known liquid mixtures are used in the machines which are best calculated to avoid freezing in cold weather. The machine can be placed in all positions, horizontally, vertically or at any angle. It is also especially adapted to cantilever work, as it can be suspended by means of a cable and brought easily in position where required. Its light weight permits its easy transportation, even in the case of bad roads. The dynamo is operated by three switches: the dynamo switch, the working switch and the safety switch. The dynamo switch admits the current to the motor field; the working switch, which is worked by hand and automatically, admits the current to the brushes and to the rheostat; the safety switch, worked automatically, causes the rheostat to act, and short-circuit the motor, in case the working switch should fail to work. The current admitted by the dynamo switch goes through the motor field and to the working switch, then to the rheostat, and then to the dynamo switch, or directly to the latter. The speed and direction of rotation of the motor are automatically controlled by the working switch connected to the rheostat. The points and spring contacts of the former are so arranged that the total resistance point is in the middle. The switch, when turned on either side of that point, gradually cuts off the resistance, causing the motor to revolve in either direction. But after the full speed is attained, all the current will automatically pass through the rheostat. The machine has been improved by adding the safety switch by which the motor is automatically short-circuited at the end of the stroke of the plunger, thus using the resistance of the motor to stop the rivet punch. By this combination mistakes in switching become impossible. Great rapidity of operation is attained, even as high as 120 rivets per hour.

CHAPTER VII.

TRANSMISSION AND REGULATION OF THE ELECTRICAL CURRENT.

MODERN practice in the transmission of the electrical current has come to a point where perfection depends more on the mechanical construction of apparatus, and on care in making installations of machinery, than on any new principles that are being advanced. There have been some experiments recently made, however, by two experts, under the auspices of the two largest companies in this country, looking to the employment of electricity in a much wider field than heretofore occupied. These two experts are Prof. Nikola Tesla, of the Westinghouse Company, and Prof. Elihu Thomson, of the General Electric Company. Especial significance attaches to the experiments of Tesla, owing to the fact that the Westinghouse people have gone extensively into the practices advanced by him, and are now making a great part of their apparatus under his patents and according to his system. The two noted installations employing the Tesla high-frequency alternating system are the great incandescent lighting plant that served the Columbian Exposition so magnificently, and the long-distance high-frequency alternating light and power plant at Niagara Falls, now in course of construction. It has been admitted by many that both Professors Thomson and Tesla were doing some very interesting work in their special directions; but by these same people, who ought to be well posted, it was denied that the experiments were of any practical value, or that they were likely to prove of value in the future beyond the demonstration of certain scientific principles. That this class of people have been mistaken, the splendid work of the Westinghouse people abundantly shows. In addition to the class of work above alluded to, it will be proper, under this head, to treat of that class of devices and apparatus employed as auxiliary to the transmission of the current; such as subways, transformers, switches, rheostats, lightning arresters, etc. Wire cables may be more properly considered under a special head.

WESTINGHOUSE ELECTRIC AND MANUFACTURING COMPANY.

POWER TRANSMISSION APPARATUS.—The Tesla poly-phase system consists, practically, of two or more alternating currents used to produce a rotating magnetic field. The advantages of this system may be briefly classified as follows: The use of alternating current for transmission at high potentials; the ability to economically obtain both alternating and direct current by the use of transformers; the ability to supply incandescent lamps, arc lamps, alternating-current motors of two types and direct-current motors of every type from a single generator and a single transmission circuit; efficiency for every kind of service; variety in types of motor. In addition to the Tesla motors of the rotary field type, the system is adapted to two-phase motors of the synchronous type. By means of a commutating machine, direct current is also readily obtained; and this of course may be utilized for any kind of direct-current constant potential motor. The company showed in operation the most extensive plant of the poly-phase system which has been erected up to the present time, either for exhibition or for practical operation. They have showed their faith in apparatus constructed under Tesla patents by building and installing an operative plant on a very large scale. The power for operating the machinery here exhibited was obtained from the lighting circuits supplied by the plant in Machinery Hall. Two-phase currents from this source were employed to operate two Tesla motors, one a 500-h.p. machine, which took the place of the turbine or engine which would be used in practical operation to drive the generator, the other a 5-h.p. motor, direct connected to a small continuous current machine which furnished the field excitation for the generator. Neither of the above mentioned Tesla motors was intended to constitute a part of the exhibit proper, but served to furnish the power necessary for operating the plant and to prove that the two-phase alternating current of sixty periods per second supplied by the installation in Machinery Hall, was adapted, not only for supplying incandescent lamps and arc lights, but also for the operation of motors. The apparatus to make this exhibit comprised, in the generating station, one 500-h.p. two-phase a.c. generator, one 5-h.p. direct current exciter, a marble switchboard with instruments, and raising or step-up transformers which received alternating current at a potential of 400 volts from the generator and delivered current of 1200 volts to the transmission circuit. In practical transmission, great distance would of course

115 H.-P. Edison Motor Driving Line Shafting.
Machinery Hall.

necessitate the use of much higher potentials, but it was not considered advisable to introduce extremely high voltages in an exhibit. The conductors were carried from the step-up transformers to a pole, and thence to another pole located near the switchboard of the receiving station. The wires were supported upon special glass insulators designed by the Westinghouse Company. No oil is used in these insulators, which are similar to those in successful use in the transfer of energy from the power station in the San Antonio Canon to the city of San Bernardino in southern California, a distance of 28 miles. The step-down transformers were located behind the switchboard in the distributing station, where the potential was reduced to 400 volts. From this switchboard a part of the alternating current was conveyed to a small rotary transformer delivering current at a potential of about 50 volts, intended for electrolytic purposes but used for the operation of the Schuckertt light. The remainder of the current was conveyed to a large 500-h.p. rotary transformer, used to supply direct current for operating a Dorner & Dutton railway truck, equipped with two 30-h.p. reduction motors or to drive a 60-h.p. direct-current motor mounted upon an 18-inch Ingersoll-Sergeant air compressor, and also to supply direct-current arc lamps of the constant potential type arranged in sets of nine across the 500-volt circuit. The rotary transformer was also used as a motor, and, by means of a belt made to drive a Worthington pump and an alternating-current arc lighting machine.

As stated above, the transmission exhibit began properly with the two-phase generator, which was driven by a motor supplied with current from the lighting circuits, and was identical in every respect, save one, with the system which has been recently adopted by the Cataract Construction Company for their large installation at Niagara Falls, the one difference between the two systems being in the rate of alternations employed, which, instead of being 30 periods, as used in this exhibit, has been reduced to 25 periods per second for the Niagara Falls installation.

Exhibit of Tesla Apparatus for High Frequency Experiments.—In one of the Westinghouse spaces was erected a room about 25 feet square, in which were daily shown experiments illustrating the remarkable results obtained by Mr. Tesla in the use of high-frequency, high-potential currents. The current was taken from the lighting circuit connected with the plant in Machinery Hall, and transformed up to an exceedingly high voltage by transformers situated back of the room. The rate of alternation was increased from 7,200 per minute to 400,000 per second, by the use of condensers and an arc-breaking device.

Within the room was suspended two hard-rubber plates covered with tin foil. These were about fifteen feet apart, and served as terminals of the wires leading from the transformers. When the current was turned on, the vacuum bulbs or tubes, which had no wires connected to them, but lay on a table between the suspended plates, or which might be held in the hand in almost any part of the room, were made luminous. These were the same experiments and the same apparatus shown by Mr. Tesla in London about two years ago, where they produced so much wonder and astonishment.

Over the entrance to this room there was a most interesting effect. A glass sign, bearing in its center the name "Westinghouse," was supplied with a current of high-potential, one terminal of the transformer being connected to the letters by means of invisible wires, while the other terminal was connected to tin foil on the back of the glass plate.

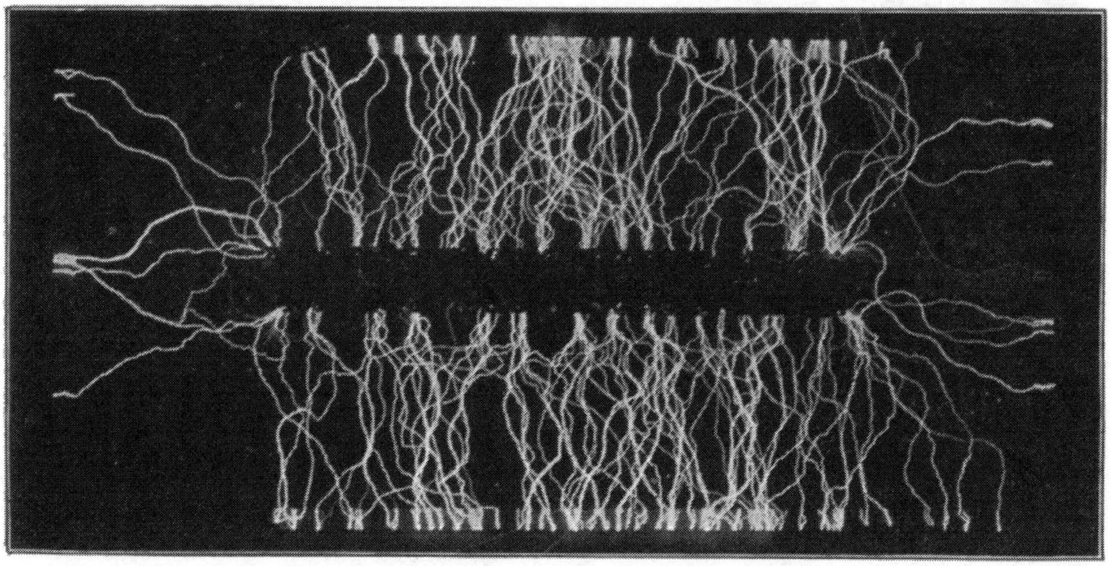

HIGH POTENTIAL DISCHARGE.—TESLA.

When the current was turned on, the electricity was discharged from the letters across the surface of the plate, and over its edge to the tin foil on the back, producing the effect of a modified lightning discharge, and accompanied by a similar deafening noise. This was probably one of the most novel attractions in a sensational way seen in the building, as the noise could be heard anywhere within Electricity building and the flash of the miniature lightning was very brilliant and startling.

Outside of the booth were exhibited a collection of the various forms of motors and other apparatus employed by Mr. Tesla in the discovery and development of his poly-phase system. Here, also, were seen mechanical and electrical oscillators, the subject of a lecture by Mr. Tesla to the American Institute of Electrical Engineers. The lecture was delivered in August, and attracted a great deal of scientific interest.

TWO THOUSAND VOLT TRANSFORMER.

On a table near this room was a Tesla ring, which illustrated the principle of the rotary magnetic field. The ring consisted of an iron core wrapped with two coils of wire, through which a two-phase current was passed. The effect produced upon metallic bodies placed within the influence of this field was practically the same as if a magnet were rapidly carried around the axis of the coils; that is, the metallic bodies were set rotating.

WESTINGHOUSE TRANSMISSION SPECIALTIES.—In addition to the polyphase transmission exhibit, the Westinghouse Company showed a great variety of protective and safety devices, transformers, lightning arresters, etc., of which the following will indicate the general character:

TRANSFORMERS.—The exhibit of transformers comprised a number of standard transformers shown in Electricity building, and also transformers aggregating something over 5,000 kilowatts capacity used through the grounds and buildings of the Exposition for lighting. There were, also, special converters insulated with oil.

Lately the company has introduced the Wurtz non-arcing lightning arrester in their transformer. Claims are made for these transformers for their high efficiency at all loads, regulation, moderate temperature, safety, cheapness, method of fusing, and protection. The drop, or loss, of potential at the secondary terminals of the transformers, between the limits of no load and full load, is, when the primary transformer is supplied with constant potential alternating current of proper frequency, exceptionally slight. The question of temperature has frequently been overlooked in building transformers; this allows the efficiency to be readily increased at the price of excessive heating. All transformers tested by the Committee on Awards were operated under load for a sufficient time to make it certain that they obtained their maximum temperature, and the special award given the Westinghouse Company makes it evident that this and similar points have been carefully considered in their converter design.

Safety is provided by extremely thorough insulation of the secondary from the primary coil by the method of separately winding and separately insulating each. Only primary fuses are used, and these are removed and replaced without danger of contact with the circuit. The transformers are supplied with independent lighting arresters of the Wurtz non-arcing metal type, where this is desired.

LIGHTNING ARRESTERS.—The standard form of lightning arrester used is known as the non-arcing arrester. It consist of a series of gnurled cylinders arranged side by side and separated by $\frac{1}{64}$ inch air gap. These cylinders are made of non-arcing metal, which is a patented alloy. The two outside terminals are connected to the line, and the third terminal, at the center, is connected to the ground. When lightning strikes the line, it jumps the small air gaps between the cylinders and passes directly to ground. On account of the peculiar, yet positive, action of the metal, the dynamo current is unable to follow the lightning, as would be the case were any metal other than the non-arcing alloy used. The claims for this arrester are as follows: It has no moving parts; it is always ready for a discharge; it operates so quickly that no short circuit of large rush of current takes place; it requires no attention. The form of arrester used for direct-current arc light circuits consists of a coil which holds a small weight supported by magnetism. When the lightning jumps the air space between the carbon electrodes within the enclosed chamber, the dynamo current follows, cutting out the coil and allowing the weight to drop on an extended portion of the arms which hold the electrodes, thus causing them to swing out and breaking the arc. The current then flows through the magnetizing coil, and the weight is again lifted, making the instrument ready for another discharge.

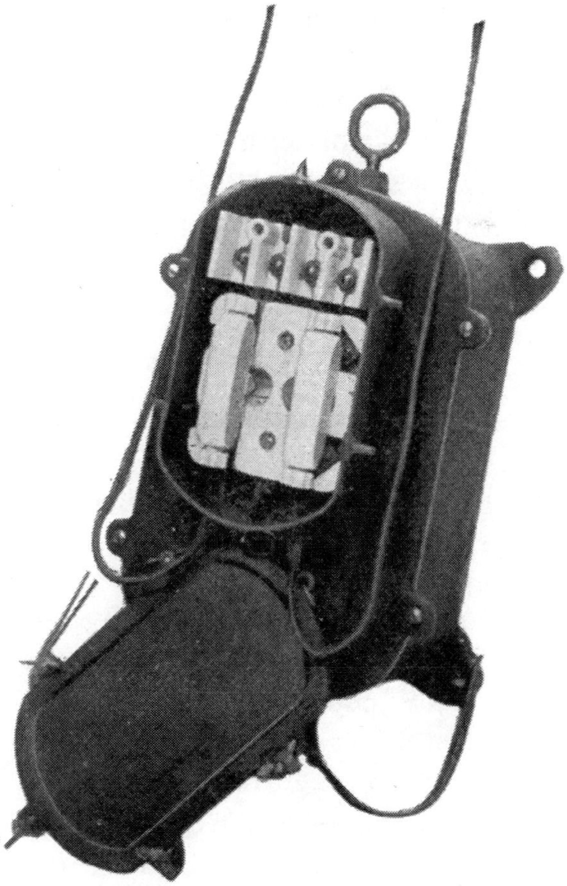

SPECIAL CONVERTER, 1000 OR 2000 VOLTS.

CIRCUIT-BREAKERS.—The automatic circuit-breaker is an instrument to take the place of the inconvenient and unreliable fuse. The instrument opens the circuit instantly and exactly at a predetermined value or strength of current. After the circuit has been opened, the switch may be again closed, and it is immediately ready for action. The instrument is mounted on a solid marble base. It consists of an electro-magnet, below which is an armature connected with a trigger which holds the switch closed. On the sides of the switch are carbon plates and carbon pencils so arranged that they will continue contact for a short period after the opening of the metal jaws. A spring catch below the breaker receives the lever after it is thrown back by a spring. The instrument is placed in series with the machine which is to protect, and consequently, the whole current flows through the solenoid shown near the upper part of the breaker. The current in the solenoid magnetizes a soft iron core, which, when the current has increased sufficiently, becomes strong enough to attract its armature. The lifting of the armature raises the trigger, which releases the jaws, and they are carried out of contact by a powerful spring. The instruments are made standard in all sizes up to 1800 amperes. These switches are very simple and durable in construction, and have been designed with the purpose of making all details of the system as complete and satisfactory as the larger apparatus. For high potential circuits a dynamo-changing switch is used. For connecting supply circuits with the alternating mains, cables and plugs are used.

FUSES.—Each style of fuse has its advantages for some particular line of work. Fuse No. 5 is used for high potential circuits, such as 2000 volts, and is wonderfully satisfactory and efficient. It consists of a lignum vitæ block in two parts, between which is laid an aluminum fuse. When the current becomes abnormal the fuse melts and the arc is blown out—through a hole seen on the face of the block—by the expansion of the air within.

THOMSON HIGH-FREQUENCY TRANSMISSION EXHIBIT.

In a portion of the space occupied by the General Electric Company was the personal exhibit of Professor Elihu Thomson. In the center was a piece of apparatus for producing high voltage discharges, by which the current was increased through a series of step-up transformers, condensers, and special induction coils from 75 volts to about 1,5c0,000 volts. This device produced a discharge through 64 inches of dry air, the potential differences of same having been calculated by Professor Thomson to be about 2,000,000 volts. In addition to its interest in the study of high potential discharges, it is valuable in demonstrating the high insulating qualities of oil. At one side of the above apparatus were two tables, one supporting an alternating-current horizontal repulsion coil, and the other a similar, but vertical coil, with rotating discs of various forms, solid copper rings showing the repulsion between induced and inducing currents, magnetic resonators or sound discs, induction coil with lamp in series for

Farmer Motor of 1849, and Painting of the Inventor.

exploring the magnetic field, and other forms of interesting experimental apparatus for use in connection with the repulsion coils. On the other side of the high potential apparatus was a table, on which was an automatic regulator for keeping constant the pressure on the secondaries of a transformer system. It consisted of a re-active coil of variable resistance, operated by a small alternating current motor which cuts in or out the secondary winding of the coil according as the potential rises or falls due to changes in the lamp load. A special form of photometer was shown, having a floating coil and lamp, the former suspended on one end of a balance, with a dampening device on the opposite end, in order that the lamp may burn at constant brilliancy. It may be adjusted to a standard of lights and then used itself as a standard. A three-carbon arc lamp, which may be used in connection with this photometer, was shown also. It takes about 30 amperes at 110 volts, and throws a powerful light, almost always in one direction. On stands in the rear of the above experimental exhibit was a collection of very interesting historical apparatus, selected from a list of several thousand specimens in the laboratories of Professor Thomson, at Lynn, Massachusetts.

THE GENERAL ELECTRIC COMPANY.

The exhibit this company made of special transmission devices was complete, extending in all lines of work.

UNDERGROUND SYSTEM.—The three-wire underground system was shown in model form. There were two 1½-k.w. dynamos, driven through a countershaft by a 3-k.w. motor, all of which were controlled from a slate panel switchboard. The dynamos were connected to a three-wire system of underground tubes of many different capacities and in various stages of manufacture, showing method of insulation. Above these racks were cabinets, showing cross sections of the many styles of wires and cables manufactured by the General Electric Company. Adjoining the underground exhibit was the "Patent Exhibit," showing models used in evidence before the United States courts by the General Electric Company in suits for infringement, in which its patents on the three-wire system and the feeder system were involved.

TRANSFORMERS.—Upon two graduated step stands were shown the company's oil transformers, in all sizes from 6 to 60 lights capacity, the latter size having an efficiency of 97.65 per cent. Upon these stands were also exhibited reactive coils of various sizes, some connected in circuit with lamps to show their effectiveness in operation.

SWITCHES.—A number of non-arcing switches were shown, some with multiple breaks, some with magnetic blow-outs, and one with an air jet to blow out the arc.

MISCELLANEOUS.—On fourteen large display panels were arranged switches; cut-outs, fuses, sockets and all manner of station appliances and parts of dynamos. The switches shown were beautifully finished, of kinds and of capacities ranging from the small 5-ampere snap switch to the immense knife-blade switch for 1200 amperes.

THE MATHER ELECTRIC COMPANY.

A list of specialties was shown by this company, embracing the following: Voltmeter switches, triple-pole single-throw dynamo switches, triple-pole, double-throw circuit switches, double-pole single-throw circuit switches, double-pole cut-outs, connected into the power and lighting circuits. On a switchboard were mounted all of the rheostats for the dynamos. For a further protection to circuits there were automatic circuit-breakers, adjustable from 100 to 600 amperes, and suitable for any voltage up to 600 volts. These circuit-breakers act within 10 amperes of the quantity for which they are set, breaking off current absolutely, thus affording reliable and safe protection for apparatus and circuits.

THE JENNEY ELECTRIC MOTOR COMPANY.

RHEOSTAT.—The motor starter, or rheostat, embodies new ideas. The principal features are found in the switching and releasing device. This device is such that the switch arm cannot be moved to close the circuit through the armature of the motor till the main line switch has been closed, which throws the current on the field circuit. When the motor is in operation the rheostat lever is held in position with all resistance cut by means of a small electro-magnet in connection with a lever and pawl, the said magnet being connected in the field circuit of the motor. When it is desired to stop the motor the main line switch is opened, and as soon as the motor has nearly stopped, the magnet referred to releases the arm and pawl and allows the main switch lever to be thrown back and armature circuit to be opened by means of a spring provided for that purpose. The construction is such that the main switch cannot be left on intermediate sections. Should the main circuit be interrupted by accident or otherwise, the action of the device will be the same as though the main line switch is opened. The automatic action of this device, in cutting the motor out of circuit under such circumstances, prevents injury to motor and a "short circuit" on the line when the current is again thrown on.

THE HOPE ELECTRIC APPLIANCE COMPANY.

There were shown in this exhibit a variety of cut-outs. There were constant-current arc cut-outs; cut-outs for shifting the current, 100 amperes capacity, porcelain insulated, in storm-proof iron boxes. The shifting mechanism is a shaft fastened to a lever which, carried by the center, is thrown fully over by an effective spring in combination with links of brass so placed, and so fastened as to assist and make positive the action of the spring. Constant-current arc cut-outs were shown intended for small central station use, or for cutting out any given number of arc lights from a circuit. Mechanism substantially the same as above, capacity 500 amperes. Alternating and direct current cut-outs; these cut-outs enclosed in substantial iron boxes, protected from moisture, and ventilated, are for use between the main line wires and the transformer, and are made with and without fuses. Their capacity is from 1000 to 4000 volts, and 5 to 50 amperes without arcing or sparking. Bi-pole or trolley-circuit boxes, made substantially as the double-pole primary circuit box, for 500 volt currents with 150 amperes were also shown.

THE ELECTRIC APPLIANCE COMPANY.

TRANSFORMERS.—The Elkhart Transformer exhibit consisted of the transformers in actual operation supplying the current used for illuminating and power purposes in the exhibit. The converters were so arranged that their close regulation, high efficiency, and cool running could be readily shown; and being in actual operation it made a very practical converter exhibit.

SWITCHES.—An interesting sample table showed all sizes and styles of New England double-pole snap switches and the switch parts from the cover to the porcelain base, in capacities ranging from 10 to 100 amperes. Jack-knife switches were shown in all finishes, shapes and capacities.

LIGHTNING ARRESTERS.—The swinging ball lightning arrester was an important factor in this exhibit.

ELECTRICAL SPECIALTY COMPANY.

SWITCHES.—The exhibit was of the Triggs Standard Single Push Button Flush Switches for controlling incandescent light. These switches are in the form of a single push button and so arranged that by pushing a button the lights or a set of lights may be turned on, and by pushing the same button the lights may be turned out. The switch stands flush with the wall instead of projecting as others do. A set of lights may be controlled from an unlimited number of places which is the only single push button apparatus for doing this work. The switches are made single-pole, double-pole and three-way. The making and breaking quality is worthy of mention as the light is on till off, and off till full on.

E. S. GREELEY & COMPANY.

FUSE PROTECTOR.—The W. B. B. Protectors are mounted in various ways to accommodate different circuits. This protector will fuse at its rated capacity, to within one-tenth part of an ampere. They are made to fuse from three-tenths of an ampere up. They are made of a piece of fuse wire on one end of which is fastened a small coil of German silver wire; the other end of the German silver wire forms the end of the fuse which is ordinarily made three inches long. Thus the fuse operates by the hot wire coil principle. Any excess of current over the rated capacity will instantly blow the fuse. The fuses are mounted on vulcanized fibre bases for any number of circuits. The value of such a device is readily seen when one remembers that an ordinary fuse will generally carry twenty-five per cent. to fifty per cent. more than its rated carrying capacity depending upon the temperature. For protecting the telephone from the dangerous sneak current is its primary object, but it is used to protect all circuits where magnets are used, as in telegraph or fire-alarm work.

EMPIRE CHINA WORKS.

INSULATORS.—The exhibit was of highly vitreous hard rubber porcelain insulators and ware for insulating purposes,—insulator knobs, bushings, cleats, circuit breakers, push buttons, cut-out bases, switch bases, buttons or tips for electric lamps, etc. The goods are liquified by subjection to the operation of baking to a snowwhite heat whereby the ware is made non-conductive throughout and becomes the best of hard porcelain whether glazed or unglazed. This porcelain ware was in use in the Exposition in the switches and cut-outs furnished by the Bryant Electric Company of Bridgeport, Connecticut.

THE INDIA RUBBER COMB COMPANY AND THE GOODYEAR HARD RUBBER COMPANY, NEW YORK.

This exhibit consisted of electrical supplies made of hard rubber,—sheet, rod, tubing, cells of different forms and capacities, bushings, insulators, separators, telephone handles, receivers, transmitters, and many other articles made for other parties as parts of their apparatus.

O. S. PLATT.

NEW ENGLAND SWITCHES.—This exhibit consisted of double-pole switches, and their parts unassembled, showing the construction of, and the workmanship on these goods, as well as the material used, which was of the best quality and selected especially for the different functions each piece had to perform. The insulator is secured to the cross-bar or working bar, and strengthened by means of a metal band, to which is attached the copper and spring metal contacts, thus preserving the strength of solid metal and gaining a perfect insulation. The "cross-bar" which holds the contacts firmly in position and liberates them instantly, is a cam, the most positive and quickest mechanical device; it is so constructed that when turned it forces the cross-bar down to the base of the switch, and in locking allows it to recede slightly, and the contacts take their place on a clean surface, made so by the wipers while going down. The switch as above constructed, is supplied with a suitable cap or cover, a handle, and convenient means of wiring up. This exhibit consisted, in addition to the above, of samples of switches with very wide break, for use on high voltage circuits, and can be made to stand as high as one thousand volts, fifty amperes. These switches are made of interchangeable parts and can be assembled without the application of technical knowledge.

THE STANLEY ELECTRIC MANUFACTURING COMPANY.

Transformers.—This exhibit consisted of a number of latest type transformers. In these transformers the iron loss has been reduced to as low a point as possible compatible with good regulation; the idea being to keep the all day efficiency of the transformer at as high a point as possible. The method employed in insulating these transformers is worthy of mention. After being wound the coils are boiled in a bituminous compound, while in a practical vacuum, the result being that the wires are completely embedded in the dialectic, and a cross section of a coil presents the appearance of a cross section of a telephone cable, each wire being entirely encased by the insulating compound; the coil is thus rendered absolutely moisture-proof. The weight of the transformer has been reduced by using an iron with the least possible hysteresis loss. The fuses are used only in the primary circuit, and the fuse boxes are made entirely of porcelain. The fuse proper is attached to a porcelain plug which can be used interchangeably in all sizes of transformers, and although the fuse box is "double-pole" each fuse is in a separate apartment, so that the possibility of an arc being formed through the vapor caused by the melting of a fuse is precluded. The arrangement for attaching the Stanley transformers to cross arms or to the side of buildings enables them to be erected with a minimum of trouble.

N. C. GAULT & COMPANY, WINONA, MINNESOTA.

Lightning Conductor.—This firm made an exhaustive exhibit of rods and cables for the conduction of lightning. One kind of construction deserves mention. It is composed of a series of small hollow cables

Motor for Door Operator.—Box Open.

made by twisting twelve galvanized steel wires around a small copper tube, making a very smooth and flexible cable. These cables are then covered with copper as they are wound or twisted around a large hollow copper wire which composes the center of the finished cable. By a peculiar device of the machine making these cables, the seams of the copper coverings of the small cables are all carried to the center, thus rendering them impervious to water and the whole having the appearance of a cable made from solid copper wires. Yet, with all its apparent complexity, the mode of construction of this cable is so simple that a coil of two hundred and fifty feet can be made in twelve minutes. The machine for the manufacture of these cables was also shown. The simplicity of construction of this machine together with the perfection and rapidity of its work attracted much favorable attention.

GENERAL ELECTRIC COMPANY, LIMITED, LONDON.

SWITCHES, CUT-OUT, FUSE BLOCKS, ETC.—The main part of the exhibit of this company was the excellent switches, and fittings for installation work which are the outcome of experience, necessity and rigid rules of the English Board of Fire Underwriters. The switches, which are highly finished and carefully designed, were shown for all capacities common in practice; however, all have the same fundamental construction, namely, a link movement, which gives them a double-locking and quick-breaking action. In the larger ones the contacts are adjustable and have large contact surfaces and comparatively long break. The smaller ones of china in various colors and highly ornamental, designed to harmonize with the wall papers, furniture or surroundings, are exceedingly well worked out and disclose in every point careful consideration in preventing trouble. Some of these switches, while mounted on a china base, are provided with a bronze or enameled shell of ornamental design and very solid structure. Among the fittings the most important part are undoubtedly the fuse blocks, cut-outs, wall sockets, ceiling rosettes, etc., etc. This special make is known to the trade as the "H.I." (high insulation) system of electric fittings manufactured by this company. All of these devices are constructed with a view of obtaining highest possible insulation, and are thoroughly proof against fire, short circuits and leakage. The conductors are either screwed into the china itself or fitted to surface walls. Contrary to American practice no metallic parts such as screws, or nuts are to be found at the back, so that dampness of the wall against which they may be placed or other similar causes can have no effect upon them. Neither screws nor covers, nor handles, etc., come into contact with any metal parts of the device carrying current or in some way forming part of the circuit and all flexible cord is especially protected by dividing it at the base by means of insulating wall. The fact that the flexible cord or, better, the copper becomes brittle, after the current has passed through it for some time, has caused a further modification of construction especially in ceiling rosettes, etc. to prevent the weight of the incandescent lamp and shade to be carried by either the attaching screws or the wire conductors alone. A china wall inside, in one piece with the base, supports the terminals of the leading-in wires. The latter are passed through holes in the wall just named retaining the insulating covering which is only removed at the part to be placed in contact with the terminals. To prevent their becoming loose the two wires are twisted before being attached to their respective screws. Evidently this construction prevents all arcing from one terminal to the other and the short-circuiting by some of the fine wires forming part of the flexible cords, which is a common occurrence in many other designs. To manufacture these goods the English porcelain lends itself particularly well, especially as not only the coarser threads by which the covers are attached to the base can be properly made but even the finer threads for the small screws can be made in the porcelain so that all conducting metal parts can be screwed on from the front, thus leaving the back a perfectly smooth insulating surface. An inspection of a section of these small threads in the porcelain discloses that they are as sharp and clean cut as they are accustomed to be found in metal. The cut-outs are well worked out and embody all the improvements named. The so-called "A" terminal for all currents up to 25 amperes, consists of a brass tube, the whole of which is in a straight line with the leading-in wire holes in the china base, and has a set screw at the side, so that the wire may be firmly clamped in the terminals. Particularly noticeable were also the many artistic fixtures ánd electroliers which disclosed considerable taste and novelty.

ALLGEMEINE ELECTRICITATS GESELLSCHAFT.

Regulating Switches and Safety Devices.—The automatic regulator serves for maintaining a constant pressure in electric installations, especially in small ones, where it is often desirable to employ a special attendant for regulating the voltage by inserting resistances or changing the number of elements of the battery as required, and the tension or current is thus kept constant. This ingenious device may be used, on the other hand, for controlling the engine driving the dynamos. In the second place were seen fluid resistances for starting motors, metal-resistances and cut-outs for the exciting circuit in shunt-wound machines, controlling resistances for incandescent lamps, resistances for arc lamps, safety-fastenings for arc lamps, and switches and cut-outs of different kinds. The most important task for the electrical engineer is the manufacture of reliable and practical installation fittings with a view to increasing the safety of installations and diminishing their cost. The Allgemeine Electricitats-Gesellschaft has made satisfactory progress in this difficult field of work, producing an independent and finished system, to which the rapid extension of electric light in and out of Germany owes a great deal. A sufficient proof of its good qualities will be found in the fact that it has been copied on all hands. To give a detailed description of the house-connections with their accessories, distributing switchboards, and various installation fittings, would lead too far. To show the care which has been taken to exclude all possible accidents, the safety switches, made of fire-proof material may be referred to. Mistakes in inserting plug-fuses of different sizes are ingeniously obviated by using plugs and screws of different heights, because inserting too big a plug into a weak circuit would make no contact. Lightning conductors serve for additional safety against thunder storms. The cut-outs are of solid and durable construction, capable of resisting most external influences.

GAULARD & GIBBS.

TRANSFORMER.—This was one of the first instruments combined by Gaulard of the type that he exhibited in Turin (Italy) in 1884, and with which he started the industry of alternating current transformers in London in 1883.

VIALET CHABRAND.

MAGNETIC SWITCHES.—The lamps of an apartment can be lighted or put out either from any of the doors or from either side of the bed, that is to say, from as many points as may be desired; the control is effected by means of pushers very easy to fit up, one of which serves to light and the other to put out the lamps; the ability of putting out the light in all the rooms however short, may be the time one has to be away whereby there is not only no danger from fire but also a great saving both with respect to current and lamps, which thus become cheaper than any other mode of lighting whatsoever; the staircase lamps can be either lighted or put out on opening the front door on the ground floor, on each landing, or even from the porter's lodge; on board ship, the cabin lamp can be lighted from the door and from each berth, a most comfortable thing for passengers and a saving of both current and lamps; this last advantage arises chiefly from the fact that a passenger prefers not using the lamp when the latter becomes tiresome and not wanted and also from the fact that all the cabin lamps can be put out from the main switch board by rapidly moving the levers of the switches controlling the battery circuits, whereby all the cabin lamps are extinguished when there is no one in the cabins, as at meal times, for instance. These magnetic switches intended to be of use everywhere for electric lighting purposes, were combined in order to allow of both lighting and extinguishing the lamps from as many points as may be required and with the greatest possible ease, so that the lamp might be turned out when not wanted, in thus reducing the working of the lamp to what is strictly necessary, and the cost of light and current to the very lowest point. For each place where the electric light is used, the magnetic switch type depending on the number of lamps, comprises any number of ivory or porcelain cases, in which are placed two push buttons, one of which serves for lighting and the other for extinguishing the lamps. When the one is depressed the current is sent into the apparatus, which closes the circuit of the lamps of the apartment; and when the other is depressed, the direction followed by the current is changed in the switch in opening the circuit and, consequently, in extinguishing the lamps. On account of the operation of the new apparatus each of the two button cases does exactly the same work as the switch of a group of lamps in thus suppressing the current in all the wires of the apartment, but with this essential difference that the fitting up of the two button cases all of which are connected together by means of a three wire cable, can be most easily done, whilst it would be absolutely impossible to arrange a number of switches from each one of which all the lamps could be controlled.

L. SELIGMAN.

Samples of Gutta-Percha.—The samples shown were collected in a mission performed by Mr. Seligman, engineer of the Postal Telegraph department of France, in 1881-2, to Singapore, Sumatra, Malacca, etc. They represented various species of gutta-percha collected in those countries. This mission has given the impulse to the different enterprises started during the recent years for the culture of gutta-percha.

M. CLERAC.

Carbon Resistance.—Resistance was made with a carbon paste dried and placed in ebonite tube. The tube is closed at its two ends by metallic fittings provided with binding posts. The resistance of the carbon mixture varies according to its composition, and to the pressure exerted upon it by the metallic fittings.

SUBWAYS AND CONDUITS.

THE ELECTRICAL CONDUIT COMPANY, NEW YORK.

This company's system obtained the only award at the Exposition for underground construction; the system has also been adopted officially for the cities of New York and Chicago. The exhibit showed a section of subways with manhole, feeders and various sized branches complete.

Manhole.—The company adopted iron manholes as being more durable and substantial, and preventing the admission of moisture. A great advantage in the form and plan of the structure of this manhole is its being made in sections or rings of varying widths starting at the bottom with a solid iron pan, and by simply placing rings in succession, one on top of the other, until the structure thus made has reached a height in its rise towards the surface on a level with the grade at which the conduits lead into the manhole from the several directions. This ring, or section of the manhole, is called the spigot ring, and is formed in segments. In order to accommodate any size conduit, the spigots that are to be placed in position of any size aperture (for the conduits leading laterally therefrom) are all of uniform size on the outer edges of the plate, fitting exactly into the spigot ring, taking the place of one of the segments that construct the spigot ring. By this mechanical construction a larger or smaller spigot can be substituted at any time. Surmounting this spigot ring is a hood made of sufficient strength to bear the street traffic, tapering to a proper size opening at the street for the admission of a man.

Conduit.—The conduit system consists of an iron box composed of two troughs, one placed upon the other, with the longitudinal divisions at the sides, a groove being cast on the edge of the lower trough and a tongue cast on the edge of the upper trough, so made and exact in size that when put together must necessarily result in an exact fit. These sections are five feet in length, as are all the varying sizes of internal capacity. To insure joints being water tight, prior to the laying, all grooves, recessed clamps and abutting ends are filled with what is known as plumber's putty. The more it is immersed in water the harder it becomes. Its entire reliability for making a tight joint to resist the passage of water has been ascertained by experience and careful observation in conduit structure in continuous operation for over eight years.

Distribution.—The question of distribution of wires and their easy handling is the one vital question underlying the entire subject and prob-

Subway Used at the Exposition.

lem of placing electrical conductors underground, and this point is the one to which this company has given the greatest attention, and assert that they have achieved and worked out to a state of completion a mechanical means that enables them to handle wires underground, from point to point with perfect facility and convenience. In the course of laying the conduits, they place a handhole section in front of each building. These handholes are formed in the top section, and consist of an opening large enough to give free access to the wires in the top ducts or compartments. This opening is covered with a tight and secured plate, removable at will, so constructed that when a connection is desired to be made, with any house, an excavation is made immediately over this plate, its location being known by marks upon the curb-stones, or measurement from the nearest manhole plate (careful records always being kept) in case the handhole is not brought to the surface. The workmen go into the cellar of the building with which connection is to be made, remove a single stone in the foundation, and by means of a system of dirt-augers and drills, a hole is bored beneath the sidewalk and curbstone to the excavation over the handhole. An ordinary gaspipe of the size and capacity to accommodate that particular building (in the end of which is forced a conical wooden plug) is started in the hole thus bored from the cellar, and driven through to the handhole opening. When arriving in sight it is very little work with crow-bars to raise or lower or bend the pipe sufficiently to allow its proper admission into the handhole spigot provided for its reception. The handholes are provided with an extra piece, which is called the handhole spigot, which is placed in position at the time the house connection is made. It extends out from the line of the conduit structure about six inches, forming a bell shape. Within this recess the connection or splice made with any wire is entirely outside of the line of the duct from which it is taken. The handhole plan provides also for a separate and distinct opening into any of the ducts on the other side of the conduit structure in respect to the top ducts, as well as to those on the bottom, the idea being to give a separate and distinct duct leading from the conduit to the house for electric light wires and a distinct duct for telephone and other services.

THE STANDARD UNDERGROUND CABLE COMPANY.

This company presented a practical demonstration of the methods of suspending cables in the air, and the installation of cables through underground means for the conduction of electrical energy for telephone, telegraph, electric light and power transmission. The company also showed many samples of carefully prepared joints, tees, and branches, in which were presented the various methods of joining sections of cables one to the other, the looping out of wires for service, or the taking off of branch lines. In connection with the cables as a necessary adjunct, they presented various means of terminating the cables in appliances known as terminals; whereby the end of the cable is hermetically sealed, and yet presenting an easy and effective method of reaching the ends of the wires for tests or connections. Under this general head of terminals, may be included distributing boxes, by means of which a main cable may be cut into, the wires for various services may be taken out without disturbing the working ability of other existing lines through the cables. This treatment of the cables was practically shown by the bringing out from the manholes and up the pole a section of underground cable; and it in turn being shown as connected through a terminal to the aerial cables for distribution above ground.

THE NATIONAL CONDUIT MANUFACTURING COMPANY.

This company presented for exhibition a complete installation of underground conduit and manholes. The process of installation was shown, and the chief feature of their manufacture is best described as a cement lined sheet iron pipe, made in such lengths as to produce the greatest flexibility of application, if impediments in the path of the conduit work are encountered, and designed to facilitate the work of construction. The conduits are laid in a matrix of cement, the whole mass, when finally completed, being a veritable continuous tube of imperishable rock. They are also manufacturers of pipe for water works and corresponding systems; also manufacturers of multiple conduits. More than 15,000,000 feet of their product has been placed in service in the various cities of the United States and Canada. They are also owners of patents covering manhole construction, which affords unequaled facilities for the accommodation of cables and their protection.

THE FIBRE PIPE COMPANY.

The Fibre Conduit Company exhibited a conduit specially adapted to underground work and composed of spruce fibre and a preservative. The fibre is mixed with water and formed into a pipe or tube in "the white" by machinery. It is then placed in a kiln until perfectly dry when it is taken out and immersed in a bath of material to preserve it and render it water proof. The finished pipe is strong and not only water proof but is unaffected by either acids or alkali and can therefore be laid directly in cement if necessary. The finished pipe can be readily turned in a lathe and joints made absolutely tight.

THE INTERIOR CONDUIT AND INSULATION COMPANY OF NEW YORK.

This exhibit consisted of a system for the installing of wires in buildings, for the carrying of electric current. The system is that of tubes of a light but fireproof material, laid in the plaster of the walls with proper junction boxes, elbows, etc., giving a continuous duct air tight at the joints, throughout the building, in which are drawn the electric wires carrying the current. There were two large sample boards showing in detail the methods of installing the conduit, and also the tubing plain and brass covered, of various tints, junction, feeder and main line boxes, for the cut-outs, elbows, tee's, couplings, cut-outs, pendants, Johnson Switches, dead beat ammeters, and all the tools used in reaming, cutting and fitting the conduit. The brass covered conduit has double lap seam and the couplings are cemented on, so that the tubing when installed becomes a continuous insulated brass pipe throughout the building, with soldered joints rendering the wire protection from dampness, fire, or any action of the destructive alkalies of cement. The fittings of this conduit are so arranged as to match in color and design almost any interior decoration so that the system may also be used successfully in old buildings, where it would be placed on the surface of the walls.

THE H. W. JOHNS MANUFACTURING COMPANY.

Insulating Materials.—The electrical exhibit of the H. W. Johns' Manufacturing Company, displayed insulating compounds in great variety. These were Asbestos in its various forms, adapted for electrical purposes; Vulcabeston, which is a composition of Asbestos and India Rubber; Moulded Mica, which is a compound of flake mica and insulating gums; and "Monarch" insulation, a compound consisting of Mica and other materials. The facility with which Vulcabeston may be manufactured in various shapes and sizes was evidenced by the numerous samples exhibited. These ranged in size from the smallest bushings and washers used in arc lamps and rheostats, to the large insulating rings used in the Westinghouse 750 k.w. generators (used for lighting the Exposition), field-spools for the 250 h.p. Eddy motor or generator, dynamo and street-car motor spools and bushings of the General Electric Company, and many other pieces. The street-car controllers of the General Electric and Westinghouse Companies are insulated throughout with Vulcabeston. A method of insulating armature cases was illustrated by a Gramme armature ring of a 100 arc light dynamo of the Excelsior Electric Company. This ring or core has a section about six inches square, with an outside diameter of about thirty inches, and is covered with Vulcabeston in the form of an envelope, one-quarter of an inch thick, moulded and vulcanized in place. This method of insulating an armature core is noteworthy. The insulating envelope is unbroken, except at those points, in the interior of the ring, which are provided for the mechanical attachments with the feet of the spider ; thus producing an armature, the insulation of which is integral with the core, and practically perfect. Vulcabeston has been prepared to endure a temperature of 500 degrees F. without losing its insulating and mechanical properties. As ordinarily manufactured, for practical requirements, the material is sufficiently vulcanized at temperatures between 300 and 400 degrees F. Moulded Mica is, as above mentioned, a patented composition of flake mica and resinous gums. The material is moulded in dies under pressure, possesses superior insulating qualities, and is practically waterproof. Metal pieces may be moulded within the material or filled with this insulating compound. Samples of the company's trolley line insulators were displayed, prominent among which were the "Round-top," "Swivel," "Salem bell," "Iron Clad" and No. 1 bell hangers, with pull-offs, span and guard wire insulators, globe strains, the new "Giant" strain, insulated with both sheet mica and moulded mica ; ears, mechanical clips, and various other pieces included in a complete list of trolley wire insulators. The exhibit also included samples of insulated screw-drivers and pliers, switches, rheostats, thermostats, controllers, arc lamps, etc., illustrating the application of the materials mentioned above and their practical ability and utility as insulators either when applied in moulded masses, such as blocks and other forms of the materials, or when associated with metal parts moulded within and insulated by them. It was the purpose of the H. W. Johns Manufacturing Company, to demonstrate by their exhibit their ability to furnish insulation for all purposes.

THE STANDARD PAINT COMPANY OF NEW YORK.

Insulating Materials.—The Standard Paint Company, manufacturers of the P. & B. products, made a display of their goods which are very extensively used for electrical purposes. The exhibit included P. & B. electrical compounds, insulating tape, armature and field coil varnish, insulating papers, preservative paints, the P. & B. pipe coating compound, etc. From the nature of the products it was not possible to advantageously show all the articles, but a large and interesting exhibit was made. The P. & B. insulating compounds have, as a base, a practically indestructible residuum, which is dissolved by a very powerful solvent, which acts as a vehicle to deposit the body of compound on surfaces to which it is applied. It is claimed for the P. & B. compounds that they become part of the material to which they are applied almost immediately. They become perfectly dry and hard within an hour after application. The uses of the compounds are many and varied, as they are perfectly water, acid, and alkali proof, and are unaffected by earth or sewer gases. They are used extensively by electric light companies, street railways, manufacturers of insulated wire, cable, etc. Among other purposes for which they are especially adapted are insulating connections, painting moldings and wires when laid in damp places or subjected to acid or alkaline fumes, or under cement; to make cut-out boxes waterproof for coating circuit and switchboards, lamp-hoods, frames, brackets, and insulator blocks, converter boxes, dynamo and motor bases. The compounds are also especially adapted for coating lead and other cables, for underground and submarine uses. These compounds were entered for competition and received highest medal and diploma. The P. & B. armature and field coil varnish—the name of this material implies its use, being especially adapted for armatures and field coils. It will not soften at a temperature of 300 degrees F., and has very elastic and adhesive properties which prevent it cracking or flaking off. It is perfectly moisture proof, a good insulator and will not be cut by oil.

THE P. & B. INSULATING TAPE.—This, like the other P. & B. products, is perfectly water, acid, and alkali proof. The base of the tape is a good quality muslin, which is thoroughly saturated and coated with the P. & B. compound. The features of this tape are that it will not vulcanize with heat nor crack or harden and become defective by exposure and use. It is claimed that joints properly made with it will outlast, under ordinary circumstances, any wire covering. The P. & B. motor cloth has as its base a heavy jute burlap. This material is saturated with a compositon of the P. & B. compound, making it perfectly waterproof, very pliable and elastic. It has been extensively used on rheostats with great success, as sharp corners can be turned without injuring the coating. The P. & B. insulating papers have, as their base, a good quality of manilla fibre paper, thoroughly impregnated with the P. & B. compound. Every fibre of these papers is absolutely waterproof, acid and alkali proof, and they are under all circumstances a perfect insulator. They are not affected by heat up to a point of carbonization of the fibre. Several grades of these papers are manufactured by the company; they are known to the trade under the names of Giant, Hercules, and Universal and P. & B. They were entered for competition and received two medals and diplomas for waterproof and insulating qualities, and were especially commended for superiority over papers coated with wax, tar, or oil. The World's Fair cold storage house was insulated throughout with P. & B. paper. The P. & B. pipe-coating

H. W. Johns' Exhibit of Asbestos.

compound is an absolutely pure bitumen, prepared with great care, possessing the especially valuable properties of great adhesiveness, and of being entirely odorless and tasteless. For coating pipe, both wood and iron, more particularly for conveying water, it is far superior to any mixture of coal tar and asphaltum, or of any grade of either of these materials. It is applied by dipping pipe into a vat containing the coating compound in a liquid condition. A considerable exhibit of the efficiency of this coating was the line terminating at the Fair Grounds of the Waukesha Hygeia Mineral Spring Co., over 100 miles long, this pipe being coated inside and out with the P. & B. pipe coating compound, and a report made to Mr. Higinbotham, president of the World's Columbian Exposition, Jackson Park, Chicago, by the chemists of the Michigan University, Rush Medical College, and Northwestern University schools of medicine and pharmacy certified that the water was delivered at the Fair grounds as pure and tasteless as it left the grounds. The P. & B. paint was given a prominent space in the exhibit. Although the uses for P. & B. among the electrical fraternity are many and varied, it is largely in demand in many other industries—in breweries, distilleries, tanneries, cider and vinegar factories, for railroad and bridge uses, in acid, reduction, chemical and dynamite works, fertilizing works, gas companies, packing and cold storage houses, coal mines, water companies, etc. Neither salt nor fresh water, alkaline or acid solutions make any impression on surfaces coated with P. & B. paint. It is fire resisting. It is extensively used by street railway, electric light companies, and others for coating and preserving all kinds of iron and wood work, such as butts of poles and posts, and as a preservative for iron or wooden conduit placed underground. It does not dry out and will effectually preserve iron from rust and wood from decay. The liquid which is used as a solvent for the paint, evaporates almost immediately after application and causes the base of the paint to penetrate deeper into the pores of metal, wood or brick than any other coating; in consequence of this the expansion and contraction and the ordinary cracking and breaking of other coatings does not occur with the P. & B. paint. The P. & B. paints and compounds are all put up in liquid form of the proper consistency for applying with a brush or by dipping. The peculiar character of this paint also makes it valuable for application to roofs, and to all surfaces exposed to the weather. The P. & B. products occcupy a special and unique place among manufactured goods, and are so recognized, both in the United States and foreign countries. They are manufactured under a number of United States patents.

WIRES AND CABLES.

The dossibilities of transmission of the electrical current depend in no small degree upon the insulation and protection of the medium of transmission, that is the wires. Some of the brightest men in the electrical profession have given their time and talents for a long while to the improvement of the insulation of copper conductors. The best results of their work were shown at the Exposition. A large variety of wires for all purposes was shown, from the bare copper wire to the most perfect of submarine cables, and in sizes ranging from 1-2,000 of an inch in diameter up to ¾ of an inch diameter, solid and flexible cables insulated and incased in a lead covering and two inches in diameter capable of transmitting immense quantities of current. A variety of forms of simple wires were shown insulated in various ways. Wires with an insulation of rubber impregnated tape, wires with a double covering of same, wires further protected by being braided, rubber covered wires simple and braided, tape and braided and further covered with a thin covering of lead for use in damp places. Aerial cables were shown containing conductors ranging in number from 6 to 178 for telegraph and telephone use, these bound with tape, others braided and still others armoured with steel wires for submarine use. For underground use there were shown these same forms encased in lead, and for underground telephone use lead covered cables containing a large number of conductors insulated with paper and with a further air space insulation. Several kinds were shown which resisted interior heat even to the melting point of the wires without destroying the outside insulation.

KERITE INSULATED WIRE AND CABLES.

W. R. Brixey exhibited Kerite Insulated Wires and Cables manufactured by him at Seymour, Connecticut. This exhibit consisted of Kerite Insulated Wire covered with various colored braids and wound upon handsome polished oak reels of graduated size and piled in pyramids. Huge polished oak reels upon which were displayed ærial underground and submarine cables, were appropriately scattered over the space. One reel showing a 334 conductor ærial telephone cable, was particularly noticeable, while another showing an 8 conductor signal cable, 300 miles of which had recently been furnished the New York Central Railroad Company for their signal service between Albany and Buffalo ; another showed an 18 conductor submarine Osmond telephone cable similar to one recently laid across the North River for the Metropolitan Telephone Company.

These, together with a great variety of special and regular cables, made up an exceedingly attractive and instructive exhibit. Three handsome plate glass show cases contained short samples of ærial and submarine cables made up into paper weights, samples of electric light feeders and special wires, also samples of Kerite tape, and original autographic testimonials from prominent electricians and users of Kerite crude rubber, crude Kerite, cleansed rubber, and finished Kerite compound were shown on tables by themselves. Four large rubber trees together with a liberal display of bunting and pictures, put the finishing touches upon this very interesting and attractive exhibit. The exhibit was made upon a raised platform enclosed by a handsome polished brass railing made to represent a cable. Convenient gates enabled the visitors to make a closer inspection of the Kerite products shown within the enclosure and have access to the handsome booth or cabin erected in the center. This booth was built of sycamore handsomely finished on the inside in the natural wood in such a manner as to represent a ship's cabin, and was luxuriantly furnished with upholstered chairs and lounges and provided with electric light and fans and all the necessary auxiliaries to ease and comfort. The outside was finished in white and gold presenting an artistic and harmonious center to the surrounding wire and cable exhibit. Kerite is the trade name given to an insulating compound first made by Austin Goodyear Day about the year 1865. Mr. W. R. Brixey, proprietor of the business, was superintendent of the factory for many years before he became its owner, and under his able management and experience the quality of Kerite has been maintained and improved until it stands to-day a standard of high grade insulation for electric light wires and cables.

Under the direction of the judges of award of the Exposition, after break down tests up to 38,000 volts, a continuous test at 6000 volts to the point of destruction and a soak test covering a period of 100 days, in ammonia, and sulphuric acid solutions, Kerite No. 14, $\frac{5}{32}$ B. and S. insulation, received the only award given to insulated wires, and a detail of these tests and their result is incorporated in the award cards made to the Commission of Awards.

THE OKONITE COMPANY, LIMITED, NEW YORK CITY.

The exhibit of this company consisted of seven reels of cables, composed of Okonite wire, showing the different styles in which cables are made up for submarine, electric light, telephone, telegraph, and railway use, lead-covered cables for street railway work, underground and pair cables for telephone and telegraph uses, both underground and overhead. They were manufactured after the most approved patterns, and were superior articles in their line. One peculiarity of the Okonite products is that they improve instead of deteriorate with age, and are said to stand a higher test after they have been in use for some time than when they first leave the factory, although subjected to very severe tests before they are put on the market. A proof of the high estimation in which Okonite products are held, is found in the fact that the Ferris Wheel, the greatest modern engineering achievement, was wired with Okonite. The Libby glass works in the Midway Plaisance was also wired throughout with Okonite, and the feeding cables of the Intramural Railway were furnished by this company, and the constant and heavy wear and tear to which they were subjected, proves them to have been of the highest quality.

THE SAFETY INSULATED WIRE AND CABLE COMPANY.

LEAD COVERED CONDUCTORS FOR UNDERGROUND SERVICE.—The Safety Insulated Wire and Cable Company of New York furnished as a working exhibit about ninety-six miles of lead-covered underground cable for the arc-light circuits of the Columbian Exposition grounds and buildings. These cables were placed in the ducts, joints, and connections made and maintained from the beginning to the close of the Exposition by The Safety Insulated Wire and Cable Company. In the construction of the foregoing, No. 6 B. & S. gauge, soft copper wire well tinned was used; the copper being the best Lake Superior, having a conductivity of 98 per cent. of pure copper. The insulation was the "Safety" rubber underground compound of which pure Para rubber was the base. This was placed around the conductors in seamless form to a thickness of $\frac{4}{32}$ of an inch. It was then spirally wrapped with rubber tape and covered with lead with an alloy of three per cent. of tin, to a thickness of 2½ thirty-seconds of an inch. These cables were subjected to a severe practical test as to their durability and reliability from the opening of the Fair until its close. In the manufacture of the cables the rubber compound is applied in a plastic state over the wire, without laps or seams of any kind. The compound is then vulcanized at a temperature of about three hundred (300) degrees Fahr. to a moderate degree of hardness, but amply sufficient however, to prevent decentering of the heaviest conductors which it insulates, and yet it retains a high degree of flexibility, a valuable quality in underground cables.

RUBBER COVERED CONDUCTORS FOR ARC LIGHT CIRCUITS FOR AERIAL USE.—About 200 miles of this cable were furnished in the construction of this wire; No. 6 B. & S. gauge best Lake Superior copper was used, having a conductivity of 98 per cent. of pure copper, insulated with "Safety" Para rubber compound to a thickness of $\frac{4}{32}$ of an inch, with a heavy braid thoroughly saturated with a water-proof compound placed over it. This was not lead-encased.

RUBBER COVERED CONDUCTORS FOR TELEPHONE CIRCUITS.—About three million feet of No. 18 B. & S. gauge, soft copper wire tinned were used. This wire was insulated with "Safety" rubber compound to a thickness of $\frac{4}{32}$ of an inch, and was then made into cables of from five to fifty conductors; the conductors being laid up in pairs. The cables thus laid up were served with jute, two layers of tape wrapped in reversed directions placed over all. This company did the construction work in connection with this contract, made all the connections and maintained them from the opening to the close of the Fair.

RUBBER COVERED CONDUCTORS FOR FIRE ALARM AND POLICE TELEPHONE AND SIGNAL SERVICE.—This company furnished about fifty miles of No. 14 B. & S. gauge for the fire alarm circuits. The conductor consisted of No. 14 B. & S. gauge, soft copper wire well tinned, covered with the "Safety" Para compound to a thickness of $\frac{6}{32}$ of an inch, and with a strong substantial braid over the insulation. For the police and signal service there were furnished about fifty miles of No. 14 B. & S. gauge, soft copper wire well tinned, and insulated with best "Safety" rubber compound to a thickness of $\frac{6}{32}$ of an inch and braided. They were then twisted in pairs.

THE NEW YORK INSULATED WIRE COMPANY.

The New York Insulated Wire Company, of New York, occupied space of about 500 square feet in the southeast gallery of Electricity building. They showed eight pyramids of five reels made of wild cherry wood, on which were symmetrically wound their different classes of telephone, telegraph, electric light and railway feeder wires and cables. One pyramid was composed of bare tinned copper wire stranded conductor, showing the manner in which they lay up their cables before putting on the insulation. The compactness and flexibility of one of the cables equivalent to four hundred thousand (400,000) Circular Mils, was particularly noticeable. They also showed four hundred thousand Circular Mils cable insulated with Grimshaw Insulation, which consists of a layer of white non-oxidizing compound known as "White Core," over which is served a layer of black rubber, three thirty-seconds of an inch thick, and a wrapping of Grimshaw Tape and finishing compound, showing a smooth and highly polished moisture-proof surface. There were pyramids of single-coated Raven Core Wire with braided covering, both solid and flexible conductors; also several reels of high grade railway feeders, known as competition wire, as this company is a strong advocate of rubber-covered wire for pole lines. There were three easels, made of cable twisted brass, on which were placed gold lettered glass signs denoting the class of goods manufactured. One of the easels held a highly polished oak

Westinghouse 15000 Light Alternator.

plaque on which was a central disc of polished copper bearing the words "TRADE MARK," encircled with Grimshaw Wire, with white braid, surrounded by several rows of Grimshaw Wire in black braid, on which were fixed in nickel letters the words "GRIMSHAW WHITE CORE WIRE," thus producing an exact fac simile of their well known trade-mark. In displaying their Vulca Electrical Wire Ducts, a very unique old "Virginia Cabin Home" was made with the material and attracted much attention. The New York Insulated Wire Company obtained the contract for installing the incandescent plant for the World's Columbian Exposition Company, from switchboard to lamp, using many million feet of Grimshaw White Core Wire for the secondary and primary work, and also thousands of feet of Vulca Electrical Wire Ducts and tons of Grimshaw Tape. Notwithstanding the entire work was done during one of the most severe winters, and under many adverse circumstances, when the current was turned on, May 1st, 1893, no trouble was found to exist from defective insulation, thus demonstrating the high quality of material used. Voltage used on primary, two thousand volts; and on secondary, fifty volts.

THE NORWICH INSULATED WIRE COMPANY.

The exhibit of the Norwich Insulated Wire Company, comprehended in its various forms the use of paper specially prepared for the insulation of electrical conductors. Not the least of the claims made for this insulation is its lasting qualities. Protected from moisture its life is practically endless as is illustrated in the life of paper as used in other ways, as in books and in documents. In its use as applied to electrical conductors for high potential it is bound to the conductors in so dense a form as to render it non-inflammable. In an actual experiment a current was applied to this cable so intense as to completely melt the conductor yet the insulation remained entirely intact. This result is of course accomplished not alone through the density of the paper in its application, but in its treatment thereafter. It is claimed for this cable that it is capable of sustaining a pressure of twenty-five thousand volts.

In cables for telephone purposes, paper has been the means of so reducing static capacity as to bring about its general use. The system of insulation is mostly applicable for conduit service.

STANDARD UNDERGROUND CABLE COMPANY.

Samples of wire on reels were shown that exhibited the various methods in use by this company of insulating wires for electric light and transmission of power. Feeder cables, which are designed to be erected on pole lines, insulated wires for electric lighting, both incandescent and arc, samples of lead covered cables for fire alarm and police signal work, were shown both in the anti-induction form in which each wire is surrounded by lead forming a cable of wires under a united sheath of lead, samples of bunched wire for telegraphic purposes, in which case the wires are laid up in concentric layers but not twisted in pairs. A complete set of tools necessary for removing the lead casings and the insulation from the wires was also shown, indicating the progress of the art and the necessity of facilitating quick work in this direction. The Standard Underground Cable Company installed and maintained within the grounds of the Exposition 390,000 feet of feeders and mains, which were used in the distribution of the alternating currents for the incandescent lighting of the Exposition grounds and buildings. The work of installing this amount of cable was carried on through the most trying circumstances, having been performed through the winter and spring, when nature interposed every conceivable obstacle in the path of continuous and perfect work; notwithstanding all of the annoyances experienced, the installation was a perfect working system, subjected to a normal strain of 2,000 volts alternating current pressure. The use of alternating currents for long distance transmission of electricity for lighting and power, was shown by the exhibition of two cables known as "Duplex," encased in one lead sheath side by side. These cables possess the peculiar faculty of reducing to the minimum the effects of self-induction, and are used most exclusively for the alternating current, although the same types of cables are many times used for arc-lighting currents, where an economy of duct space is considered. The company showed, as another special feature of their exhibit, a bunched wire cable for telegraphic use, which is lead-encased and each of the copper conductors is insulated by the use of paper, and the conductors are twisted in pairs and laid up in concentric layers, forming the cable. The especial feature of the type of cable so constructed is that of unusually low electrostatic capacity, combined with high insulation resistance.

WESTERN ELECTRIC COMPANY.

Wires and Cables.—The exhibit of insulated wires and cables comprised a full line of magnet wires of various sizes, with silk and cotton insulation; office and annunciator wires; flexible conducting cables; "Underwriters" and "Weatherproof" wire; and underground and aerial cables for telegraph, telephone and electric light service. In addition to the finished product, machines were shown in operation for covering wires with silk, cotton and paper, and the cable exhibit comprised terminals, model splices and branches and a complete cable-laying apparatus for underground cables. The "dry core" cable now almost exclusively used for telephone work was advocated by the company for several years before it was generally accepted. The cable now made for telephone uses is of 100 pair of No. 19 B. & S. gauge wire, having an outside diameter of 2 1/8 inches with lead and tin composition covering 1/8 inch thick. With these dimensions the static capacity per mile does not exceed .075 microfarad. Telegraph and electric light cables were also shown. Terminals, pole boxes, junction and branch splices were shown as models, and also directly connected to cables as in actual service. A complete cable-laying outfit was exhibited, showing a couple of brick manholes sunk below the floor, with main conduits and branches and with cables in position and splices in different stages of completion. A wagon with a full kit of tools stood between the manholes, and a reel of cable on a stand was in position to be drawn into a duct at one manhole, while a capstan and all the necessary apparatus was at the other. This company furnished the underground telephone cables entering the exchange in the American Bell Telephone Company's building and connecting that exchange with the main office in the city nearly eight miles away. Another interesting part of the cable exhibit was the cable used for connecting multiple switchboards. This is composed of wires insulated first with silk, then with cotton and laid up into a core covered outside with lead tape and a braiding of cotton saturated with a fireproof paint. This cable has an oval cross-section, the reason being that the cables are piled up back of the boards so that no cable can exceed in vertical diameter the thickness of the strip of jacks to which it is connected; horizontally there is more room; so to get the insulation as thick as possible the cable is made with a cross section having a minimum (vertical) diameter the same as the thickness of a strip of jacks, and a maximum (horizontal) diameter of about 50 per cent. greater.

CABLE AND WIRE MACHINERY.—In this exhibit were also found spooling and insulating machines, used in the manufacture of annunciator and magnet wire as well as telephone cables. The cotton yarn used in covering the wire is wound upon spools and then put into revolving heads, where it is wound upon the wire which is in turn wound upon spools for shipment. Magnet wire is covered in the same manner, silk being used in the place of cotton, and the heads being made to take in any size wire from the smallest up to any size desired. Here was also shown the first process of insulating wire for telephone cables, paper being used in the place of cotton. The paper is wrapped loosely upon the wire in two layers, the design being to leave as much air space as possible between the wire and its covering, in order to reduce the static capacity of the cable. These wires are then twisted in pairs, after which any number of pairs are put into lead pipes and used for cables. In this space was also a magneto armature winding machine.

SIMPLEX ELECTRIC COMPANY.

The wires exhibited included rubber covered as well as braided wires for indoor and for line work. The "Plain Caoutchouc" wires are said to have a perfectly cured and very pure Para rubber coating, which in the "Braided Caoutchouc" wires is protected by several closely worked braids. These braids are well saturated with a compound which not only helps to resist abrasion but also protects the caoutchouc from the action of certain chemicals found in wet plaster or in subways. An extra thickness of these protective braids is used on the Simplex Mining Wire, which also has a heavier rubber coating, this wire being intended for use in tunnels, mines and chemical works. The same compound is used on the "T. Z. R." or line wire, which is claimed to stand remarkably severe tests in all kinds of weather. Various sizes of this were shown, including some made up of a number of heavy cables stranded together. There were also large specimens of leaded and armored cables for underground and submarine work, most of these being made with heavy caoutchouc insulation so as to stand high voltages.

THE INDIA RUBBER AND GUTTA PERCHA INSULATING COMPANY.

The goods represented by this exhibit were for house wiring, aerial and underground. The quality of the insulating material employed is of the best used.

THE ELECTRIC APPLIANCE COMPANY.

This exhibit consisted of Paranite wires and cables. The center of the space was occupied by a large cone about five feet in diameter at the base and eight feet high covered with Paranite wires, running from heavy submarine cable at the bottom to No. 18 wire at the top. In addition to this there were four pyramids of fancy reels, and several display tables were covered with coils of Paranite wires and cords. "O. K." weatherproof wire was also displayed.

THE NEW ENGLAND BUTT COMPANY.

WIRE BRAIDING MACHINES.—An interesting exhibit was the machinery for insulating electrical wires shown by the New England Butt Company, of Providence, Rhode Island. Many new and novel features were shown in some of this machinery, as also the adaptation of other machinery for the purpose. The exhibit was a working one, the power for operating it being derived from an electric motor and transmitted by belt to a line of shafting in the rear, from which each machine was operated. One of the machines was a six-spindle flyer winder, a machine of delicate parts, of great capacity, simple in construction, and of fine workmanship; its purpose is to cover the smaller sizes of electrical wires with fine cotton or silk, the thickness of the covering required being the smallest possible fraction of an inch, gauged as low as two and one-half thousandths on No. 40 wire. There was also a double winder, a simple and inexpensive machine containing novel features. This machine was designed for office and annunciator wires. There were also a few of the many sizes and kinds of braiding machines, the first two representing the single forms of machine, in sizes sixteen and twenty-four strands, with different kinds of take-up for the different kinds of wire desired to be covered. The principles of these machines are the same in all the braiders, but there are many sizes and combinations designed for different classes of work. Another machine was the braider in double form for putting on a double-layer of braid, with scored rolls under the machine so that they may come in contact with the bare wire and thus prevent cutting or squeezing the braided wire, which the rolls are apt to do when feeding the larger sizes of wire. There were a number of other machines in this line, such as cabling, stranding, polishing, twinning, weaning, taping, paper covering, etc., of which the company make a specialty.

FELTEN AND GUILLEAUME, GERMANY.

This firm exhibited telegraph cables in great variety with sections visible and otherwise. The cables are insulated with India Rubber, with Gutta Percha, with Impregnated Fibre and with Paper and Air Spaces. Besides the usual Round Wire Armor or Sheathing there were also cables sheathed with flat-shaped wires which are lighter and possess the additional advantage that the cable in drawing through conduits has not a tendency to twist. Others were sheathed with locked coil wires to keep out water and acids and, at the same time being incompressible, the conductors cannot be damaged from external causes. For lead covered cables, flat or ribbon iron sheathing is also employed. For those interested in tracing the origin of the Telegraph Cable were found a number of specimens of cables composed of several cores, each having its own sheathing, or rather several cables stranded together. These are nearly 40 years old. Torpedo Cables and cables for blasting in mines are made with one or more conductors according to requirements. The greatest importance is attached to lightness combined with strength to withstand rough usage and handling. For this purpose such cables are preferably sheathed with fine steel wires of great tensile strength and for submarine purposes these are hemp covered to make the cable lighter. The insulation is either Gutta Percha, India Rubber or Okonite. Telephone Cables

THOMSON-HOUSTON POWER PLANT.—MACHINERY HALL.

were exhibited in much the same variety as telegraph cables, the same remarks applying to both except that the insulation is more confined to fibre and paper. The latter, especially in conjunction with air spaces, has proved a most efficient insulator for telephone cables. Specimens were shown both for single wire and metallic circuits. Electric Light Cable, single, concentric and bi-concentric, with India Rubber, Impregnated Fibre and Paper Insulation were represented by numerous specimens. For alternate currents and high tension, paper has proved an excellent insulator; the cable is reduced in diameter and can be produced to withstand the highest tensions. Some specimens of electric light cables insulated with paper and air spaces were also exhibited. The compartment devoted to joints contained specimens in various stages of completion both in the core and the cable. The mining ropes, two in number, were also shown, the one embracing the constructions more commonly known, while the other was representative of the new and improved construction. The latter included the Patent Lock Coil Rope with interlocking wires. The advantages claimed are the smooth and uniform surface, hence less friction; wires are obtained with the greatest regularity, and range from 50 to 175 tons per square inch, according to the material employed and the process of manufacture. This firm has taken up the manufacture of Phosphor, Silicum, and other Bronze Wires for telegraph and telephone lines, and produce bronze wires having a conductivity of 60 per cent with a breaking strain of 75 kilogr. per square millimetre, also another type with $\frac{85}{90}$ per cent and a breaking strain of $\frac{52}{65}$ kilogr. per square millimetre. Felten & Guilleaume also exhibited compound wire having a steel core for strength and a copper mantle for conductivity; also patent double bronze wire, in which the core is of bronze with high tensile strength, and the mantle of bronze with high conductivity. This wire, in consequence of the perfect manner in which the two metals, being different in their properties but of equal metallic base and therefore of even expansion and contraction by heat and cold, are united, constitutes the climax of the art of producing bronze wire, and is especially remarkable for its extreme ductility. Since telephony and electric lighting have been brought into public notice, Felten & Guilleaume have been most active in constructing the special cables requisite for these two branches of electricity. They have been particularly successful with their anti-induction telephone cables, which have been proved to be practically free from induction and retardation, and can be used both as underground and aerial cables, and are also specially adapted for tropical climates. The most important exhibit of the firm was that of their patent system of air space cables, in which, by the use of paper partitions, they succeed in practically surrounding the conducting wires with air, thus insuring the least possible capacity, together with excellent insulation. These cables have been an entire success. The same system of insulation insuring the lowest possible capacity is also applied by Felten & Guilleaume to the manufacture of long distance submarine telephone cables, rendering speech possible over distances which it was considered impossible to reach hitherto; also to telegraph cables and heavy electric light cables. Felten & Guilleaume also exhibited electric light cables and leads insulated with Gutta Percha and India Rubber, both vulcanized and not vulcanized; also with "Okonite," an improved class of vulcanized India Rubber which is exceedingly durable and heat resisting.

ALLGEMEINE ELECTRICITATS GESELLSCHAFT, BERLIN.

Conducting and Insulating Materials.—The products of the Insulated wire Rubber, and Stabilit works of this firm, formed an effective decoration for the pillars on both sides of the exhibit, showing by a simple arrangement various patterns of conducting wires, flexible cords, cables and rubber goods for electric and other technical purposes, such as required for the insulation of cells, brushholders, etc. There were also patterns of a new insulating material invented by the firm called "Stabilit." This material combines the advantages of vulcanite fibre with those of ebonite without possessing their disadvantages. The Stabilit can be worked in the same way, screws may be cut on it; and it insulates just as well as ebonite and is able to endure higher temperatures. As compared with vulcanite fibre it is entirely proof against dampness and is also more solid. This material may be supplied in the form of plates, rods, tubes and pieces of various shapes.

SUBWAYS FOR ELECTRIC WIRES USED AT THE COLUMBIAN EXPOSITION.

When plans were first formulated for the lighting of the grounds of the World's Columbian Exposition, it was decided that all electric wires and conductors, not only for arc and incandescent lighting, but for power transmission, police signals, fire alarm, telephones and telegraph lines should be placed underground out of danger to the public and yet be accessible. For this purpose the electric subway was designed and constructed. The original plan was to build such a subway of solid brick, but as this was found unnecessary for temporary service, and expensive, the plans were changed to a wooden frame-work lined with cement, plaster and with concrete floors. The plans that were approved and executed called for a subway to connect the electrical plant in Machinery Hall with Mines and Mining, Electricity, Manufactures, Government and Fisheries buildings. The main subway starting from Machinery Hall was 15 ft. 8

in. wide by 8 ft. 4 in. high and was divided in the center by a fireproof partition, making two divisions 6 ft. 6 in. square on the insides. This subway was run to within 50 feet of Electricity building where, from the west division, extended two branches 8 ft. 4 in. high by 6 ft. 1 in. broad—one to Mines and Mining and the other to the Electricity building. These were so arranged that all the wires on the west wall turned west to Mines and Mining, and those on the east wall were run directly into the Electricity building without crossing. The east division at the 50 foot point turned east to the bridge at the southwest corner of Manufactures building, where it widened out into a fan shape the width of the bridge and the wires were carried across on supports placed between the bridge girders. From the bridge the subway extended 100 feet east to the west loggia of Manufactures building where it turned north, going the entire length of the building, but as all the wires on the east wall turned into the building at branches placed near the southwest corner and center, the size of the subway changed just north of the west center to a section the same size as the branch to the Mines and Mining building. At the northwest corner of the Manufactures building the subway turned east to the north center, where it again changed to a section 5 ft. 9 in. wide by 6 ft. high, reducing the capacity by one-third. From this point it turned north, running under the Government building and across the north inlet bridge to the Fisheries building, where the subway ended. In Machinery Hall there was a large double subway commencing opposite the Thomson-Houston switchboard of the power plant, and running 825 feet east under the south aisle, where it connected at the east entrance with the main subway and at the west end with a duct trunk line. The general construction of all the subways consisted of a framework of 3 inch by 8 inch material placed 1 foot centers, and covered on top, bottom and sides with 2 inch matched planking. The inside was lined with standard metal lath and Acme cement mortar on top and sides, the bottom being covered with 4 inches of sand and 6 inches of concrete. The cross section of the main subway at Machinery Hall showed two bodies of wires in each of the two divisions, supported by arms projecting from the wall. These cross-arms, twelve in number, two feet two inches long, were held in position by cast iron uprights, lagged to the frame work of the subway. Each cross arm supported five pins and insulators, making a total capacity of 240 insulators in a cross section. These uprights were placed at distances of about 30 feet apart through the entire subway and consisted of six different types, known as types 4, 6, 8, 8 a, 12 and 12 a; the number indicating the number of cross arms each supported. Types 8 and 8 a, 12 and 12 a, were used in going around the corners. The cross arms projected into the subway 2 ft. 2 in. from either side, giving a clear passage way of 2 feet in the center. In all there were about 6,000 cross arms and 30,000 pins in use throughout the subway. For access to the subway and for convenience in pulling in wires, manholes were placed at distances of about 150 feet apart. These consisted of a round cast-iron box 20 in. in diameter and 20 in. high, resting on the framework of the subway and supplied with a cast-iron cover. The first contract, which called for the construction of the main part of the subway, was awarded January 23rd, 1892, to T. C. Brooks & Co., of Jackson, Michigan, for the sum of $35,-094.49. Work was begun the first day of February, 1892, and was to have been completed by April 15th, 1892, but owing to frozen ground and rainy weather it was greatly delayed. The bottom of the subway being only a little above datum, considerable difficulty was experienced in putting in the concrete and flooring on account of water, but this was overcome by the use of an electrical portable pump. This contract did not include the subway running east and west under Machinery Hall, the portion under Government building, the branches under Manufactures nor the bridge approaches. This latter work was done by the Exposition Company, except the plastering on the subway under the Government building, which was done by Wm. Pickland & Co., for the sum of $1010.00. The east and west subway under Machinery Hall was put in under many difficulties, chiefly because of necessity to follow the building aisles and on account of steam and water pipes. The total length of subway, including east and west subway under Machinery Hall, all the branches and approaches to the bridges, was 6,195 feet. Wiring of the subway was begun in the middle of February, 1893, and continued for about six weeks. It was found that there were so many arc wires for Manufactures building which had to be placed on the east wall of the east subway that it was necessary to run two wires on one insulator, and for that purpose a special two-wire insulator was designed giving a perfect glass insulation between the two wires. Wires were also arranged so that no two wires of different potential would come on the same insulator. The wiring of the subway required 4,000 of the special two-wire insulators and 20,000 of the regular single glass insulators. The subway contained 25 2-10 miles of power, 28 7-10 miles of incandescent and 51 miles of arc wires, making a total of 104 5-10 miles of wire for lighting and power transmission. Besides these, were telephone and telegraph cables, fire alarm and police signal wires.

For the convenience of drainage the profile of the floor of the subway was so arranged that all water which might collect would flow to four points, namely, north of Machinery Hall, south of the Electricity building, west end of north railroad bridge and north center of Manufactures building. At the bridge, an opening was made into the lagoon so the water would flow out, but at the other three places marine pumps were installed and the water pumped to the nearest catch basin. In this way the subway was kept comparatively dry except on one or two occasions when a water pipe burst, completely flooding everything. The subway was lighted with 225 110-volt 16-c.p. Edison incandescent lamps placed at distances of about 30 feet apart throughout the entire length. The lamps were placed five in series and supplied with current from the 500 volt power circuit. The lamps of different circuits alternated in location so that in case one lamp burned out it did not leave the subway in total darkness. For the convenience of keeping an exact record of the position of every wire throughout the entire course of the subway, a system of card cataloguing was arranged. The cards were printed showing the exact position of all the cross arms and insulators supported by one set of uprights. Each insulator on a cross arm was numbered and each cross arm was also numbered. On each card was marked the position of the wire on the insulator, and the circuit number of every wire, at any given point of uprights. The uprights also all being numbered a card was made out for every set of cross arms and arranged consecutively in a file. By this means it was always possible to tell the exact position of any wire at any point in the subway. Wires which ran north from the Fisheries building were carried from the subway into a duct trunk line which ran east to the Intramural railroad and thence north, following the line of the road around the Montana State building. This trunk line was 2,250 feet long and contained 15,270 feet of pump logs. A clause in the contract of the Intramural railroad reserved the right for the Exposition Company to carry light and power lines along the structure underneath the road-bed, on extension insulators, and lines were run in this way wherever they were desired along the route of the road.

SERVICE TRANSMISSION OF POWER FOR THE COLUMBIAN EXPOSITION.

The work for permanent power service was taken up on January 1st, 1893, and contemplated power transmission as follows: Mines and Mining Building, 400 h.p., from Mather machines. In the Transportation and its annex, 400 h.p. was to be provided for. The center of distribution was to be under the gallery at the center of the building on the west side; main to be run all around under gallery, suitable for tapping 200 h.p. on a basis of evenly distributed power. In the Agricultural Building arrangements were to be made for 300 h.p. in the annex and 100 h.p. distributed over the main building. Arrangements were to be made for a complete circuit through the building to connect by the route of the Intramural Railway back to the generators,

WILLANS VERTICAL ENGINE COUPLED DIRECT TO 125
LIGHT GENERATOR.

allowing for the following capacities to be taken from the Intramural railway loop: Krupp Gun Exhibit, 50 h.p.; Forestry Building, 50 h.p.; Dairy Building, 30 h.p.; 100 h.p. to be evenly distributed for Stock Pavilion; maximum loss in conductors with full load for above service 50 volts; current to be taken from Eddy generators. Another installation of conductors was to be made which would make a circuit from Machinery Hall on the structure of the Intramural Railway and returning by the way of the subway for the following capacities of electric power service: Service Building, 20 h.p.; Horticultural Building, 30 h.p.; Midway Plaisance, 200 h.p.; northeast corner of the park, 100 h.p.; Government Building, 50 h.p.; Manufactures Building, 150 h.p. The main in the Manufactures Building was to be run all around the building under the gallery, and to be suitable for tapping the above power on a basis of even distribution. The 100 h.p. called for at the northeast corner of the park was intended to be evenly distributed power from the California Building to the end of the railway loop. The Midway, likewise, was to have an even distribution of power. This service was to be taken from the Eddy generators and a loss of 50 volts allowed between brushes of the generators and the motors on full load, for all power except that at the north end of the park and Midway Plaisance, where a loss of 80 volts could be allowed. Distances of less than 3200 feet were figured for carrying capacities, and beyond this distances were figured for drop. The amount and size of wire figured on the above basis and for the various locations as hereinbefore mentioned were as follows: 0000, 157,640 feet; 000, 17,500 feet; 0, 10,000 feet; No. 1, 22,000 feet. Ten per cent. was added to this and specifications were drawn in accordance, all wire to be rubber covered. The reason rubber-covered conductors were specified was that the most perfect insulation was desired and it was deemed advisable not to deviate from the rules as already adopted, namely; to allow nothing but rubber-covered wire to be used throughout the entire installation of the electrical plant. Better results would follow from the use of such wire and the maintenance be thereby reduced to a minimum. Bids were advertised for and finally awarded to the United States Wire & Cable Company; on the 2nd day of February, 1893, a contract was made with the company for the following amount of wire at the prices named below:

No. 0000 Stranded,	180,000	feet at	$187.32	per	1000	feet.
" 000	"	19,000	"	155.71	"	"
" 0 Solid,		12,000	"	101.27	"	"
" 1	"	24,000	"	73.85	"	"

The amount of wire used for power work exclusive of exhibitors' service wires was as follows: 181,640 feet 0000, 19,860 feet 000, 29,710 feet 0, 24,074 feet No. 1, 1,500 feet No. 6, and 5,400 feet No. 4.

There were used approximately 100,000 feet of wire in service for exhibitors. Below is given the rules, as laid down by the World's Columbian Exposition, which governed the supply of electricity for power service:

RULES.

1. The generators and main conductors will be supplied, operated and maintained by the World's Columbian Exposition. The service conductors will be furnished at consumers' expense and at the rates hereinafter mentioned. The motors and appurtenances must be supplied, operated and maintained by and at the expense of the consumer. The service connections to the motors shall be installed, maintained and owned by the World's Columbian Exposition.

2. The motors shall be suitable to operate on a 500 volt constant potential circuit. The rheostats shall be constructed wholly of non-combustible material. The main line switch shall be of the "Knife Blade" type, and suitable for working on a 500 volt constant potential circuit.

3. No service will be rendered for less than $20.00. Fractions of a horse power will not be considered, except for motors for less than 1 h.p. Charges will be made for service connections to the main lines at the following rates:

For ½ h.p. and less	$10 00	
For 1 to 5 h.p.	15 00	per h.p.
For more than 5 h.p., not exceeding 10	12 00	" "
For more than 10 h.p.	10 00	" "

4. Charges for service will be based on the maximum electrical h.p. delivered to the motor, irrespective of the class of work to be operated by the motor, at the following rates:

For ¼ h.p. and less	$20 00	
For more than ¼, not exceeding ½ h.p.	40 00	
For more than ½, not exceeding 1 h.p.	75 00	
For more than 1, not exceeding 2 h.p.	70 00	per h.p.
For more than 2 h.p.	60 00	" "

5. Special service for motors not exceeding 2 h.p. can be furnished from the regular alternating incandescent circuits.

6. The above rates contemplate continuous service, or service on demand, at any time during the hours of the Exposition, from May 1st to November 1st, 1893, inclusive. Consumers requiring power service for a specified number of hours only will be charged at the rate of 5 cents per electrical h.p. hour; the specified time of day in which this service can be rendered to be determined each day by the chief of the department.

All power circuits had a working potential of 500 volts, with the exception of that for the Administration Building, which was 220 volts.

The standard of 500 volts was adopted after consideration, as that is the potential most generally in use by all central stations throughout the country and is in conformity with what is considered to be the best practice of the present day. Especially was this of advantage from the fact that it was necessary to cover a wide area in distributing power to all parts of the Exposition grounds, including the Midway Plaisance, and also due to the fact that regular street railway generators could be used in connection with commercial motors. A contract was made with the Crane Elevator Company to furnish electric elevators for use in the Administration Building. The motors furnished for the operation of these elevators were 220 volts and this necessitated the making of an extra contract for 250 volt power generators. This contract was made with the C. & C. Electric Motor Company on September 28, 1892, for two 80 k.w. 250 volt generators. This was the only deviation made from the 500 volt standard. It might be well here to state that the Exposition had 103 motors in permanent service on the grounds with a total h.p. of 1810, and in addition to these there were 212 exhibitors' motors which used 538 h.p., making in all 315 motors with a total h.p. of 2348. These 315 motors comprised 28 different makes of machines. On all exposed power circuits lightning arresters were placed, 24 being used on the different circuits throughout the grounds and at the switchboards. There was placed a lightning arrester on each circuit that left the board. The Westinghouse double pole lightning arrester was used throughout. The first permanent power work commenced and finished was that for the Administration Building, for which 200 h.p. was provided. This circuit operated eight passenger elevators, each elevator being operated by 1.25 h.p. Eikemeyer motor. In each pavilion was placed a cut-out box which governed the two motors in that pavilion. Each of these four cut-out boxes was fed by an independent circuit directly from the main cut-out box, this circuit consisting of two No. 0000 wires drawn through vitrified sewer pipe from the main or distributing box to the motor cut-out box which it fed. In addition to these eight elevator motors which were situated on the ground floor, there were in each pavilion and on the top floor four 2 h.p. Eddy motors, each of which operated a 48-inch ventilating fan, used in exhausting the vitiated air from the pavilions. Each of these motors had an independent circuit run to it from the cut-out box directly below it and which governed the elevator motors in that pavilion. A tap of No. 6 wire was taken off this circuit and run over to the tramway near the southwest corner of the Mines Building, to operate a 5 h.p. 220 volt T. H. motor, used in running the exhibition ore tramway on the outside of the Mines Building. This circuit was approximately 2000 feet long.

TRANSMISSION OF THE ELECTRICAL CURRENT.

In order to avoid the unsightly appearance of overhead conductors from the elevated structure to the different buildings, where power was to be supplied it was decided to run such wires under ground; several methods were considered but the scheme as finally adopted consisted in running iron pipes varying in diameter from 1½ to 3 inches. From a manhole box located at the foot of one of the structure supports to nearly the top, the wires were carried from the cross-arm to these pipes and thence directly into the manhole box, from which point they were drawn in ducts leading to the various points of distribution in the buildings. This method was used in all cases, whether for the main circuit itself, or for taps at the point where they left the Intramural. All conductors running into the various buildings first entered a cut-out box lined with asbestos in which was placed a double pole switch of the "Knife Blade" pattern large enough to carry all the current for that building. Each circuit that left this box was provided with a similar switch of sufficient carrying capacity; also two single pole fuse blocks fused up to protect the wire that left them so the entire current could be cut off from the building by simply opening the break-down switch, or each circuit could be independent or absolutely cut off at the cut-out box. All the conductors inside the buildings were carried on glass knobs and in non-combustible tubing where they passed through walls or inaccessible places. At each and every place that a tap was taken off for an exhibitor's motor, two single pole magnetic blow-outs were installed and fused up to protect the wire leading from them; also, wherever the wire changed from larger to smaller size, a single pole cut-out was placed in each leg of the circuit. Located in the power plant, which occupied the entire space south of the main south aisles of Machinery Hall and Annex, were the power generators contracted for by the Exposition Company, and also in connection with these were installed the generators for furnishing power to the several exhibitors' spaces in the Electricity Building.

POWER GENERATORS.

EXPOSITION.

MAKE.	NUMBER.	K.W.	VOLTAGE.	TOTAL K.W.
Mather. - - - - - -	2	225	500	450
Mather. - - - - -	2	120	550	240
C. & C. - - - - - -	4	80	550	320
C. & C. - - - - -	2	80	250	160
Eddy. - - - - - -	4	186½	550	746
Westinghouse.	1	373	550	373
				2289

EXHIBITORS.

MAKE.	NUMBER.	K.W.	VOLTAGE.	TOTAL K.W.
National.	1	80	500	80
Jenney.	1	40	500	40
Western Electric.	2	137½	250	275
Wood.	1	120	500	120
Edison.	1	800	150	800
Edison.	1	150	150	600
Edison.	1	42	150	42
Siemens Bros.	1	180	120	180
				Grand total, 4426

making a total generating capacity of 2289 k.w. or 3068 E.H.P. for Exposition use, and 2137 k.w. or 2864 E.H.P. for exhibitors' use, or a grand total generating capacity of 4426 k.w., or 5932 E.H.P. With the exception of two 80 k.w. machines which supplied the power for the elevators in the Administration Building, all the Exposition generators were regular street railway generators, wound for a potential of 500 volts, with a guaranteed E.M.F. of 550 volts, full load. With each separate type of machine was furnished a switchboard complete with all switches regulating and controlling devices, the board being erected in close proximity to the machines to be controlled. The switchboards as installed were of white polished marble, with the exception of one which was made of dark slate. In connection with the wiring of the power switchboard, it was deemed necessary to install relay circuits between each board to be used in case of emergency, so as to cause as little delay as possible in transferring the load from one to another set of machines. For this purpose a 600 ampere triple pole switch was installed at each power board and connected to the bus bars by ¾-inch copper rods. Between each switch a circuit of two No. 000 wires was run, together with an equalizing connection made up of two No. 000 cables with the Westinghouse generators. It was necessary to run a separate equalizing cable from the machine to the transfer switch as installed on the board, owing to the fact that this switchboard was not supplied with an equalizing bus bar, not being run in multiple with any other generator. Due to this relay system, it was possible to run any set of machines in multiple with any other; this arrangement turned out to be invaluable; in fact, the successful operation of the entire power plant depended upon this system, because of the manner in which the power plant was laid out, as each separate machine was in itself a small central station and intended to operate independently. This arrangement was due, however, largely to the fact that the power generators were entered as exhibits and independence of action was demanded by all manufacturers. Thoroughly appreciating the difficulty, the thing to do was to arrange as complete a system of connections between these separate central stations as was feasible under existing con-

W. R. Brixey's Exhibit of Kerite Wire and Cable.

ditions. The outcome was the relay circuits as described above. The inflexible system originally contemplated was by these simple connections, rendered elastic and capable of as many combinations as could be desired.

ELECTRICAL SERVICE IN ELECTRICITY BUILDING.

The special service for Electricity Building was not in a general way far divergent from the plans followed at other parts of the Exposition. The rules made special for that building, and a statement of the general plan, may properly be made at this point, while the transmission of power is the topic. It was determined at an early day in the organization of the department to except the Electricity Building in the matter of contracts for electrical service. Most of the large companies having operative exhibits in the building were contractors under the Exposition Company and had either power or lighting plants in operation in Machinery Hall. By unanimous consent it was agreed that each company should have the privilege to increase its Machinery Hall plant enough to allow each company to do a certain amount of lighting and power supply in the Electricity Building for purposes of exhibition. It was determined that such light and power should serve the Exposition, in the Electrical Building, and the contracts were made in accordance with the following rules, which will not only serve to show the complete system of electrical equipment, but also the terms upon which service was rendered:

RULES GOVERNING LIGHT AND POWER IN ELECTRICITY BUILDING.

Exhibitors in the Electricity Building will be divided into two classes: First, those who contribute to the service of lighting and electric power transmission for the building; second, those who do not so contribute.

Contributors will be again divided into two classes:

(1) Those who furnish generators connected to prime movers in Machinery Hall.

(2) Those who contribute to the service lighting of Electricity Building through motor power taken from circuits in the Electricity Building.

Contributors of either class will be required to enter into formal contract for furnishing and operating machinery offered, in accordance with the following general terms:

No steam power being available in the Electricity Building, all generators provided by contributors of class 1 must be installed in Machinery Hall.

Generators provided by American companies will be located in the regular service plant in Machinery Hall. Generators provided by foreign companies will be located in the spaces severally allotted to these countries in Machinery Hall.

Power circuits will be led from the generators of each contributor of class 1 through existing subways from Machinery Hall to the Electricity Building. The several exhibitors offering generators for this service are expected to furnish and install, at their own expense, subject to the supervision of the World's Columbian Exposition, the several power circuits before mentioned.

The aisles of the Electricity Building will be illuminated by arc lights, which will be considered the service lighting for the building. Arc lights in addition to these, as well as all incandescent lights and power for keeping exhibits in motion, will be considered as special service rendered to individual exhibitors.

Power for operating generators needed for service lighting will be furnished free of cost to the exhibitors installing this light. In consideration of such contribution to the service lighting of Electricity Building, exhibitors participating in this service will be granted such amount of extra power and light for the operation of their own exhibits as may be mutually agreed upon in contract.

Arc and incandescent lights may be taken direct from the several power circuits leading from Machinery Hall, or they may be furnished from generators driven by motors taking current from the power circuits before mentioned, motors and generators in this latter case being located in the Electricity Building. Arc lights for all-night service are an exception to this rule, and will be supplied from the regular service plant already contracted for by the World's Columbian Exposition.

Contributors of class 2, who use current from the power circuits, will be expected to provide the necessary motors, generators and shafting, free of cost to the World's Columbian Exposition. Exhibitors contributing to the arc lighting, necessary in the service of the building, will be expected to wire, hang and maintain such lamps as may be assigned to them, free of cost to the World's Columbian Exposition. Such arc lamps, incandescent lights and motors, as may be installed within the contributing exhibitor's space, will be wired and hung by said exhibitor at his own expense.

In order to provide for lighting and power service to those exhibitors who do not contribute to the service of Electricity Building, it is proposed to assign to each exhibitor contributing to said service a specified portion of the Electricity Building in which to furnish special lighting and power. All work of wiring, hanging, and maintainance of lamps, etc., done for other exhibitors at the following rates:

(1) For each incandescent lamp, including the first lamp and socket with installation on a plain cord or pendant, $3.50 per 16-c.p. lamp.

(a) Special fancy wiring will be furnished at an extra cost, to be paid by consumer, under special agreement.

(b) Fixtures and shades will be furnished and maintained by consumer.

(c) Lamps of other capacities, fancy or colored lamps, will be furnished in accordance with special agreement. Lamp renewals must be furnished free of charge by contributor. Breakages must be purchased by consumer from the exhibitor supplying his exhibit, at the usual market price.

(2) For each arc lamp, including ceiling block, suspension rod and lamp complete, $35. Lamps must be furnished with opal globes. All care and attendance to lamps and circuits must be furnished without extra charge by the contributor maintaining the circuit.

(3) All wiring necessary for the installation of motors within the space assigned to each contributor shall be done by said contributor, at the expense of exhibitor desiring said power, at the following rates:

½ h.p. and less - - - - - -	$ 8 00
1 to 5 h.p., per h.p. - - - - - -	12 00

More than 5 h.p. and not exceeding 10 h.p., per h.p.	$10 00
More than 10 h.p., per h.p. - - - -	8 00

Fractions of a horse-power will not be considered, except for motors of less than 1-h.p. In all cases the rating of the motor will be the standard rating of the manufacturer.

Service connections, as well as all lamps and sockets installed at contributor's expense, shall be maintained and owned by said contributor. All alterations of original installation shall be made by said contributor, after written agreement with the exhibitor desiring change.

All installation and operation shall be subject to the inspection and approval of the Chief of the Department of Electricity. Installation must conform to the National Code of the Board of Underwriters, subject to the inspection and approval of the Director of Works of the Exposition.

NOTE.

Service will be provided only from May 1, to October 30, 1893, inclusive, and during Exposition hours, except by special agreement.

The service rendered will be entirely at the risk of the consumer. The World's Columbian Exposition will not be responsible for stoppages from any reasonable cause.

RULES GOVERNING THE SUPPLY OF LIGHTING AND POWER TO EXHIBITORS IN THE DEPARTMENT OF ELECTRICITY.

Electric power will only be supplied in the Electrical Building.

The aisles of the building will be illuminated with arc lamps without cost to exhibitors; additional lighting to be paid for by exhibitors at the following rates:

Arc lamps, service May 1, 1893, to October 30, 1893, inclusive, during Exposition hours, $65 per lamp of 2000 nominal c.p.; lamps suspended from ceiling, with opal globes and globe nets, special fixtures or globes, or special ornamental lamps, will be subject to an extra charge.

Incandescent lamps, service May 1, 1893, to October 30, 1893, inclusive, during the hours of the Exposition, $8 per 16-c-p. lamp, including installation of first lamp and socket and lamp renewals; breakages must be purchased by exhibitor of the contractor lighting his space, at the usual commercial rate. Lamps will be hung on plain cords or pendant; special or fancy wiring will be furnished at an extra charge. Fixtures and shades will be furnished and maintained by consumer. Special colored or fancy lamps, or lamps of other than 16-c.p., will be subject to special agreement. Installation of 500 lamps or more will be subject to special discount.

MOTORS.

Exhibitors requiring motive power will furnish necessary motors, rheostats, main line switches, insulating base, etc., complete, together with all belts, countershafting, and other means of connecting motors to operating machinery. Motors should be suitable for operation on class of circuit furnished in exhibitor's location.

The rheostats must be constructed wholly of non-combustible material. The main line switch shall be of the "knife-blade" type, of ample carrying capacity, and suitable for breaking current at the potential used. Motors shall be erected in position at the expense of the exhibitor by the contractor furnishing power in the location of his exhibit, service connections being installed, maintained and owned by the said contractor.

Charges for service connections with the main line will be as follows:

For ½ h.p. and less - - - -	$ 8 00
For 1 h.p. to 5 h.p., per h.p. - - - -	12 00
For more than 5 h.p., not exceeding 10 h.p., per h.p.	10 00
For more than 10 h.p., per h.p. - - -	8 00

Fractions of a horse-power will not be considered, except for motors of less than 1 h.p.

Ratings of motors will be the standard rating of the manufacturer.

Charges for service will be based on the maximum electrical horse-power delivered to the motor, irrespective of the class of work to be operated by the motor, at the following rates:

For ¼ h.p. and less - - - -	$15 00
For more than ¼ h.p., and not exceeding ½ h.p.	30 00
For more than ½ h.p., and not exceeding 1 h.p.	50 00
For more than 1 h.p., and not exceeding 2 h.p., per h.p.	45 00
For more than 2 h.p., and not exceeding 3 h.p., per h.p.	42 50
For more than 3 h.p., per h.p. - - -	40 00

A limited amount of power may be furnished to exhibitors free of charge, to simply turn over an otherwise inoperative exhibit.

Each company operating a plant in Machinery Hall, and desiring to participate in the electrical equipment of the Electricity Building, was given certain territory as nearly as possible in the neighborhood of its exhibit. A certain percentage of the light and power brought into the building by each was allowed for the operation and lighting of each exhibit proper, the amount being fixed by agreement, and the balance of the light and power service was extended to neighboring exhibitors who had no primary plants of their own. In this way the building was equipped for service arc lighting and with power for the operation of motors practically without expense to the Exposition, the smaller exhibitors were enabled to get their lighting at a cost to them of about 30 per cent. less than the Exposition charged in other buildings, and the larger companies operating primary plants were enabled to show their particular systems and their specialties, they being reimbursed for their expenditure to a certain extent by those exhibitors whom they supplied. This plan was entirely satisfactory and, as the service rendered by each company contributing was understood to be the best possible service under that company's system, it will readily be seen that the very best of service was secured.

CHAPTER VIII.

MEASURING INSTRUMENTS.

METHODS of measurement of the electrical current are no small part of the business of electrical scientists of the day, and the display made at the Exposition of instruments of precision was exhaustive inasmuch as the principal manufacturers of dynamo electric apparatus all showed various types of instruments used for commercial purposes in measuring the output of their apparatus. The principal instruments used for measuring were the voltmeter, two kinds, for direct and alternating currents, which showed the tension of a current; the Milli-voltmeter used for measuring minute fractions of potential; the ammeter used for measuring the quantity of current, and the Milli-ammeter for measuring very minute quantities; the Wheatstone Bridge and Galvanometer for determining the resistance offered to the passage of the electric current; the Wattmeter, which combines the functions of the voltmeter and ammeter and gives directly the commercial units used, and the Clark cell, which is the standard used for potential. In their finest and most finished form, scientifically and mechanically, all these instruments were shown, most of them in operation or in condition to operate for the benefit and interest of visitors.

QUEEN & COMPANY, PHILADELPHIA.

The electrical exhibit of Queen & Company included in general a complete line of high grade resistance boxes, standards and bridges; galvanometers of every style and grade for the most delicate and accurate as well as for the more ordinary classes of work; batteries, condensers, keys, induction coils, dynamometers, electrometers, hot wire instruments, station instruments, photometric outfits, Geissler Crooks and spectrum tubes; a display of cheaper electrical apparatus intended principally for school demonstration purposes, and, in addition to the above, a complete

Night View of Grand Court and Agricultural Building From Electricity Building.

line of balances, electrometers, etc. As a result of the great amount of experimental work which this firm has been carrying on for the past two years, their exhibit presented very many novel feasures.

RESISTANCE MEASURING OUTFITS.—The display of high grade standard resistance sets, Wheatstone and slide wire bridges, resistance testing sets of all ranges, and standard coils, was thoroughly complete. Platinoid wire is employed throughout in all the sets, the spools being wound and allowed to stand for several weeks before the final adjustments are made. This ageing of the coils, as it is called, is necessary whenever strictly first-class results are to be attained, for the reason that the resistance of wire changes slightly for the first weeks after having been wound on the spools. The temperature co-efficient of each spool of wire is carefully determined and only those spools having the same temperature co-efficient are placed in the same box. The woodwork for all of the resistance boxes is of highly polished mahogany and the sets are in addition provided with handsome quartered oak carrying cases in place of the cheaper cases usually supplied. There was a standard resistance set with Wheatstone bridge arrangement (Anthony form). This style of box comprises five sets of ten coils each, all the coils of a single set being of the same ohmic resistance. The values of the different sets are one-tenth ohm coils, units, tens, hundreds and thousands, making a total resistance in series of 11,111 ohms. This arrangement here adopted has many advantages over the "Dial Pattern." The coils of each denomination are arranged in rows so that any or all of them may be connected in series or multiple. Thus in the row marked "unites" a plug to the right of Block "0" and to the left of Block '10," gives the ten unit coils in series, or ten ohms, while placing plugs to the right of Block "0," left of Block 1, right of Block 2, etc., gives the same ten coils in multiple or one-tenth of an ohm. The bridge arms consist of one, ten, one hundred, one thousand and ten thousand ohms on each side, giving a ratio of one to ten thousand ohms, and a total range of measurement from one millionth of an ohm to one hundred and eleven megohms, by taking advantage of the multiple arrangement. For determining the temperature of this set, instead of employing a thermometer as is ordinarily the custom, a temperature coil is provided which may be thrown in the bridge arrangement as the unknown resistance to be measured. The resistance of this coil, as measured in ohms on the bridge itself, corresponds to the exact mean temperature of the box in degrees centigrade. The adjustments of all the coils on this, as well as on many of the other high grade resistance boxes, are to within one one-hundredth of one per cent, and a certificate is furnished guaranteeing the accuracy to one-fiftieth of one per cent.

The standard subdivided megohm box consists of ten coils of 100,-000 ohms resistance each. The terminals in this box are all mounted upon corrugated hard rubber pillows 1½ inches high. The inside of each pillow is cut out so as to enclose the terminal rods, to which the coil terminals are soldered, without touching them excepting at the top, thus making leakage practically impossible. The coils may be connected any number of them in series or multiple. Aside from these there was on exhibition a complete list of the company's standard subdivided resistance boxes of various ranges. Of standard coils there were on exhibition, aside from the old British Association form of standard, several forms of Queen's new oil immersion standard coil. One of these standards is the most perfect yet gotten out, embodying features adopted by the German Reichenstalt, to which important modifications and additions have been made by the firm. The standard consists, in short, of an exterior brass case, double walled and filled in between with asbestos. This contains an open cylinder upon which is wound a resistance wire termed the heating coil, the terminals of which connect to binding posts mounted on the hard rubber top. The standard coil and the temperature coil are also wound upon cylinders, the former being of platinoid and the latter of both platinoid and copper wire, of such resistance and relative proportions that its resistance changes one-twentieth of an ohm, measured in terms of the standard, for a change of 1° C. of both the standard and coil. A sufficient quantity of resin oil is employed to cover the heating coil contained in the outer chamber, also to cover the heating coil contained in the central chamber. On passing current through the coil, heat is transmitted to the oil, is equalized over the surface of the box and communicated on through the oil to the inner parts. Of the other forms on exhibition one was without the temperature coil and the other without both temperature and heating coils. One form is especially adapted to the determining of temperature co-efficients of samples of wire. The line of shunt boxes exhibited for use in varying the sensibility of galvanometers was very complete, including all styles of the ordinary and compensated shunt boxes in cylindrical brass cases with hard rubber tops, some being of the regular style of mounting and others being mounted on high corrugated pillars. The compensated box contains in addition to the regular $\frac{1}{9}$, $\frac{1}{99}$ and $\frac{1}{999}$ shunts, an arrangement of resistance such that the total resistance in circuit remains unchanged when the different shunt coils are used, thus increasing the accuracy of the galvanometer. The commutator arrangement is especially intended for use in making extremely accurate resistance measurements and for determining temperature co-efficient of wires by the Carey-Foster method. The arrangement of parts is perfectly symmetrical and the workmanship is of the best throughout. All binding posts as well as the bars are of copper and the base on some is of white marble and on others of highly polished wood. There are copper mercury cups provided and arranged so that two standard coils may be placed in multiple on either side if desired. The commutator bars, fitting as they do loosely in the rubber platforms, readily adjust themselves to the bottoms of the cups in which they rest, thus securing good contacts. The two standard accessory coils are wound on the same spool and inclosed in a brass case and may be quickly changed for coils of other values if desired. With this commutator either the standard cylinder bridge or the compensated meter bridge may be used. The wire on the cylinder bridge is wound in a spiral groove of fifty turns, in the hard rubber cylinder. Contact is made with this wire by means of a small platinized wheel moving horizontally as the cylinder is revolved. Alongside this sliding wheel is a graduated scale. At the end near the handle is a circle graduated into one hundred equal parts, each indicating accordingly one one-hundredth of a revolution. Each one of these spaces may readily be separated into ten equal parts with accuracy, thus enabling the bridge to read 50x100x10, or 50,000 parts. Since the resistance of the wire is but one ohm, resistance as low as .0002 ohms can be measured and, by the addition of a shunt, readings can be reduced to .000002 ohms. In the meter bridge the frame is of brass provided with hard rubber feet for insulation. A metal frame being used there is but little change in tension on the stretched wires due to change of temperature. The working wire is of platinoid, No. 17 B. & S. gauge, and in addition there are two shunts of the same material which may be thrown in or out as desired. The contact is made by means of two platinized wheels which press continually against the working wire, one above and the other below, and roll along, always making good, firm contact at a well-defined point. A separate key serves to make and break circuit. The carriage is capable of either a fast or slow movement and a vernier reading to $\frac{1}{20}$ m m is provided. An extremely compact form of the slide wire bridge and Carey-Foster attachment was on exhibition. The working wire is twenty centimeters long and accompanied by a single shunt. A millimeter scale with vernier attachment is provided. The usual symmetrical arrangement of heavy copper bars and mercury cups is employed. By turning the large commutator through an angle of 180°, the positions of the standard and unknown coils are interchanged while, by turning the smaller one through 90°, the direction of the battery E.M.F. is reversed. Thus only two reversals of the commutator are required, an important feature not possessed by other forms of the apparatus. The new Queen conducting bridge was quite an attractive feature of this exhibit. It was designed for the express purpose of measuring the conductivity of heavy conductors, only short pieces, from one to two feet long, being employed. The set is handsomely and substantially made and occupies a rectangular mahogany case 23⅜x11½ x8⅜ inches.

GALVANOMETERS.—In all high grade galvanometers, of the Thomson type, graded coils are now used. The grading is such, in accordance with Maxwell's theory, as to give the maximum magnetic effect for a given length of wire, five changes of wire being employed. The coils in all of the instruments are arranged so as to be easily interchanged; for suspensions quartz fibres are invariably employed. Of the Thomson type of galvanometer, four designs were on exhibition. Two pairs of graded coils are supported on four corrugated hard rubber pillars 6 inches high. The connections being made by means of platinized contacts, renders the coils perfectly interchangeable and, by means of a switchboard arrangement, the coils may be thrown in any combination of series or multiple, and may also be used differentially. The astatic system consists of two sets of magnets, made of pieces of fine piano wire, arranged parallel on small circular discs of mica, which, in turn, are cemented to a quartz rod. At the center of this rod is placed the mirror, just back of which is a mica vane moving in an air-tight compartment for damping purposes. The control magnet used in this and in many of the other galvanometers, consists of a hard steel cylindrical magnet with threads cut on each end. Two nuts of soft iron turn on these threads so as to approach to or recede from each other as the magnet is rotated. By means of this arrangement the sensibility of the instrument may be varied throughout a very wide range.

The galvanometer having probably the widest range of usefulness is the new horizontal magnet d'Arsenal. The magnet for this instrument consists of a number of discs stamped from sheet steel. No iron core is employed and the shape of the straight scale is exactly in proportion to the current measured. The coil, wound usually of No. 40 copper wire, is long, narrow and circular in cross section, and is contained in an aluminum tube which forms a part of the system most effectively. The aperiodicity of the system may be made anything desired by properly slotting this tube. For ballistic work the coil is wound on a boxwood frame, and the system is brought to rest by means of a short-circuit key. In either case the system is contained in a long brass tube which may be slipped into place between the poles of the magnet. The connections being automatic, a number of systems furnished with the instrument may be quickly interchanged, this being a most desirable feature. Phosphor bronze strip is employed for both the suspension and the spiral spring at the bottom. A clamping device is provided, so that injury to the system during transportation is absolutely impossible. In addition to the instruments already described, there were on exhibition innumerable smaller pieces, such as go to make up a complete apparatus. Among these should be mentioned the Carhart-Clark standard and one-volt cells, condensers, telescopes, lampstands and scales, keys of various types, etc. Also, there was a complete line of students' apparatus of a cheaper grade, consisting of voltmeters, galvanometers, resistance boxes, condensers, etc.

Special mention should be made of the magnificent display of Lord Kelvin's well known instruments, including a complete line of his well known balances, station and multicellular electrostatic voltmeters, and the like. This display was included in the exhibit of instruments of Queen and Company's own manufacture.

THE WESTON ELECTRICAL INSTRUMENT COMPANY.

The Weston instruments were adopted by unanimous consent of the jury as standards for all electrical measurements at the World's Fair. Prof. Ayrton, of England, who was himself identified with the manufacture of electrical measuring instruments for many years, made the proposition, although excellent instruments of English, German and French make were at the disposal of the jury. The exhibit of the Weston Electrical Instrument Company, in which one of each of the many different types and ranges of instruments manufactured by them was represented, afforded the best illustration of the accuracy and care with which the manufacture of these instruments is carried out, even to the minutest details. This was especially true of the exhibit of sample boards containing all parts, down to the smallest screw, which enter into the construction of Weston instruments Although quite a number of these instruments were shown in operation at the exhibit of the company, their behavior under actual working conditions might be studied to better advantage on a number of switchboards in Electricity Building and Machinery Hall. In some cases they were in disguise, as, for instance, on a very handsome white marble board in Machinery Hall and on sundry small ones in Electricity Building, where the instrument was covered by a circular brass case with a plain white dial, in which an opening is made sufficiently large to show the illuminated dial. A very neat novelty in the portable instruments was a voltmeter, placed in a mahogany box with lock and key. A separate compartment is provided in which are contained four "lamp adapters," one for each of the principal lamp sockets in use at the present time. The addition of this ingenious device is apparently a great convenience to inspectors, since, in order to make suitable connections, it is only necessary to insert the respective adapter in place of a lamp and to

French Government Exhibit of Scientific Instruments.

make connection with the contact plug which fits all the different adapters. The laboratory standard instruments were designed for the most rigid requirements of accuracy. The principle involved in their construction is the same as that used in the portable direct current instruments, only the dimensions of the apparatus are considerably increased, so that a scale 14 inches long is obtained. The scale itself is of peculiar construction. A number of concentric circles divide the scale marks into five equal parts. A diagonal line is drawn from the top of one division mark to the bottom of the next work of interpolation, since by means of the very narrow pointers the $\frac{1}{10}$ part of one division can be read with accuracy by passing the eye along the index to a point where index and diagonal intersect each other. The variety of uses to which this type of instrument may be put is almost endless. With two such instruments, one with a high resistance coil for voltmeter purposes and one with a coil of low resistance for currrent measurements by the fall of potential method, together with the necessary resistances and shunts, a most complete laboratory outfit can be obtained and measurements can be made with an accuracy of $\frac{1}{10}$ of one per cent, from a minute fraction of a volt or ampere up to several thousand. By means of Mr. Weston's special alloys without temperature co-efficient a remarkably close compensation for temperature is obtained in these instruments, a result which has so far not been attained in any other type of instruments.

The direct reading wattmeter for alternating and direct current circuits involves essentially the same principle in its construction as that employed in the Weston portable voltmeter for alternating and direct current circuits, with this modification—that the main current passes through the stationary field coil of few turns of comparatively heavy copper wire, while the movable coil, in series with an appropriate amount of external resistance wound non-inductive, takes a very small portion of the main current, and is connected in parallel with the object to be measured. Wattmeters were exhibited with ranges varying from 150 to 15,000 watts over the whole scale. One instrument is especially intended for the test of incandescent lamps, and has a range from 0 to 150 watts in single watt divisions. The maximum current capacity for this wattmeter is given as two amperes, and the maximum voltage allowable, 150 volts. Electrically, these instruments are so dimensioned that errors due to self-induction are inconsequent.

The Weston ground detector was shown in operation at the exhibit. It is really a modified voltmeter, and so constructed that it quickly indicates the existence of grounds, and also shows the insulation resistance of the circuit whenever consulted. It can, therefore, be trusted as a guide to determine whether the insulation of a circuit is deteriorating, and to indicate when the circuit needs attention. In order to determine the actual insulation resistance of a circuit with this instrument, a simple calculation would be necessary; this calculation may be dispensed with, however, by means of a very ingenious and simple mechanical calculating device which is furnished with the instrument.

Alloys with negligible and negative temperature co-efficients formed probably one of the most extraordinary exhibits of its kind. When Mr. Weston gave to the world in 1885 for the first time the discovery that it was possible to make alloys which possess the above mentioned properties the fact was received with a good deal of incredulity. Since then the observations have been verified by other investigators here and abroad, and it has been found that the material possessing these remarkable properties has moreover the great advantage that it is, with proper treatment, absolutely free of the annoying qualities of german silver and platinoid, which cause a continual change of their resistance with time. These alloys are used by the company not only for the compensation for temperature of their instruments, but also in their portable and laboratory standard resistance boxes, of which a complete set was on exhibition.

E. S. GREELEY & CO., NEW YORK.

This company made an exhaustive exhibit of measuring instruments of all kinds, prominent among which was a sensitive Thomson's Reflecting Galvanometer. The coils of this galvanometer are hinged so as to allow easy access to the delicate astatic system. This system consists of a ⅜ inch mirror, on the back of which are the magnets. The mirror is suspended by a silk fibre obtained from a silk spider found in South America. The Frey Portable Galvanometer for street subway testing is a sensitive reflecting galvanometer having a device by which the system is released and the galvanometer allowed to swing free. The galvanometer swings on a universal joint, thus always being level. The instrument is also locked and the system lifted to take the strain off the fibre by turning a key on the base. The instrument is locked and can readily be transported from place to place without breaking the system. The Frey Compensating Shunt used with these galvanometers is an improvement over the Kempe compensating shunt. Various styles of test sets, tangent galvanometers, electrometers, and divided meter bridges were shown. The Frey Reversible Bridge used for comparison is built on the Wheatstone principle and has a double coil which can be reversed, thus determining whether any error exists in the bridge itself. There were also arranged Western Union Tangent Galvanometer, Differential Double Shunt Galvanometer, Zinc and Tangent Galvanometers, Condensers, Resistance Boxes, of various kinds, and various forms of contact and discharge keys.

THE WESTINGHOUSE ELECTRIC AND MANUFACTURING COMPANY.

METERS.—The Shallengerber meter for alternating currents is a recording instrument which registers upon its dials the total flow of current in ampere hours. The meter consists of a coil or coils in series with the circuit in which current is to be measured. Within this coil, and at an angle of about 45 degrees, is placed another coil short-circuited on itself. In practice this second coil consists of stamped rings of bare copper. Within the inner coil is a soft iron disc supported on a shaft which is at right angles to the plane of the coils. The iron disc serves as an armature and its shaft is properly connected with gears operating the indicating dials. When an alternating current passes through the main coil, magnetic poles are induced in the disc and an induced current is set up in the short-circuited coil, which also results in setting up magnetic poles in the disc that are at an angle of about 45 degrees from those induced by the main coil. The induced current has its phase later than the phase of the inducing current, and as a result the poles induced by the former current are a little later in time than those produced by the current in the main coil. Consequently there is alternately set up and dying away two magnetic fields or lines of force that are practically independent, differing in angular position and also in phase or time. As any set of poles in the disc persists after the dying away of its current, each set has a tendency to be attracted into the field being maintained by the other coil, and thus rotation is produced. As the torque of an electromotive device of this character is not proportional to the amount of current flowing, but the tendency of rotation increases approximately as the square of the current, a retarding device is required whose resistance shall increase as the square of the speed, in order that the resulting speed of the armature shall be directly proportional to the current flowing in the main coil. This retarding device consists of fans of aluminum placed upon the shaft; thus the speed of the motor is directly proportional to the current and the number of revolutions is indicated upon the dials, which are properly arranged for registering ampere hours. No commutator is employed and all electrical connections are solid. The mechanical friction of the rotating element is limited to that at the lower end of the spindle, as the upper end simply passes through a guiding ring. The spindle is of hardened steel and rests upon an agate bearing. A drop of watch oil is held in the bearing. The registration is simply in ampere hours. The meter is of course usually employed on constant potential circuits and this being so it is unnecessary to read in watt hours. A shunt coil facilitates the starting of the meter but introduces a constant loss and a certain degree of additional complication, both of which are open to objection. These meters may be used for circuits carrying practically any potential, the rate charged per ampere hour being of course varied proportionately. The rotating elements are extremely light and wear is reduced to a minimum. The consumption of energy (drop) in the motor is altogether insignificant.

GROUND DETECTOR.—This is an instrument for instantly and accurately determining a ground on alternating current generator or supply circuits. The instrument consists essentially of two transformers which have their primaries connected in series with one another across the generator or supply circuit. The secondary circuits of the transformers are individually closed upon themselves each through an incandescent lamp. From the wire connecting the two transformers in series, connection is made to the ground through a push-button, shown at the bottom of the detector.

STILLWELL REGULATOR.—The Stillwell Regulator is an instrument by means of which the potential of the circuit to which it is attached is increased or decreased without altering that of the dynamo supplying the electrical energy. It consists essentially of four parts, viz.: one or more converters, a multi-polar switch, a reversing switch and a preventive coil. The converters are built with secondary coils adjustable in length, i.e., connections are brought from different parts of the secondary coil to the multi-point switch. The primaries of the converters are in parallel with each other and with the circuit, while the secondaries are in series with the supply circuit. The fundamental principle of the regulator is that of a converter which shall increase or decrease the primary potential by an amount determined by the length of the secondary coil. This length is increased or diminished by loops carried from the various coils of the secondary to the multi-point switch. The arm of this switch sweeps over these contacts, and by this means connections may be made with any given length of the secondary coil. The instrument is designed to obviate any short-circuiting effects while throwing from one contact to the next.

COMPENSATOR.—The different types of compensator made by this company, are all on the same principle, the difference in design being made merely to suit the local conditions. The instrument is used, in connection with the station voltmeter, to enable the attendant to accurately maintain a constant potential at the distant end of the line, notwithstanding variations in load and consequent loss or drop in the line. The compensator does not of itself either increase or decrease the electromotive force on the circuit, but it acts on the voltmeter in such a way, that the pressure on the circuit must be increased as the load increases, in order that the voltmeter may give its normal indication. The instrument consists of a converter with a primary of an adjustable number of turns connected in series with the supply circuit. The secondary also has an adjustable number of turns and is connected to the voltmeter. As the load on the circuit increases, current through the primary increases and the action of the secondary on the voltmeter is appreciably increased. When in use, the plug must be in one of the contacts along the top of the instrument. Which one it should be in depends upon the amount of current flowing in the circuit. Semi-circular lines of contact, near the bottom of the instrument, indicate the points of contact of the switch for the varying percentages of loss which it may be desired to allow in the supply circuit.

VOLTMETER.—The voltmeter used with the compensator described above is known as the type "E" voltmeter. The small coil connected with the compensator is at the upper end of the coil, to the left of the center of the instrument. The voltmeter is mounted on a marble base and enclosed in a plate glass cover, so that it is attractive in appearance and at the same time non-combustible. The necessary resistance is secured by the use of flat coils placed next to the marble base. In series with this resistance is a German silver wire coil, which, by means of a sliding contact, is made to serve as a variable resistance, so that the instrument may be adjusted to indicate any voltage within its range. The main coil, which actuates the plunger connected to the lever of the indicator, is placed to the left of the center of the instrument. A plumb-bob is used for setting up the instrument. The instrument may be calibrated by varying the weight on the balancing arm at the top of the instrument.

AMMETER.—The alternating current ammeter is similar in appearance to the voltmeter. Upon a marble base is mounted a solenoid connected in series with the main circuit. The solenoid actuates an iron core connected with the pointer, which, by the movement of the core, is deflected across a scale at the lower part of the instrument. A plumb-bob is used for setting up the instrument. Connections are made back of the marble base. The ammeter is made in various sizes.

ANTHONY ELECTRICAL INSTRUMENT COMPANY.

The exhibit consisted of instruments for measurement of electric currents and for the determination of electrical resistances. The apparatus exhibited was as follows: magnetic vane ammeters, large size, and mounted in polished brass cases, reading from 0 to 150, 0 to 300 and 0 to 500 amperes; magnetic-vane voltmeters, large size, and mounted in polished brass cases, reading from 0 to 150, 0 to 250 and 0 to 600 volts; magnetic vane ammeters, small sizes, and mounted in polished brass cases, reading from

Corona Lighting.—Manufactures Building.

o to 50, o to 100, o to 150 and o to 200 amperes; magnetic vane voltmeters, small size, and mounted in polished brass cases, reading from o to 150, o to 250 and o to 600 volts. The ammeters and voltmeters have fixed and movable pieces of very short sheet iron which are both magnetized by the same coil and act by mutual repulsion against a non-magnetic flat spiral spring. By this arrangement, the use of a permanent magnet is avoided, and the power developed is very considerable for a small consumption of electrical energy. On a switchboard were attached instruments, switches and cut-outs.

THE EDISON MANUFACTURING COMPANY.

KENNELLY STANDARD AMMETER AND MILLIAMMETER.—These instruments have no iron, steel, or magnetic metal in their moving parts, and consequently will not alter or demagnetize under any reasonable treatment. They are not affected by the earth's magnetic field, nor by the stray field of dynamos beyond a few feet radius; the scales are nearly uniform, and very distinct, being read off with mirror to avoid parallax; the movements of the indicator are practically dead-beat; the instruments are calibrated and standardized at the Edison Laboratory. They have very low resistance and inductance, so that they possess great electro-magnetic power, combined with very little influence upon the circuit whose current strength they measure. The resistance of the 500 milliampere instrument is 0.3 ohm. and its inductance 5 microhenrys; they are compact and portable, the outside dimensions of the instrument being 6¼ in. x 8¼ in. x 2¼ in. high; their weight in wooden case, six pounds; there are no adjustments, levelling screws or clamps for shipment, so that the instruments are always ready for operation, and can be used in any position. The binding posts are large and strong. They will take a large wire by penetration, or a small wire by clamping.

J. C. VETTER & COMPANY.

The Vetter Standard Direct Reading Milliampere Meter, especially designed for physicians' use, is of the upright pattern, as two distinct scales for low and high readings, extending full from left to right, can be seen at a distance. The staff is of hardened steel, and pivoted in ruby bearings. The Vetter carbon current controller is an instrument for controlling with the greatest nicety a current of high voltage and small quantity from off or no current to the full amount. In combination with electro-medical apparatus, this controller is far in advance of any rheostat switchboard or cell selector. The Vetter Incandescent Current Adapter and Tap is a new and simple device for utilizing the incandescent current in electro-therapeutics; to charge storage batteries, to operate electric motors for fans, dental work, sewing machines, etc; to extend a drop light (without removing the original lamp), and to adapt the incandescent light current for the operation of various translating devices requiring different amperage or quantity currents in their employment.

GEORGE CUTTER, CHICAGO.

A new departure in measuring instruments is marked by the Cutter Non-Inductive Ammeters and Voltmeters, made by George Cutter, these being a line of hot wire instruments exhibited in the Electricity building. The measuring is proportioned to the difference in expansion between two exactly similar wires, one of which is heated by the current while the other acts as a correction for changes in temperature of the surrounding air. For the ammeters, a heavy conductor is placed in shunt with the measuring wire, while the voltmeters have a non-inductive resistance placed at the back of the instrument. These instruments give equally accurate readings with direct or alternating currents, and their range as shown in the exhibit is from a hundredth of an ampere to a thousand. The voltmeters shown were for fifty and a hundred and ten volts, both the station or wall type and the smaller portable form being represented. All of these had a long and open scale, while a larger form of six hundred and fifty volts had a dial covering nearly a complete circle. A meter for arc lamps is what was shown in the Pattee Lamp-Hour Recorder, this being a clock which registers the exact number of hours during which the current is used. One form of this had the arc light switch mounted in the same casing, so that the whole device was self-contained. Then there were other forms adapted for a time record for motors on power circuits, or for a group of lamps on low tension incandescent lines.

J. H. BUNNELL & COMPANY.

Western Union Tangent Galvanometer.—This instrument was handsomely mounted on a circular hard rubber base 7⅜ inches in diameter, provided with leveling screws and anchoring points. The galvanometer consisted of a magnetized needle ⅞ inch in length, suspended at the center of a rubber ring 6 inches in diameter, containing the coils. The coils were five in number, of a resistance of 0, 1, 10, 50 and 150 ohms. The first was a stout copper band of inappreciable resistance; the others were of different sizes copper wires carefully insulated. Five terminals were provided, the plug-holes of which were marked respectively 0, 1, 10, 50 and 150. The ends of the coils were so arranged that the plug inserted at the terminal marked 150, put in circuit all the coils; at the terminal marked 50, all except the 150 ohm coil; and so on till at the zero terminal only the copper band was in circuit. Fixed to the needle which was balanced on jewel and point, was an aluminum pointer at right angles, extending across a 5-inch dial immediately beneath. On one side the dial was divided into degrees ; on the other it was graduated, the figures of the scale corresponding to the tangent of the angles of deflection. Large testing set galvanometer and rheostat, in leather case, sensitive horizontal needle galvanometer with magnifying reading lens, 3 coils, 0, 40 and 100 ohms resistance and combined bridge (Wheatstone) and rheostat covering all ranges of measurement direct from 100 to 11,000 ohms, and with bridge having in each side coils of 10, 50, 100 and 150 ohms, admitting of multiplying measurements up to 550,000 ohms, were also shown.

THE McNEILL-TINDER ELECTRIC COMPANY, WINCHESTER, KENTUCKY.

Recorder.—This little machine consists essentially of an electromagnet and an armature; the latter is attached to a series of numbering dials. At central station a clock making a momentary contact every tenth part of an hour (six minutes), is installed and connected with a relay key which closes the circuit in which the recorders are inserted. This circuit is composed of a single iron wire, the other end of which is grounded. One such wire—the kind commonly used on telephone lines—will carry current sufficient for, on an average, of about two hundred recorders. This number will vary according to distance, etc. This line is, practically speaking, dead, as the impulses are only momentary and are given at intervals of six minutes. The recorders are put in, in series, wherever a group of lights may be. A double pole switch is used and the connections are so made that when the lights are turned on the recorder is thrown into its circuit at the same time, it also being thrown out again when the lights are turned off, although the recorder circuit is entirely separate from the light circuit. As only the time is measured the same recorder may be used for any sized group, provided, always, that when being used for any one group the same number of lights are, at all times, being used. In connection with the contract system this device is invaluable. By simply throwing a switch at central station the 9:30 customers have their recorders thrown in and the company at once see which of these are running lights over-time. The ten o'clock customers are next treated in the same way, and thus a heavy leak in the company's dividends is stopped. This machine can equally well be used on arc lights and motors. In the arc light field it has no peer, giving the exact time the lights are in use.

THE McINTOSH BATTERY & OPTICAL COMPANY.

Milliampere Meter.—The construction of this instrument is essentially that of an Astatic Galvanometer, and it combines the desirable qualities of cheapness with great accuracy and adaptability to use in electro-therapy. It is not easily liable to injury, and yet under repeated tests, in comparison with the high-priced milliampere meters, constructed particularly with reference to their accuracy as test instruments, this meter has proven itself wonderfully accurate and reliable.

CHARLES L. JAEGER.

Electric Recording Compass, Log and Chronometer.—These instruments were designed to keep a continual, simultaneous record on one strip of paper of the direction, distance and time in which a ship travels. The records are made by means of sparks which pass between certain electrical conductors placed in the bowl of the compass, these conductors being insulated from one another in certain places with hard rubber, and the atmosphere; between these conductors there is a strip of paper which is perforated by the sparks. The paper is kept continually moving by means of a suitable mechanical device. By the position of these perforations you can determine the direction, distance and time in which the ship has been moving. To keep the direction record, there is attached to the magnetic needle a light metallic spiral; near the spiral is arranged a straight conductor which participates in the movements of the ship; a strip of paper is fed across the conductor by a suitable mechanical movement. To keep a record of the distance and time in which the ship has been moving, there is an electric log and chronometer which are in electrical communication with the compass, the log completing a circuit every one-fifth of a mile the ship travels, this current perforates the paper at the lower edge. By substituting an electric cyclometer for the electric log, the apparatus can be applied to a wagon. The chronometer sends a stream of sparks every minute through the paper at the upper edge. In this manner the three indelible records are produced on one strip of paper at the same moment, without interfering with the proper working of the different instruments. The current to produce the records is taken from a Ruhmkorff Coil which requires about two and one-half volts and five amperes to operate. These recording attachments can be applied to any form of compass, whether liquid or dry; the spiral is made of aluminum and therefore is very light. It can be applied to one of Prof. Thomson's compasses as well as any other to good advantage. The apparatus is very simple and requires very little attention besides putting in a new strip of paper each morning, and does not cost much to operate. The records can be duplicated by any well-known process in virtue of the perforations.

HARTMANN & BRAUN, FRANKFURT A/M, GERMANY.

SCIENTIFIC INSTRUMENTS.—The firm exhibited a very full line of measuring instruments, some of which are described as follows: Appliances for measuring resistances of 0.01 to eleven million ohms, consisting of a universal resistance bridge with plug connections, reflecting galvanometer with reflecting magnet and reading telescope. Apparatus for measuring current strengths and E. M. F.'s by the compensation (Null) method, consisting of a Clark standard cell and ballast resistance box for main and shunt circuits, a branch resistance of 0.01 ohm, an aperiodical reflecting galvanometer with Siemens' bell magnet and reading telescope. Apparatus for measuring high insulation resistances consisting of an astatic cable galvanometer with Bruger's magnets, shunt resistances, a scale lamp for objective reading, a megohm with 10 subdivisions as standard of comparison, a battery of dry accumulators with universal pachytrope and switches. Apparatus for measuring capacities of cables, consisting of a ballistic reflecting galvanometer with Weber's magnets, a reading telescope, a condenser of one microfanad with subdivisions, together with charge and discharge keys. Apparatus for measuring small resistances, in particular, for testing the conductivities of materials, consisting of a resistance bridge with key, amperemeter and sensitive galvanometer. Range from 0.00001 to 5 ohms. The measurements are made on the principle of the Wheatstone bridge, modified, however, in such a manner that by double reading the intermediate resistances are eliminated. The result is obtained without extensive calculation by direct reading from the scale of the apparatus. By means of universal clamps conductors of any section and also carbons may be placed in the circuit. Apparatus for testing iron and steel with respect to magnetic permeability and coercive properties, consisting of a standard electromagnet, a Lenard's bismuth spiral, a bridge for determining the strength of the magnetic field, a galvanometer constructed on Deprez's principle, and an amperemeter. Apparatus for determining the resistances of electrolytes according to Kohlrausch, consisting of a Kirchhoff-Wheatstone bridge, cylindrical form, an induction apparatus for generating alternating currents, an unifilar suspension electro-dynamometer, a telephone, three U-tubes with platinum electrodes and Arrhenius' resistance vessel. Photometric apparatus, consisting of a large precision photometer with Lummer-Brodhun's comparison prism, standard amylacetate lamp, universal incandescent lamp stand, revolving mirror for determining the illuminating power of

Westinghouse 15000 Light Alternator in Course of Construction.

arc lights at various angles of emission, voltmeter and wattmeter for measuring E. M. F.'s, and energy expended in incandescent lamps.

SMALL APPLIANCES.—Resistance bridge for determining the resistances of solid conductors, e. g., insulated wire wound on a bobbin, with aperiodical needle galvanometer. Range from 0.1 to 10,000 ohms, reading directly without calculation. Small universal resistance bridge, according to Kohlrausch for determining resistances of solid conductors and electrolytes, e. g., the internal resistance of cells by means of alternating currents, generated by means of an induction coil supplied with the apparatus. A telephone is used for measuring in place of the galvanometer. Range from 0.01 to 1,000 ohms, direct reading. Wheatstone-Kirchhoff Bridge with measuring wire of 1 meter length may be extended for very accurate measurements by additional stretched wire resistances. The meter wire is calibrated by means of a second contact slider, which renders the apparatus also available as a Thomson double bridge. Precision resistance box or single standard resistance coils serve as comparison resistances. Differential Galvanometer with index for comparing any resistances; the latter was found among the exhibits in the shape of a decade crank resistance box and a series plug resistance box. Differential Galvanometer with bobbins, wound with resistances in the ratio of 1.20: by means of this instrument the resistance of an incandescent carbon filament may easily be found by Voller's method. Apparatus for determining small resistances according to Kirchhoff's differential shunt method, consisting of a standard resistance of 0.01 ohm, two precision resistance boxes and a differential galvanometer, the latter modified according to Deprez, was also shown. Nippoldt's Testing Apparatus for lightning conductors, consisting of Wheatstone bridge reading directly contained in the casing of a telephone, a small induction coil, together with cells for generating alternating currents, or if desired a simple galvanometer, the whole in a sling case, also, a folding earth plate; the resistance due to the passage from the lightning conductor to the earth plates has, on account of the polarization taking place at the plates, like the resistance of electrolytes, to be measured by means of alternating currents, hence a telephone is substituted for a galvanometer. It is useless to measure the resistances of the metallic conductors, as these can only be examined by ocular inspection.

APPARATUS FOR MEASURING TEMPERATURES BY ELECTRICAL INSTRUMENTS.—Braun's Pyrometer for temperatures up to 1000 degrees and up to 1500 degrees C., based upon the variation of the resistance of a platinum wire wound so as to obviate self-induction inside a non-combustible box and forming one branch of a Wheatstone bridge was shown. The sensibility is so great that with the highest temperatures an increase or decrease of but a few degrees is rendered apparent. The absolute accuracy is extremely great, the instrument being standardized by numerous comparisons with a porcelain air thermometer.

MEASURING APPARATUS ACCORDING TO KOHLRAUSCH.—Wheatstone-Kirchhoff Bridge. The resistance wire, which is 3 meters long, is wound upon a cylinder of white marble, there being 10 convolutions; the permanent connections are insured by brushes; the sliding contact is made by a small wheel. The comparison resistances of 1, 10, 100, 1000 ohms, and the resistances which are to be measured, are interchangeable by means of plug connections. Unifilar suspension electro-dynamometer for small currents with mirror reading: the current is led to the movable bobbin by the suspension wire and a platinum coil, acting as a damper, dipped into dilute sulphuric acid. Portable differential galvanometer with mirror reading: provided with a small magnetic mirror suspended within the oval multiplier. The amount of damping may be regulated by means of a copper sliding piece. The sensibility is rendered variable by various thicknesses of the duplex winding and by a controlling astatizing magnet. Without astatization and with a resistance of 50 ohms a deflection of 1 mm reads 0.0000004 amp. at 1 meter from scale.

ABSOLUTE TANGENT GALVANOMETER. The current passes through a copper ring which can be accurately measured: All unnecessary metal parts are avoided. The reflecting magnetometer, which is free from metal, may be removed and replaced by a needle. Portable Bifilar Suspension Variometer for Terrestrial Magnetism: tubular strongly damped magnet suspended by two fine wires placed a short distance apart. The constants are found by means of a torsion circle provided with vernier and micrometer adjustment. Changes of position are controlled by stationary mirror. Intensity Variometer for Terrestrial Magnetism: The magnetic steel mirror is, by means of four small deflection magnets, placed at right angles to the meridian, so as to neutralize the influence of variation of declination. The sensibility may be regulated to any degree. Absolute bifilar suspension magnetometer: by exact measurement of the relatively great distance of 12 mm between the wires, the deflections due to the reversal of the tubular magnet gives the product of the magnetism of the bar magnet and the terrestrial magnetism in absolute measure. Simultaneous observation made with a unifilar suspension magnetometer gives the intensity of the terrestrial magnetism in absolute measure. The advantages of the method consist in rendering the measurement of time and of the influence of temperature dispensable. Absolute bifilar suspension galvanometer: the magnet is replaced by a wire ring of many convolutions which may be measured exactly. The deflections of this ring due to a current represent the product of strength of current, area of circuit and terrestrial magnetism. This instrument may be used for absolute measurements of currents or terrestrial magnetism by using it in combination with an absolute tangent galvanometer or a single suspension magnetometer.

SPRING GALVANOMETER.—A tube of thin sheet iron sides is suspended by a torsionless spiral of many convolutions and is drawn into a solenoid by the current. The amount of the attraction is seen from a scale divided in amperes. The indications are aperiodical and independent of strong extraneous currents. The instruments are adapted for currents up to 3000 amp. Among other apparatus exhibited by this firm were the Clark Cell standard resistance of all grades, small and large compensation apparatus, telescope galvanometers, suspended coil galvanometers, dead-beat reflecting galvanometers, rheostats for industrial exact measurements, and measuring instruments of all kinds for school and educational purposes.

ALLGEMEINE ELECTRICITATS GESELLSCHAFT.

MEASURING AND CONTROLLING INSTRUMENTS.—This company has paid special attention for many years to this important branch of electrical work. The ammeters and voltmeters have a comparatively large coil, which attracts a bundle of very thin iron wires, their length and position in the solenoid depending on the distribution desired and the size of the divisions of the scale. This was seen by the 600 amp. ammeter, the case of which had been removed for inspection. The following complete instruments were shown: ammeters for continuous currents from 30 to 2000 amp., and for alternating currents from 70 to 350 volts. A special compensating voltmeter was shown, for compensation of the losses in long mains. For controlling the pressure in central stations, a "registering voltmeter" was used, by which the increase and diminution of tension were determined directly by photographing at any moment the position of the voltmeter needle. In cases where the tension has to be maintained between certain narrow limits, use is made of a signalling voltmeter, showing automatically the too high or too low pressure by corresponding bells or colored lamps. This collection was completed by indicators for currents and galvanoscopes for measuring insulation resistances, and for testing joints, earth detectors and several other instruments used for installation work. For registering electric power, a wattmeter has been lately constructed by the company, containing clock work, which is affected by the flow of the current in such a way that its speed is accelerated in comparison with its normal speed, this acceleration being directly registered on the dial. In addition there is an automatic mechanism for winding up the clock work periodically, rendering all attendance unnecessary.

RICHARD BROTHERS, PARIS.

Long Distance Transmitting Instruments.—These instruments are intended for the transmission, to any distance, of the indications of the various instruments fitted with indicating hands. They comprise two stations, the first one perceiving and transmitting the indications, and the second receiving and registering the same. At the transmitting station is an instrument either indicating or registering. The hand or stylus moves in front of a dial, which, in its normal position, embraces a pin fixed on the hand. As soon as this hand moves in one or the other direction, the pin makes a contact, sending an electric current in the corresponding direction. The said current travels along the electric wires and closes the circuit of an electro-magnet, situate at the receiving or registering station. The armature of this electro-magnet is, therefore, attracted, and causes a ratchet wheel to advance one tooth. This ratchet wheel, carrying the hand which indicates or registers what occurs at the sending station, it will easily be understood that for each contact occurring at the sending station, the hand at the receiving station will be correspondingly displaced by one division one way or the other. At the same time that the circuit is closed at the sending station, the dial is brought back to its normal position, in which it does not touch the pin. Should the hand of the instrument at the sending station continue to move in the same direction, the hand of the pin, coming again into contact with the dial stops; another current is sent to the receiving station and the receiver is made to advance one further division. A contact in the opposite direction at the sending station causes, through the same agency, the hand of the receiving instrument to travel backward one division. It will, therefore, be seen that the hand of the receiving station instrument moves correspondingly to that of the sending station. The two instruments, therefore, are working in an absolutely synchronous manner. This arrangement can be applied to all descriptions of instruments provided with a hand or pointer, be they merely indicating or registering, such as pressure gauges, vacuum indicators, water level indicators, speed indicators, pyrometers, thermometers, etc.

Universal Controlling Chronograph.—The Self-Registering Chronograph writes a stroke on a card each time a contact is established, thereby causing the closing of the electric circuit of a battery connected with the apparatus. It is composed of an electro-magnet carrying a pen which traces a continuous line. Every time the process under observation closes the circuit and establishes a current, the pen traces a vertical line and the duration of the process, as well the exact moment at which the same occurs, are given by the diagram so produced. Amongst the many uses which this instrument can be adapted to, it has been successfully applied in connection with the speed indicators now so extensively adopted in factories. The adaptation is very simple; a mere electric contact being placed on the wheels of the indicator enables, according to the wheel on which the said contact is placed, to register the units, tens, or hundreds, etc.

RAILWAY TRAIN SPEED SELF-RECORDING INSTRUMENTS.—Railway train speed indicating and self-recording instruments are divided into two classes: in the first one, the instruments are placed directly on the train, and indicate or record either the space traveled in proportion to the time, or to the speed; they then belong to the class of instruments known as revolution indicators, or to that of cinemometers. The second class comprises the instruments which are placed at any given point of the line in order to record by means of an electric circuit which is closed by the passage of the trains over one or more pedals. The first instrument comprises as many electro-magnets as there are pedals, the said electro-magnets being mounted on vertical frames, and severally provided with a stylus carrying a writing pen. These pens, placed on the same line, write or trace at distances of ten millimeters from each other, lines upon a diagram paper propelled by clock work and traveling at the rate of five millimeters per minute. Each of these electro-magnets communicates with a pedal and causes a transversal line to be traced when the latter is depressed by the passage of a train. The elements for the calculation of the speed of the trains are given by the time which these lines are traced, and the distance known to exist between the pedals. The second type of instrument offers great advantages as to price, readiness of installation, and particularly as to portability; it is composed of a very finely cut cog-wheel, actuated by a clockwork mechanism, and revolving at a regular rate of one revolution every 2½ minutes. When an electric circuit is closed on a small electro-magnet mounted on the frame, a penholder or stylus is thrown into gear with the said wheel and carried with it in its motion, the pen tracing on a recording cylinder an ordinate, the height of which is proportional to the time during which the stylus has remained in gear. The instrument is completed by a double electro-magnet system, the object of which is to close the circuit actuating the recording stylus as soon as a pedal is depressed by a train, and to break it as soon as the train passes over a second pedal distant from the first one a known number of meters. The working of the instrument is, therefore, easy of comprehension. Two pedals are fixed on the line, say 100 yards apart; as soon as the first one is depressed by the passing of a train, the circuit actuating the writing pen stylus is closed. The latter is thrown into gear, and in a regular motion rises along the recording drum until the train, continuing its journey, passes over the second pedal; at that

Westinghouse Block. — Machinery Hall.

moment, the stylus, being thrown out of gear, falls back to zero. The length of the ordinate is in an inverse ratio to the speed of the train, since it is longer according as the speed of the train is slower. The advantages of this instrument are manifold. It is very portable, and can be got ready for use at any given spot of a railway track in a few minutes, therefore, allowing of a perfect check being established over the engine driver without the latter being aware of the spot at which the instrument is temporarily placed.

DISTANCE SPEED INDICATORS, ELECTRO-CINEMOGRAPHS—electrically recording the speed or the number of revolutions per minute of one or several engine shafts or pulleys. In certain establishments where there are often several motors or engines, the speed of which the engineer wishes to control from his office, and without using as many cinemographs as there are motors, the electro-cinemograph renders easy the controlling at any distance of as many engines as may be desired. This instrument is a combination of speed indicator, with an escapement system actuated by an electro-magnet. The instrument is set ready for work by merely winding up the spring of the train of wheels which causes the pointers "Time" and "Number of Revolutions" to act upon one another. Its installation is effected in the following manner: On each machine a simple electric closing circuit arrangement for each revolution is adopted. To each contact are attached two wires, one of which is connected to a common return wire, the other being taken and connected to a switch placed in the engineer's office. The switches are arranged on a board, in front of which is placed on a bracket the electro-cinemograph. This instrument is connected to the switches as well as to the common return wire, and to a battery placed in the circuit. When the engineer wants to ascertain the speed of any given engine, he, by closing the circuit of the said engine with the switch, immediately sets the electro-cinemograph at work and obtains a record on the diagram paper of the number of contacts occurring at the shaft in the unit of time, that is to say, its number of revolutions per minute. The speed of every individual motor can thus be controlled; the word "speed" being taken in its absolute sense. This instrument is also used for measuring the speed of water courses, boats or yachts, by merely connecting it to contacts established on a turbine of the Wohlmann system, or to an electric log of the Fleuriais or other system.

ENGINE REVOLUTION INDICATING AND SELF-RECORDING INSTRUMENT. This instrument is cheaper than the electro-cinemograph speed indicator, and is now in extensive use for the control of the working of engines. It is composed of a pulley driven by the engine, and actuating by means of a series of intermediary mechanism (endless screw and helix) a pen bearing stylus. The motion of the stylus is calculated so as to cause it to rise 2 millimeters for every 100 or 1,000 revolutions, according to the description of engine to which it is applied. Every 7,000 or 70,000 revolutions, the pen falls back to zero, ready to start again on its ascending journey as the engine continues to work. As soon as the engine stops, the writing pen ceases to ascend, and, the drum continuing to revolve in proportion to time, the line traced on the diagram paper is a horizontal one. The diagram therefore records for every day the hour of starting, the speed of the engine (by the length of curve), and the stoppages. The total of the figures corresponding to the ascensions of the pen give the total number of revolutions effected. The instrument can be made to work electrically at a distance.

Electric Alarm Metallic Thermometer: Air Chamber Thermometer.—This instrument will be found most useful for the indication of the temperature in the carrying out of all industrial processes; in vats, drying rooms, malt houses, purgers, etc., as well as in hot houses, workshops, hospitals, sick rooms, cargo and other vessels, etc. This instrument may be connected to two electric bells, respectively indicating the high and low temperature above, and under which it is not desirable to go, or with a single bell for both cases. It can be adjusted to all temperatures from —50° to +110° C.

Electrical Alarm Thermometers, or Explorating Rods for Enclosed Spaces.—When the maximum and minimum temperatures to be controlled occur in closed spaces, into which it is impossible to penetrate, such as silos, fodder sheds, boilers, etc., or in liquids, the explorating rod gives excellent results. The part of the instrument sensitive to the heat (a metallic receptacle filled with an expanding liquid) may be forced into the contents of the enclosed space the variation of temperature of which it is desired to ascertain. It is fixed at the end of a tube of any desired length. This apparatus is very strongly built, and can, without injury, receive violent blows, or be submitted to considerable pressure, all causes of breakage or of bad working being entirely removed. The equilibrium in the temperature is established through the envelope of the instrument, which is in direct and immediate contact with the source of heat to be measured; the electric contacts are protected by a metallic envelope which keeps them in perfect working order and free from oxidization, and the conducting wires themselves are protected by the tube up to the outside of the enclosed space in which the apparatus is placed. The temperature explorating rod is absolutely indispensable in silos, fodder sheds, pressure steam pans, etc., and can be of great use in public baths and hydropathic establishments. Its form allows of its being readily adapted to bath heaters, baths, hydro-mixers, etc.

DIRECT READING INSTRUMENTS, VOLTMETERS.—These voltmeters are fine wire electro-dynamometers; they do not contain a single trace of iron, and work through the action of a circuit passing through a fine wire fixed coil upon a movable coil made of the same size wire. They naturally show no trace of hysterics. The suspension of the movable coil is monofilar. The instruments can, without inconvenience, be left in circuit. As is the case with electro-dynamometers, the indicators increase according to the ratio of the square of the tension. It is, therefore, useful to exactly indicate the desired average voltage. For this voltage the displacements of the hand are about two to three millimeters per volt.

SELF-RECORDING AMMETERS, VOLTMETERS, WATTMETERS.—These instruments give an uninterrupted ink diagram of either the tension or the intensity of the current generated by dynamo or magneto-electric machines, batteries, accumulators. The variations of the current are written by the apparatus upon a band of paper covering a drum, which generally makes one complete revolution in 24 hours. The drum may be, if desired, made to revolve in 1 to 12 hours, or in 8 days, etc. Drums are also supplied to revolve at various speeds, in order to obtain curves of large dimensions, so as to facilitate their study. In certain instances the drum is replaced by a clock-work mechanism, called endless paper mechanism; the latter propels a band of diagram paper which passes before the writing-pen at a speed which can reach three centimeters per minute, or the rotating motion of the drum may be obtained by means of an electrical clock work similar to that used in the time meter. In this case the diagram paper moves only when the current is passing. The types currently used are the instruments with a drum making one daily revolution, and those provided with a 90 centimeter long paper band lasting a week. For the study of arc lamps hourly revolving drums are used. In self-recording ammeters, the terminals are, unless otherwise ordered, placed at the back of the case so as to come through the switchboard. They can, when intended for controling the charge of accumulators, be provided with a pole indicator and a brake (dash pot).

INDICATING AND SELF-RECORDING INSTRUMENTS OF THE CHARGE AND DISCHARGE OF ACCUMULATORS.—It is well known that during the charging of accumulators the lead loses a portion of its weight, until the moment at which the accumulators being fully charged, the current generated by the dynamo is wasted. On discharging, the lead regains its lost weight. This phenomenon is taken advantage of in the construction of instruments for recording the charge and discharge of accumulators, such instruments proving very useful in electrical installations. The apparatus comprises a kind of beam scale, to which is suspended the lead portion (negative and positive) of one of the accumulators of the battery. An indicating hand or a recording pen correctly indicates or records the variations of the charge and the discharge. The apparatus may be provided with the electrical contacts and a signal bell.

SIGNALLING INSTRUMENT; CONTROLLER OF THE PASSAGE OF THE CURRENT IN ELECTRIC CABLES.—This instrument is composed of a clock, which is started as soon as the current passing through the cable exceeds that which the said cable has been calculated to convey; it continues working as long as the excessive current passes, and stops when the passing current becomes normal. It is, therefore, possible to ascertain by a mere inspection if the maximum limit has been exceeded, and for what length of time. This control is indispensable, and, in case of damages or injuries to the cables, enables one to distinctly cast the responsibilities. The instrument is placed in circuit in the same manner as an ordinary ammeter.

SIGNALLING VOLTMETER FOR SWITCHBOARDS.— This instrument, placed on a shunt, lights one of two lamps of different colors with which it is provided when the tension of the distributed current becomes too high or too low. When the current has its exact value the two lamps are extinguished.

ELECTRICITY METERS, FOR DIRECT AND ALTERNATING CURRENTS. —The energy meters are simple, strong, and do not get out of order. Their working is based upon the principle of intermittent integration, and the totalizing, every 15 seconds, of the measurements effected by the wattmeter. The current consumed is given in Board of Trade unit hours on the dials. If desired, the meters may be set for the indication of electrical horse-power. The meter requires no attention, whatever,and is self-winding ; it is easy of inspection, and its individual consumption is less than 0.025 ampere. The correctness of its indications is guaranteed to within one per cent. All the important parts are interchangeable, and the repairs, which are very rarely necessary, can be made in sight of the inspector in one of his visits. When applied in conjunction with accumulators, the meters may, if desired, be regulated so as to totalize during the charging and subtract during the discharging periods. The percentage of loss indicates the moment when the accumulators require recharging. When the meter is intended for both direct and alternating currents, a short circuit plug, with which the instrument is provided, must be taken out whenever the meter is placed on a direct current.

JAPAN GOVERNMENT.

SEISMIC INSTRUMENTS.—The exhibit of earthquake apparatus made by the Japanese government was one of the most interesting in the Exposition. It included the Gray-Milne Seismograph and auxiliary apparatus, and charts and photographs showing damage to the country by the severest of the recorded earthquakes. The Seismograph records automatically the horizontal (E-W and N-S) and vertical components of earth motion, each magnified five times, giving an absolute measurement of the earth motion. The time of occurrence and the duration of shakings are also accurately known. The recent increase in the number of small earthquakes recorded must, in a measure, have been due to the improvements of instruments. The instrument first used for securing earthquake records was Prof. Palmieri's seismograph, which essentially consists of four upright U-shaped glass tubes containing mercury and placed in four different directions. The degree, or relative intensity, of a shock and the direction of its motion are shown by the amount of oscillation of the mercury, which will be more pronounced in some of the tubes than in others, and by the direction of such tubes themselves. Besides these, there were clock arrangements, showing the time of occurrence, and the duration of shock.

GENERAL ELECTRIC COMPANY, LIMITED, LONDON.

CENTRAL STATION OR SUB-STATION INSTRUMENTS.—The volt and amperemeters are highly finished and accurately calibrated instruments, operating on the principle of attracting an exceedingly small piece of iron placed eccentric in the field of a solenoid against gravity by the action of the current circulating through the wind-up. The instruments are made for all commercial capacities and ranges.

METERS.—For continuous current installations this company exhibited some of their Pantz—Reckenzann Meters which are manufactured for various capacities. They are comparatively small, compact and very ornamental.

The Electric Fountains.—View from Machinery Hall.

CHAPTER IX.

SWITCHBOARDS.

SWITCHBOARD mechanism on exhibition showed perfection in construction and cleverness of workmanship rather than advancement of the principles of practice. In view of the perfection to which the transmission of the electrical current has come, this will be readily understood. It will be also seen by perusal of the descriptions of the various makes of boards, that no special material has taken great precedence for their construction. There has been an evident disposition to look for a material other than slate, presumably for the reason that in all slate formations there is a large percentage of iron, many times the iron or other metallic deposit taking the form of veins. These veins may be made the means of carrying the current out of its legitimate channel and consequently of doing much damage. Marble or stone seems to be the growing fashion, largely, it may be presumed, for the same reason, inversely applied, that slate boards are not so popular as they used to be. Some makers of switchboards are disposed to use hard wood, veneered and ornamented, instead of other, more costly material.

WESTINGHOUSE ELECTRIC & MANUFACTURING COMPANY.

The principal exhibition switchboard of this company was erected upon a double gallery against the south wall of Machinery Hall. The floor of the first gallery was about 10 feet above the level of the space occupied by the engines and alternators, while the floor of the second gallery was 8 feet higher. The board was made up of marble panels, which were supported by an iron frame-work, consisting of angle iron supported upon floor beams of the galleries, and securely braced. The uniform width of all panels was 20 inches. The panels, most of which, themselves, constituted the bases upon which the instruments were mounted, extended into the framework of angle iron and were secured by bolts and vertical metal strips. These nickel plated strips were of semi-circular section, and extended from the floor line to the top of each board, covering the abutting edges of the marble panels, and assisting to hold the panels in position. A narrow frieze of marble was carried along the top of each board, and a vertical strip, 5 inches wide at each end, served as a decorative finish. The board on the lower gallery was the dynamo and exciter board; that on the upper gallery was the supply circuit or feeder board. The circuits from the alternators were carried to the lower board through lead covered cables, and thence through switching devices to the upper board. Instruments for measuring the current and potential of each set of the two-phase alternators and of the two single-phase machines were provided on 13 panels, one panel being allotted to each of the 12 double machines, and the thirteenth panel being provided for the two single-phase machines. Provision was made for quickly substituting a relay or reserve dynamo for any one of those in operation. Two-way switches were used for each dynamo circuit, and the terminals at one end of these switches were always kept charged by the reserve alternator, so that by simply throwing a switch, the load was easily and instantly transferred from any alternator to the reserve machine. The board was so arranged as to make it impossible for a careless attendant to accidentally throw two alternators together. Multiple arcing the machines was considered inadvisable, and the board was so designed as to make this practically impossible. Four panels were provided in the lower board for exciters, and the fifth for instruments measuring the combined out-put of the exciters when in multiple. One of these panels was not in use, as three exciters proved ample for the number of machines required. The upper board comprised 40 panels and provided for a like number of supply circuits. Each panel contained a switchboard converter, a compensator, voltmeter, ammeter, Stillwell regulator, switch, fuses and plug receptacles for the connection of the supply circuit to any of the 14 alternators. Each circuit was protected from lightning by the Wurts non-arcing arrester. The specifications under which the contract was awarded made it necessary to provide for supplying any circuit from any alternator, and, inasmuch as the two-phase type of dynamo was selected, it became necessary to design a board which would permit any one of several hundred combinations to be made. The generating plant was equivalent to 26 single-phase machines, and provision was made for supplying each of the 40 distributing circuits from any one of these 26 sources; at the same time the arrangement was such that mistakes were impossible, and the wiring of the board symmetrical and comparatively simple. The panels were each lighted by etched lamps suspended from "goosenecks" from the top of the board, and, in addition to these, four highly ornamental arc lamps, of the constant potential, alternating current type, were hung from arching wrought iron posts at the ends of the boards. Access to the lower board was secured by an artistic spiral stairway of iron. The upper platform was reached by means of two stairways from the lower gallery. Handsome ornamental railing of wrought iron inclosed the front and ends of both galleries. The operation of this board proved satisfactory in the highest degree. The distributing circuits throughout the grounds of the exposition were limited to 200 kw. each, and it is worthy of note that a board capable of distributing five, or even ten times as much as energy—that is to say, 40,000 or 80,000 kw.—embodying the sectional features of this board—can be constructed to occupy practically the same space. This board, which was probably the largest ever constructed, measured in the upper gallery 71 feet in length and 9 feet 6½ inches in height. The lower part was 32 feet 6 inches long, and 11 feet 3½ inches high. The claims made by the Westinghouse Company for this most interesting piece of electrical apparatus are as follows: safety against fire risk, being built entirely of non-combustible material; safety against possible injury to attendants, as no high tension circuit or terminal is exposed on the front of the board, hence, contact with dangerous circuits is impossible; safety against accidental connection of two alternators, secured by the arrangement of the board itself; flexibility in that any machine can be made to supply any circuit or set of circuits; capability of extension; the sectional construction adopted renders it easy to extend the board to accommodate additional machines without materially changing the wiring; sectional construction; the entire face of the board being built up of small units, transportation and erection are facilitated; accessibility, as all the wires are exposed and terminals readily accessible back of the board; uniformity and simplicity. Though there is a large amount of wiring on the board, the panels are alike, and hence, uniformity and simplicity are secured.

GENERAL ELECTRIC COMPANY.

The switchboard from which all constant current arc lamps were controlled was a 6-circuit, 6 machine arc plug board, in which the machines were connected to horizontal bars and the circuits to vertical bars about three inches in rear of horizontal bars. No cables were used except when connecting more than one circuit to a machine, the dynamo and line terminals being connected by a metal plug with insulated handle which connected the horizontal and vertical bars. All contacts being made in the rear of the slate board, it was impossible for the operator to accidentally place himself in circuit. The board was completed by supplementary panels, also of slate, with 6 ammeters and 12 lightning arresters.

In connection with other generators there were exhibited the standard slate panel switchboard of the General Electric Company, containing all necessary switchboard apparatus, mounted on iron floor standards. Connected to four belt-driven alternating incandescent machines was a Standard Skeleton 4-machine, 4-circuit switchboard, on which were mounted the usual station instruments as installed by the General Electric Company. In the center of this board was a safety slate board, with switches by means of which any circuit might be thrown on any machine, or any machines thrown in parallel.

WESTERN ELECTRIC COMPANY.

Besides the power plant board, a description of which will appear under the heading "Power Plant," two sample switchboards, one for arc circuits and one for incandescent circuits, intended to show the distribution of current for the various systems employed by the Western Electric Company were exhibited. There was also a testing device for measuring the resistance to ground of circuits while the current is running. All the circuits were run into the switch-board in Electricity building, which was similar to that in Machinery Hall, being composed of brown Tennessee marble 12 ft. long and 6 ft. high. It was designed to be as perfect a piece of switchboard distribution as it is possible to construct. The back of the board is finished in an elaborate manner, all brass and copper trimmings being polished and copper plated. The positives and negatives of the incandescent circuit are connected directly to bus bars, from which distributing circuits run through double poled knife switches and ammeters directly to the lamp circuits. One end of the board was devoted exclusively to the distribution of arc circuits, there being 150 arc lamps of the Western Electric Company's manufacture in Electricity building.

THE C. & C. MOTOR COMPANY, NEW YORK.

An important part of the exhibit of this company was a large marble switchboard having fifteen circuits which served to distribute current to various machines. This switchboard was divided in two sections, the right hand section having six circuits carrying current at 500 volts, and the left hand section nine circuits at 125 volts. Each section had a pair of round copper bus bars running horizontally across the back of the board and from which were carried the distributing circuits. To the bus bars on the right hand section were directly connected the 500 volt power mains from Machinery building. A 50-h.p. standard C.& C. constant speed motor, operated from the 500 volt circuit, drove a 40-k.w. standard C. & C. compound dynamo giving current at 125 volts. The mains from this dynamo were run to the switchboard and connected to the bus-bars on the left hand section, from which were carried the circuits for feeding the electric lights and 125-volt motors of the exhibit. There were ten circuits in all, feeding as many motors, each circuit being provided with an automatic starting box and double pole switch. These starting boxes were bolted to the back of the switchboard and the switches placed directly on the face of the same. The starting switches were all double pole and intended to be used without any additional switch. The double pole switches above mentioned were provided in order to show the action of the automatic switch when the circuit is opened at a point distant from the starting switch. The switchboard was also provided with ammeters and voltmeters for measuring the consumption of current and the pressure in the various circuits.

BRUSH ELECTRIC COMPANY, CLEVELAND.

The Brush Company showed the only line of arc and incandescent switchboards in the building. The arc boards were of new design, made of enameled slate, built up in sections of 4 dynamos and four circuits each, these panels being set in a frame, side by side. The slate was bushed with hard rubber (no current being in contact with the slate). Instead of long cables, permanent bus wires were placed on the back of the board. By means of these the interchanging of circuits or dynamo can be easily made, the bus wires being used to transfer the circuits from one part of the board to another. The plugs of this were of a very good design. A shield of hard rubber is arranged so that should the operator by mistake pull a live circuit, the shield covers the plugs completely, making it impossible for him to get a shock or burn. In fact, the entire front of this board was arranged in such a manner that it was impossible to get a shock, as hard rubber covered all exposed parts. A ground detector was used with the large boards, by means of which a ground might be located definitely between any two lamps, a bank of incandescent lamps being used as a voltmeter to measure the electromotive force between any two points.

JENNEY ELECTRIC MOTOR COMPANY.

The switchboard shown by this company was complete in every form, it being a polished oak frame with marble panels and trimmings of polished brass, and the greatest pains had been taken to make it a model board, both in design and workmanship. There are two sets of bus wires; one set for each dynamo, or one for the dynamo and the other to be connected to the street mains; and by use of the double throw switches the load can at will be changed from one machine to the other,

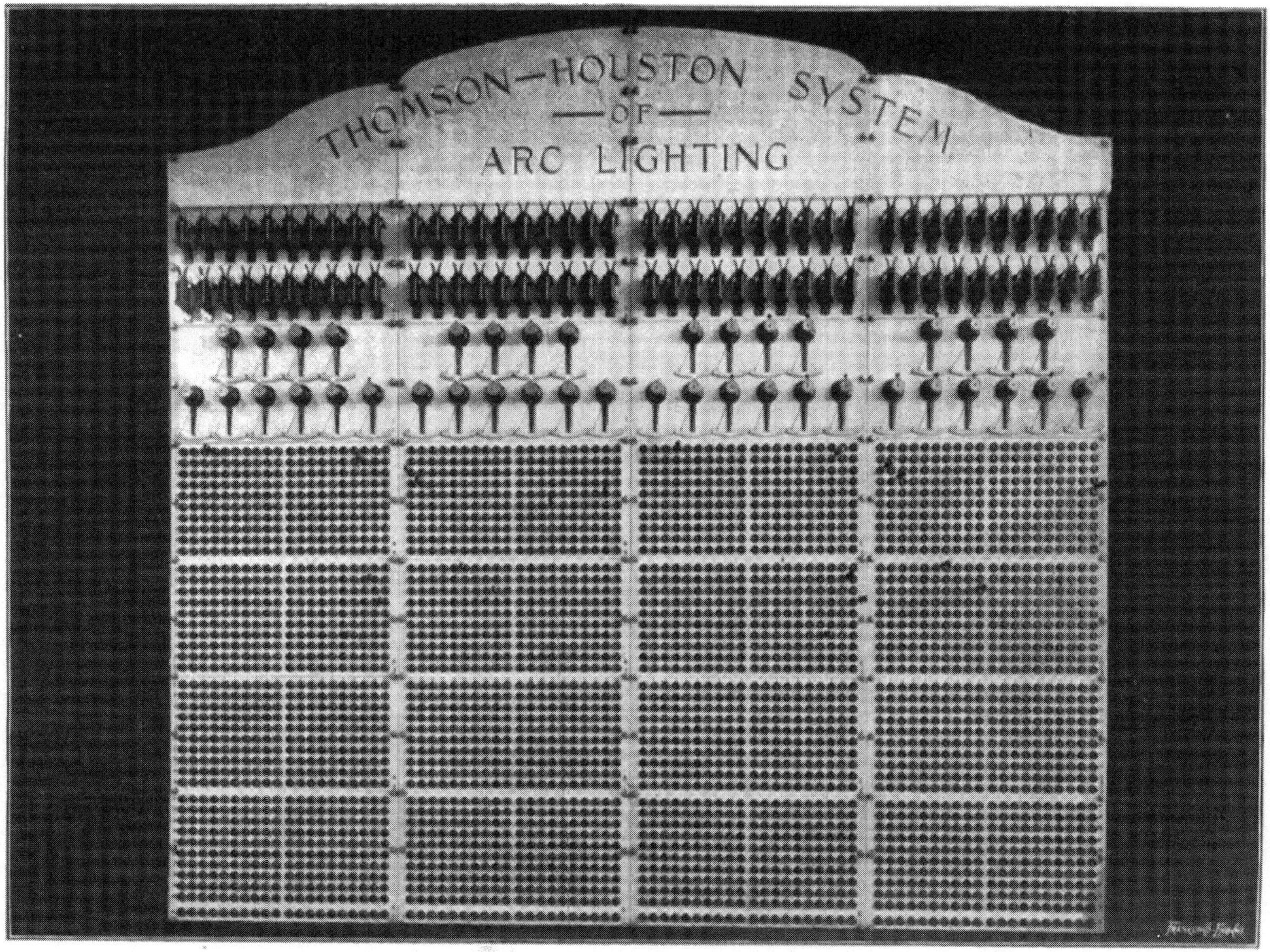

THOMSON-HOUSTON SWITCHBOARD.—MACHINERY HALL.

or, in case of a sudden break-down, street mains can be connected to the light circuits, thus avoiding total darkness. The instruments used are of the best makes. There is one ammeter for each generator or main circuit, and by means of a sliding contact switch one voltmeter can be used for each machine or circuit. From the bus wires the current is taken through the switches to a smaller set of the bus wires, to which is connected the circuits for lights, and they are controlled one at a time by means of small switches or all at one time by means of the main switch. The field controllers or rheostats are set back of the board and are regulated from the front by a small hand wheel. A ground detector completes the complement of instruments.

THE STANDARD ELECTRIC COMPANY.

SWITCHBOARD INSTRUMENTS, ETC., exhibited by this company were as follows: One six dynamo and six circuit combination plug switchboard, built of marble, mounted upon a suitable hard oak cabinet. This switchboard was arranged to accommodate the direct current ampere meter manufactured and exhibited by this company, and which, by a system of plugs, could be thrown on to any circuit in the exhibit. Double pole weatherproof non-sparking cut-outs for arc circuits were exhibited. This device permits the cutting on or off of from one to ten lamps without the slightest sparking. The contacts in this switch are peculiarly arranged so that they are slow moving, no snap action being employed. The Standard automatic arc lamp hanger and cut-out illustrated successfully a method for removing and lowering an arc lamp from a live circuit without opening the line, and at the same time cutting the lamp entirely from the circuit and leaving it free from all connections.

T. J. MURPHY, NEW YORK.

The exhibit was made in connection with the Weston Electrical Instrument Company. It was a switch-board made in eight sections, making a board 8 feet 3 inches by 10.80 inches, and was made of marbleized slate, finished in imitation of antique oak, highly polished. The imitation of oak wood proved to be quite deceptive. The shelf running across the board was another imitation; was a marbleized slate in imitation of Mexican onyx; this also was considered an excellent imitation. The translucent effect produced on onyx finish made it very attractive.

THE MATHER ELECTRIC COMPANY.

The switchboard exhibited by this company was made of two slabs, six by eight feet each, and an inch and one-quarter thick, of white Tennessee marble, mounted and encased in an ornamental wooden frame.

GENERAL ELECTRIC COMPANY (LIMITED), LONDON.

For concentrating the control of electric currents to a single point in residences a special switchboard was designed which is also adapted and frequently used for smaller isolated plants. The leads from the dynamo or storage batteries are carried to this distributing board, from which the branch circuits extend independent of one another. The metal parts are mounted on a piece of either porcelain or enameled slate, previously boiled in paraffin. Two main bars, one connected with each pole, form the terminals of the dynamo leads, which enter the teak case through well-insulated holes in the side. The two poles of the board are on separate slabs divided by an air space, and also by an insulating bar. By this arrangement, where slate is employed, any metallic vein that may exist in the material and that would otherwise make it unfit for electrical puposes, can have no injurious effect. Along the sides of the bars are located the controlling switches for the various sub-circuits, with the necessary safety plugs or fuses by means of which electrical connection is established between the circuit proper and the controlling switch.

CHAPTER X.

ELECTRICAL RAILWAY SYSTEMS AND APPLIANCES.

RAPID transit by means of electricity has received more attention during the past few years than that devoted to any other branch of the science. The special feature of improvement, moreover, has been in the direction of rapid transit in large cities. The possibilities of electricity as applied to railway service have not been questioned for a long time, but, in the installation of every railway plant, questions of economy and availability to meet special conditions arise, and these questions have brought with them, for the consideration of the electrical people, the consequent issues—simplicity of construction, lightness, durability, simplicity of control, safety, etc.

The manufacturers of railway machinery have met their part of the obligation in a marked degree, as was easily demonstrable in the light, compact and comely motors exhibited. They have also made an appreciable advance in the immediate past in the direction of high efficiency of transmission of the current by the use of insulating and protective devices, and the mechanical adjuncts of railway service have kept well in line with improvements in the electrical part of the work.

Up to two years ago, however, there was a notable failure of success in one most important direction, namely, that direction concerning the question of a substitute for that inconvenient, uncomely and highly objectionable feature, the overhead trolley. Wherever overhead trolley lines have been in operation in large cities, they have given much trouble and annoyance on account of their interference with telephone and telegraph service, owing to the inductivity of the high voltage power currents employed in them. So great has been the objection to this feature of electrical railway service that in many cities the authorities have steadfastly refused franchises for this class of rapid transit, preferring to contend with the disadvantages of horse-cars and steam until some solution could be found for the evil.

Inventors have been laboring along two directions, one to substitute secondary battery service for dynamic transmission, the other looking to the hiding of the transmission wires beneath the ground. It is all but conceded that the storage battery, under the present conditions that surround its use, is not a success where any severe service is to be supplied. The other question, that of disposing of the wires in conduits along the track bed, appears to have been met successfully, and, with a few mechanical improvements of detail, there is hardly a question of the early and complete retirement of the overhead trolley. The rapidity of the change promises to be the greater because of the fact that the patent question, involving usually enormous profits, extending over the period of the integrity of the patents, cuts very little figure, the patents being confined almost wholly to special and unimportant devices of detail.

At the Exposition, one of the most satisfactory features of service was the Intramural Railway, the elevated electrical road that carried visitors about the grounds.

WESTINGHOUSE ELECTRIC & MANUFACTURING COMPANY.

Street Railway Apparatus.—In preparing this exhibit the Westinghouse Company adopted the policy of selecting only apparatus that illustrated new and valuable features, and duplication of machinery was avoided wherever practicable, therefore, the railway exhibit was not as extensive as it might have been made. It was nevertheless sufficient to clearly illustrate the methods used by the Westinghouse Company in the construction of railway generators and motors and the method of equipping street cars with electrical apparatus. The apparatus shown comprised one 270 horse-power generator, direct-driven by a Westinghouse automatic compound engine, to which it was connected by a flexible coupling, thus constituting what is known as "The Westinghouse Kodak Outfit." This arrangement is extremely compact and substantial. Opposite the above machine was a 400 horse-power multipolar railway generator of the belt-driven type. Located along the north side of the space were exhibited armatures of both motors and generators, illustrating the simplicity of the winding and showing the mechanical excellence of the construction. The Westinghouse street railway motors of the single reduction type were fully exhibited. They were shown both complete on trucks and dismounted, and open for inspection of their interior parts. A Three Rivers truck built by the Sheffield Velocipede & Power Company, of Three Rivers, Mich., was fully equipped with two 25-h.p. motors, controllers, diverters, etc., and shown in operation. A handsome 20-foot accelerator car and truck of the Brownell Company, St. Louis, Mo., was shown equipped with two 30-h.p. motors, series multiple controllers, etc. The John Stephenson Company, Ltd., also loaned an 18-foot car with a new Tackaberry truck, which was equipped electrically and furnished with a glass floor that allowed satisfactory inspection of the motors in operation, the car having been raised from the track. In addition to this the Lamokin Car Works, of Chester, Pa., had an 18-foot car equipped with two 30-h.p. single reduction motors of Westinghouse manufacture. On an exhibition track, situated near the north end of the Terminal Station the Laclede Car Co., of St. Louis, Mo., had a 20-foot car which was completely equipped with Westinghouse apparatus. The trucks for the above exhibit were furnished by the McGuire Manufacturing Co., of Chicago.

Complete lines of switches in all sizes up to 1600 amperes capacity; ammeters capable of measuring current of 100 amperes and up to 4000 amperes; railway voltmeters, carbon point circuit-breakers, both automatic and operated by hand; lightning arresters of the Keystone and Tank types, as well as smaller details, such as trolley wheels, fuse boxes, carbon brushes, hose and wiring, rheostats, bearings, etc., were all fully shown. In Machinery Hall a 750-h.p. 6-pole railway generator of the direct-connected type was in operation supplying current for the power circuits of the Exposition. The armature of the generator was carried on the shaft of a Corliss compound engine, made by the E. P. Allis Co., Milwaukee, Wis., revolving at the extremely low speed of 90 revolutions per minute. The construction of this machine was massive in a noteworthy degree, a point illustrated by the fact that the radical depth of the commutator bars was four inches. The generator was of the same general type as all Westinghouse railway machines hereinafter described. The machine was located in a pit in order that the armature shaft might be at the proper level for the engine. There were six sets of brushes on this machine, to correspond with an equal number of pole pieces. There were three positive and three negative brushes and those of the same polarity were all connected together. In connection with this generator there was a complete marble switchboard equipped for one dynamo and four supply circuits.

Railway Generators.—The Westinghouse Company manufactures two complete lines of railway generators. One class of these machines is designed to be belt-driven, while the other class avoids the use of pulley and belt and is direct-connected to the shaft of an engine. The generators are self-oiling, self-exciting, perfectly self-regulating, and after being started, completely self-attendant. The machines are wound for a nom-

Westinghouse Switchboard. — Machinery Hall.

inal voltage of 500, but having a rheostat in the field circuit, the potential can be raised to 600 volts without undue heating, and as much as 50 per cent. overloads can be carried for a considerable period without injury to the machine. One of the noteworthy features of the generators is the fact that after the brushes have once been adjusted, further change in their position is unnecessary no matter how sudden or how great may be the variation in the amount of current supplied by the generator. The bearings are self-oiling; that is, a large reservoir below each box holds a quantity of oil which is carried upon the bearings by oil rings revolving about the shaft. In this manner the bearings are automatically oiled by the motion of the shaft and they require no attention beyond periodical examination and renewal of oil. Automatic alignment of the shaft is secured by the use of ball bearings which adjust themselves. The supports for the bearings are cast in one piece with the lower half of the field, a construction which gives great strength and rigidity to the bearings and secures the greatest possible freedom from vibration. There are as many sets of carbon brushes used as there are pole pieces on the machine. Each brush is held in an independent holder, hence any single brush can be raised from the commutator without disturbing the others, and, having its own spring, may be perfectly adjusted. Brushes of the same polarity are connected together, thus making the two terminals of the machine. The rocker arm of the brush holder is of rigid construction and the different parts are thoroughly insulated; it is designed so that it will adjust all the brushes simultaneously. Machines are mounted upon bed plates and can be run equally well in either direction by making a slight change in the connections. The largest sizes of generators have slight additions in design, due mainly to increased size of all parts. The shafts of the belt-driven machines have three very long bearings, one being placed outside of the pulley, thereby avoiding flexural strains on the shafts. A special arrangement, consisting of a worm and wheel as shown, is provided for adjusting the brushes.

THE FIELD.—The field consists of a circular yoke or ring of cast iron, having four pole pieces projecting radially inward. The pole pieces are built up of discs or plates of sheet steel manufactured to a certain formula adopted by the Westinghouse Company, which it is claimed secures magnetic properties superior to anything offered by other manufacturers. The plates are pressed tightly together and secured by iron rods riveted to cast iron plates. The pole pieces thus constructed are then cast into the yoke, which is of very high grade cast iron. Exciting coils are compound wound on metal bobbins. The windings are placed side by side, so that one can be renewed without disturbing the other. A complete bobbin, which is machine wound, is slipped over each pole piece and secured in place by bolts. Careful attention has been given to proportioning the windings of generators so that as the current supplied from the line increases the voltage rises sufficiently to make up for the loss or drop in the line, or for variation in speed, which thus enables machines to give a much more satisfactory service. The bearings and field yoke divide along a horizontal plane through the shaft, thus allowing the renewal of the upper half of the field and permitting access to the coils and armature.

THE ARMATURE.—A distinctive feature of the generator is the construction and winding of the armature. The core is built up of thin discs of soft iron punched around their circumferences with holes. The discs are then properly insulated and forced together under pressure on the shaft, to which they are rigidly keyed. The oval apertures being superimposed one above the other result in forming grooves which extend parallel to the shaft the entire length of the armature. In every groove are placed tubes of insulating material, through each of which is run a stranded cable or copper wire. The wire is thoroughly protected by insulating the material rigidly secured beneath the iron of the core, so that there is absolutely no possibility of its coming loose or making contact with another wire. The winding is of the simplest form possible, as each coil consists of but a single loop of wire. The simplicity is noticeable at the ends of the armature where the wires, instead of being overlapped and bunched together, stand out symmetrically from the core and from each other, thus allowing free circulation of air around every conductor. No wires having a large difference of potential are adjacent to one another. The necessary voltage is secured by revolving a comparatively small number of coils of wire in a strong magnetic field rather than by using a large number of coils in a weak field as is the more usual practice. The small amount of wire on these armatures accounts in a great measure for the absence of sparking, and the fixed position of the points of commutation under variations of load. No matter how great the current in the armature conductors the field is always powerful enough to overcome the effect of their cross-induction, and consequently the lead does not have to be changed. The commutator is designed on correct mechanical principles. It is long and massive, affording ample contact for the brushes. The segments are of the purest copper and carefully insulated from one another by the best quality of mica. Owing to the simplicity of construction, the orderly arrangement of the wires at the ends, and the absence of all bands or binding wires on the surface, the finished armature presents a most attractive appearance.

SINGLE REDUCTION STREET RAILWAY MOTORS.—The motor consists of a circular yoke or ring of cast iron with four pole pieces projecting radially inward. This form of construction offers the greatest mechanical strength for any given weight of material and is lighter for a given cross-section of magnetic circuits than any other form. The field yoke is supported in a cast-iron rectangular frame which extends entirely around the motor and is sufficiently strong to withstand the greatest strains and at the same time it maintains the armature shaft and car axle in perfect parallelism so that accurate meshing of the gears is assured. Easy access to the interior of the motor is allowed by the hinged arrangement of the fields, which permits either the upper or the lower half of the field, as may be desired, to be laid back, thus completely exposing the armature and field coils. The motors are series wound. The field coils are separately wound on lathes, thoroughly insulated and slipped over the pole pieces, where they are held in place by flat brass rings bolted to the yoke. The armature is of the drum type, the core being built up of thin sheet iron discs placed between thicker end-plates and the whole keyed directly to the shaft. This construction avoids the use of a non-magnetic spider, which involves elements of mechanical weakness, and is a vital objection to all forms of ring armatures. The periphery of the completed armature core presents a number of grooves or slots running parallel to the shaft. In these slots are placed lathe wound coils, individually protected by fuller board and wrapped with insulating tape. There are no crosses except of one coil upon another where the quadruple insulation offers ample protection. The windings of the wire safely separate each coil from its neighbor. As the coils are held rigidly in the grooves or teeth, no motion is possible and binding wires are practically unnecessary for holding the coils in position, but are used to secure the layer of fuller board and mica in which the entire armature is wrapped. Though the motor has a four-pole field, but two brushes are employed, which advantage is secured by the use of what is known as the two-circuit armature winding. With this winding it is impossible for the armature circuits to become unbalanced; that is, one portion of the circuit cannot carry more current than another and thereby cause excessive heating, waste of energy and injury to the insulation. The two-circuit winding allows the current to flow through all the wire on the armature by either one of two paths.

CONTROLLERS.—The controlling stand now used for street-cars by this company is known as the series multiple controller. By the use of this controller the motors are started in series with one another and in series with a resistance; the resistance is gradually reduced until all is thrown out of circuit when the motors are running in series, which gives slow speed but high efficiency. By throwing the controller handle further around, the motors are connected in multiple with one another, but in series with a resistance which is gradually cut out until the motors are left in multiple without any resistance in circuit, which gives high speed and maximum efficiency; thus by the use of this controller is secured two highly efficient speeds, one for running through crowded streets in cities, and the other for the more rapid travel required on avenues and suburban districts. The controlling handle has six notches, each of which gives a corresponding speed. The reversing is done by separate switch projecting from the side. When this switch is in its central position all the current is cut off from the motor. Throwing the switch forward allows the operation of the controlling handle to admit current in the proper direction for moving the car forward. Throwing the reversing handle backward reverses the direction of current in the armatures, and therefore the direction in which the car will move. Both controller handles have locking devices which prevent their removal in any other position than that for no current. The controlling stand is placed on the platform of the car, where it may be easily and quickly inspected. It is provided with a water-tight top and covering and made fireproof throughout, thereby insuring against injury of almost any sort. The parts are made standard and may be cheaply and easily replaced so that the controller is practically indestructible. The following are the special points of excellence which characterize the controller: Solidity and simplicity, good workmanship, ease of reversing in case of instant necessity, incombustible construction, accessibility of parts, ease of inspection and repair, absence of all interlocking devices, perfect insulation, contact tips and fingers easily removed and replaced, every part interchangeable, motion of car easy and smooth.

DIVERTERS.—In order to start the car easily and with economy it is necessary to avoid the large rush of current through the motors which would result were there not some such piece of apparatus as the Westinghouse diverter used.

The diverter consists of strips of No. 27 wrought iron and mica rolled compactly into coils and enclosed in cast-iron circular cells. The diverter is intended to be suspended beneath the car and, being strong, simple and fireproof, it requires a minimum of attention and gives practically unlimited wear.

FUSE BLOCKS AND CUT-OUTS.—These details of street car equipment have been carefully wrought out by the company and are so constructed as to insure good wearing qualities, simplicity of construction and ease of access. It is unnecessary to enter into a description or state special claims of the advantages to be found in these details.

LIGHTNING ARRESTERS.—Under this head were exhibited two styles of lightning arrester used in connection with railway work. A Keystone type arrester is enclosed in an asbestos lined, cast-iron box and used on every street car for the protection of the motors and apparatus. The arrester is of the air-blast type and consists of a marble and iron chamber having two circular openings into each of which projects an arm supported from a hinge on the outside of the instrument near its top. The ends of the two arms within the chamber are of carbon and when in proper position they are separated by a sixteenth-inch air space. The heat formed by the arc after the passage of lightning between the carbon tips expands the air in the enclosed chamber, which results in blowing out the arms and breaking the arc. When the circuit is broken the arms fall back under the action or gravity and the arrester is in readiness for further discharge. Thus it is seen that the arrester offers a simple and direct path to ground for lightning, for the reason that there is no self-induction to impede its passage. As the arrester is entirely automatic in action and incombustible, it needs practically no attention whatever after once being properly set up.

Tank Arresters.—For the protection of the railway station apparatus an arrester of an entirely different form, known as the "tank," is used. It consists of a coil of heavy wire in series with the trolley or supply circuits—and having several discharge circuits leading from different points in the coil to carbon electrodes immersed in a tank of running water which is thoroughly grounded. When lightning tends to follow the line to the apparatus within the station it is met by the coil, which interposes a large inductive resistance to its passage, while the several discharge circuits of comparatively low ohmic resistance offer an easy path to earth, so that the lightning passes readily to ground without entering the station.

GENERAL ELECTRIC COMPANY.

The railway apparatus displayed consisted of generators, motors of various types, station measuring and controlling apparatus, and overhead construction parts. There were three generators, of 100-k.w., 200-k.w. and 300-k.w. capacity each, respectively, over-compounded for 10 per cent line loss, with the Standard iron-clad armature of the General Electric Company, having no binding wires and no conductors inside the ring core. These machines were run as motors to show their performance in operation. For each respective generator there is a panel in the enameled slate switchboard containing ammeter, automatic circuit breaker, main switch, rheostat switch, lamp switch, field switch, and volt terminals or test leads. By use of these panels a switchboard may be increased indefinitely as the plant grows, and yet always remain uniform. In the line of railway motors there were on exhibition as follows: 2-w.p. 30 motors (15-h.p. each) mounted on a Taylor truck; 2 srf (16-h.p. each)

Section of the Intramural Road.

motors, mounted on a McGuire truck; 1 w.f. 30 motor (15-h.p.) mounted on a Stephenson truck; 2 w.p., 50 motors (25-h.p. each) mounted on a Robinson Radical truck; 1 s.r.g. motor (15-h.p.) mounted on a stand; 2 l.w.p. motors (125-h.p. each) mounted on a Jackson & Sharp truck; 2 g. e. 800 motors (25-h.p. each) mounted on a Bemis truck. The last named motor is the latest and most improved type put upon the market, and is remarkable for being water and dust proof, highly efficient, of great power for its weight, for the fact that only one-third of its weight is on the axle, and for the great ease with which and part of the motor may be reached or removed for examination or repair. Two of the above equipments were fitted with the Standard "Type K," series parallel controller. The Taylor truck was equipped with a Genet air brake, and the Jackson & Sharp truck with a New York air brake, the air for the latter being supplied by a General Electric oscillating electric air compresser. A platform truck, equipped with two 7½-h.p. motors for factory or other light freight service, was exhibited. Over the entire exhibit was erected a double track, double bracket trolley system, with a crossing at right angles.

SPERRY ELECTRIC RAILWAY COMPANY.

The exhibit consisted of one complete equipment mounted on a Dorner & Dutton truck, and another complete equipment mounted on a McGuire truck. Equipments are applicable to any truck. One of the equipments was provided with a car platform equipped with two dashers, steps, etc., and contained trap doors over the motor so that all parts might be inspected while running. The equipment mounted upon the Dorner & Dutton truck was supported upon bearings so that inspection was facilitated. In the southeast corner of the exhibit were armature winders actively engaged in winding regular armatures for the Sperry motor, showing all the processes applied to the armatures while winding. On the side of the exhibit were supported, in suitable racks, armatures in various stages of their manufacture, showing the bare core, plates from which the core is made, the hub, various insulating coverings, before winding; while on the other side of the exhibit were the heavy portions, such as gears, pinions, gear housings and clutches mounted in an attractive manner. There was also a motor with the upper half of the field raised and laid back. The motor is coupled in circuit and operated continuously with one field. The running of this motor with but one field, and with no sparking of the commutator, is considered quite remarkable. On one side of the exhibit were tables and cases containing all the special features, parts of the motor such as clutches, commutators, cross-connections, links, supporting bars—in fact, all of the accessory material going to make up the equipment. On a car track south of the Transportation Building a Sperry car was in operation daily, which virtually formed a part of the exhibit. On this car was exhibited controllers, trolley, and various features of the machinery which were open for inspection, with an attendant to answer any questions that were propounded in reference to the devices. The principal points in view in designing this type of motor have been to elastically support a single motor and to so connect it to the axles of the car that while, on the one hand, its torque is delivered to them, yet on the other hand the jar, vibration and hammering of the axles are absolutely prevented from reaching back to the motor. It is unquestionable that if a motor can be so protected its life will be much longer than that of one which receives all the vibrations due to bad rail joints and to railroad crossings, which are so common in this country. It has been proved by experience of this company that to insure long life and least wear and tear to the roadbed, the rolling part of the car, including wheel and axle, should be as light as possible, and should be burdened with as little as possible to increase the weight of the rails to a very large extent, in order to withstand the heavy hammering action when the motor is directly borne upon the axle itself. The motor armature is a Gramme ring with Pacinotti teeth. The field is produced by two field coils producing the same polarity in their cores, so that there are also two consequent poles, and the field has really four poles. The armature is completely closed in when running, and the top half of this cast steel box can be raised up on hinges, and the armature can be examined from the floor of the car. The field magnets are series wound, and the armature is cross-connected, so that only two carbon brushes are required. The commutator bars are hard copper and drawn up in position by the usual type of conical clutches; the bars are one inch deep. Each motor is calculated to give 40-h.p., but will in certain cases develop 50-h.p. for a short period. The whole weight of the motor is carried upon two cross-bars attached to the frame of the truck through India-rubber blocks. This arrangement is very important, as the total weight of the complete motor is 2000 pounds. The armature shaft is at right angles to the car axles, and each end of the shaft is provided with a clutch. The short part of the shaft at each end of the coupling is provided with a bevel pinion, which gears into a large bevel wheel keyed upon the car axles. Both axles are thus driven at once, and the connection between the motor and the car axles is quite flexible. The only weight directly supported upon the axles is the bevel gearing and part of the casing, the rest of the casing being supported by the cross-frames of the truck itself, while still allowing side play to the axles. By connecting one motor to both axles in this way a pull upon the draw-bar is obtained which exceeds by 11 per cent that obtainable with two motors when using the same number of watts; this is explained by the greater efficiency of the larger motor, and by the less liability to slip when all four wheels are connected together. One other point worth noting is that it is very difficult to obtain two motors so exactly alike as to give precisely the same turning effect upon all the wheels. The idea is that employment of this arrangement has the same effect as that obtained by using coupling rods upon locomotives. The controller is placed upon the platform of the car, and contains the necessary resistance coils. Only three wires pass from this resistance box to the motor, so that the connections are very simple.

THE SHORT ELECTRIC RAILWAY COMPANY.

Short Gearless Motor.—The Short 6-pole 20-h.p. gearless motor was in operation in the Short Company space, where a truck was equipped with two motors and the necessary controlling apparatus, and raised to admit close inspection. This motor has no gears, but three wearing parts and but two brushes. It has made twenty-eight miles an hour without reaching its limit of speed. Its strong points are high efficiency, immense tractive power, simple construction, ease of control and absolutely noiseless operation.

The Short Company private car in operation on exhibit track was equipped with two Short 20-h.p. gearless motors of latest design and finish. The most striking piece of apparatus exhibited was a 450-h.p. 50-inch generator of a new type. This machine delivers 600 amperes, has a speed of 300 revolutions and is wound for 10 per cent loss. The collar bearing on the armature shaft for close adjustment is one of its claims to originality. This is the latest motor designed by the company, and, aside from the construction of the shaft and the adjusting device, differs from its predecessors in several minor details that may be seen by a comparison of the working drawings with those of the well known type. In the space was shown the original gearless motor, followed by two improved forms, and finally by the perfected machine of to-day. There were also reduction motors, rheostats, etc., and small parts of machines, also a platform car with two gearless 35-h.p. Short motors. At one corner of the space was an iron frame slate switchboard with ammeters, circuit breakers, fuses, rheostats, etc.

CURTIS ELECTRIC MANUFACTURING COMPANY.

The exhibit consisted of one 30-h.p. street railway motor, series-multiple controllers, and a rheostat. The motor is entirely water-proof and is wound with such heavy wire in the armature and field spools that burn-outs due to water getting on the apparatus, and burn-outs due to overheating, are entirely obviated. The armature is of the Gramme ring slotted type. The winding is the ordinary Gramme ring winding, by which an injured coil may be replaced without disturbing the adjacent coils. The three bands which prevent the wire from leaving the slots are sunk below the surface of the armature, so that they can never be stripped by the armature touching the pole pieces. The bearings on the commutator end of the armature and the pinion end are exactly the same, and are six inches long by three inches in diameter. The pole pieces being on the side, the armature can never run in water, as the space between the bottom of the armature and the bottom of the field casting is about four inches deep. The resistance of the motor is very low, being about 1 1-10th ohms, which makes the motor have a very high efficiency. Due to its electrical design, the commutators run without sparkings, and noiselessly. The insulation in both field spools and armature consists of mica cloth, there being five or six thicknesses of this material between the wire and the iron of the motor at any point, except in the slots, where the insulation is made up of slotted mica, 1 1-16 inches thick, stuck together with shellac. The controller is of the series multiple type, very simple in construction, and of such electrical design that it does not arc, short-circuit, or ground. The contact castings on the drum are only four in number, and are sleeved on the wooden portion of the drum in a very simple way, which makes a particularly strong construction mechanically. The controller is entirely water-proof. The rheostat is made up of a long ribbon of iron bent zig-zag with adjacent sections, insulated from each other by strips of mica. This resistance material is dropped into light cast iron frames, the insulation between it and the frame being mica. Light flat bars of iron bolt the resistance material in the frame in a very strong way. Three of these sections are bolted together, and are then dropped into a cast iron, water-proof case; the whole forms a compact rheostat, water-proof, dust-proof and heat-proof. There are mica bushings between the cast iron case and the rheostat sections, so that a double ground has to occur before the electric current can get from the resistance material into the containing case.

ALONZO C. MATHER.

RAILROAD AND CANAL TRANSPORTATION.—This exhibit consisted of a complete model (one-half inch to the foot) of a "long shore" railroad, and canal boat electric installation, in active operation. The territory supposed to be the location of this installation was a section of the Erie Canal about Buffalo and Syracuse, and a road-bed along the canal. There were two complete trains in operation on the road, which had upon it two heavy grades and a draw bridge, to show the operation of the system under conditions always to be met. There were two canal boats with double overhead trolley and guide wire, upon the canal, propelled by motors. The advantages especially set forth in favor of the system are the ability of the inventor to utilize water power for the generation of the current, and the double service of the central station for the operation of road and canal boats. The plant for the operation of such a system is proposed to be located on the Niagara River at Buffalo. It consists of a large current wheel between the piers under the bridge, the power to be generated by means of link belts traversing the circumference of the wheel, and commanding a generator or generators, located on the bridge above. A third rail in the center of the railroad track is slightly elevated, so that trains running at full speed depend on it largely—on the principle of the bicycle, the outer rails serving largely as guides. The cars are equipped with pneumatic springs and steel trucks, and are constructed specially to run at high speed, the trains being vestibuled, and the axles carrying three wheels instead of the usual two. The inventor claims for the third wheel that it prevents accidents from rail spreading, broken wheels, or from trains jumping the track, also increases adhesion and reduces friction. This system of canal boat propulsion is essentially the same as that given practical demonstration on the Erie canal at Rochester, just before the close of navigation in the fall of 1893, the promoters of the demonstration having visited the Exposition for the purpose of investigating the Mather system.

ALLGEMEINE ELECTRICITATS GESELLSCHAFT, BERLIN, GERMANY.

ELECTRIC STREET RAILWAYS.—The introduction of electric street railways on the continent of Europe is due chiefly to the Allgemeine Electricitats Gesellschaft, who have carried out by far the greatest number of trolley line systems in towns. The equipment and supplies form a complete system, worked out as the result of several years' experience. As a rule, besides the insulated main conductor, a trolley conductor, consisting of a bare silicium bronze wire of 7 m.m thickness is placed over the rails, being connected with the first at certain distances by crossing wires. The rails are used for conducting the current back. The trolley conductor is supported by poles with arms or by steel ropes suspended between the poles or the houses and fixed by binding screws. The design of all the joining parts and insulators has been so chosen that they can be fitted quickly and without soldering or binding wires. For fixing the trolley conductor, insulated clamps are used, which grasp it firmly; by unscrewing these clamps, the lines can be displaced for equalizing the tension in the wires, without dropping them. Where sharp curves occur

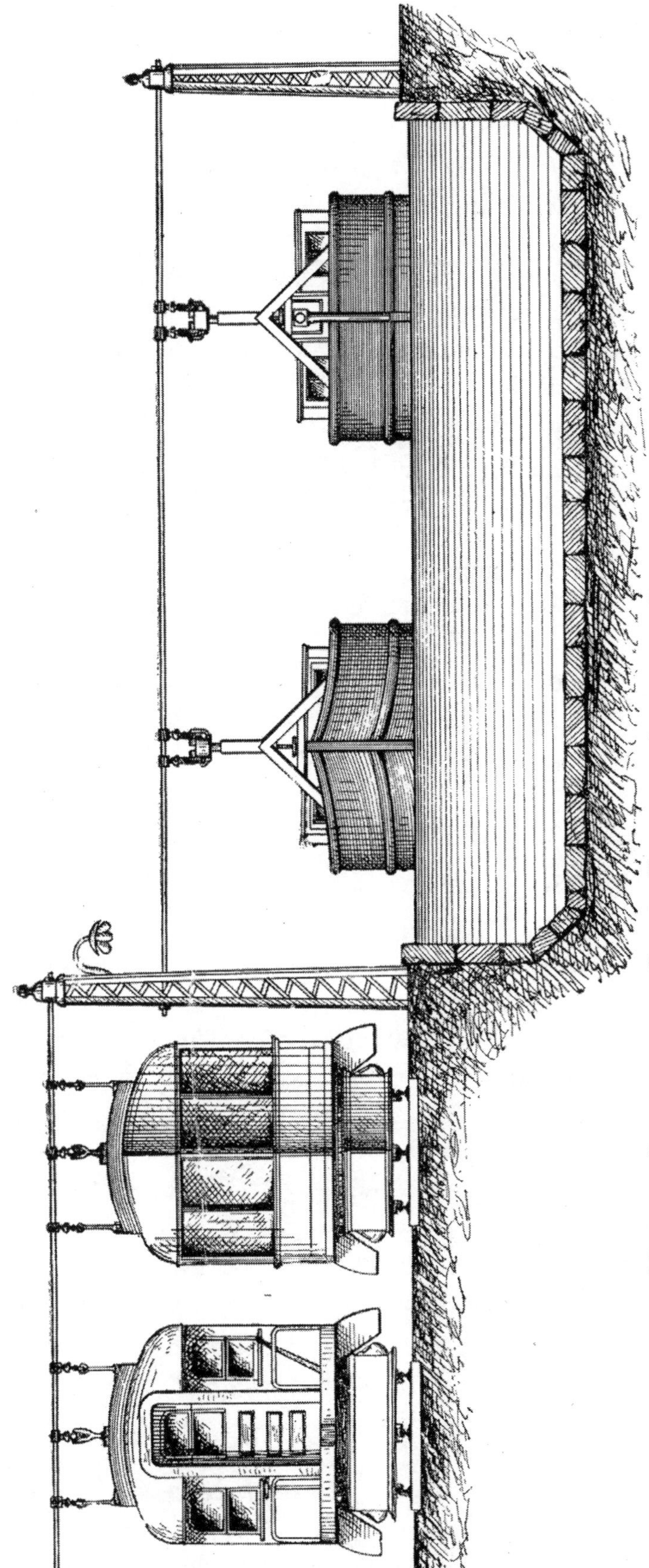

Mather Railway and Canal Traction System. — End View.

a special arrangement of the overhead wiring is provided for obtaining the greatest possible smoothness in running; this is effected by a suitably shaped instrument facilitating the speedy fixing and removal of the span-wires between the curves. The passage from one track to the other is effected by means of very light crossings. For the track as well as for the car, lightning conductors are applied, possessing no movable parts and working with absclute safety and preventing the formation of an arc. A fuse protects the motors against too strong a current; and to obviate delay during running the fuse includes a second lead strip. The lamp holders in the cars are designed with a view to being easily fitted and cleaned. They have reflectors for increasing the amount of light. There is also a new type of railway motor which shows much improvement over old methods. The armature, in the form of a toothed Gramme ring, is of great durability in consequence of the solidity and protected position of the winding. Repairs can be made even by workmen of no great skill, all the parts being easily interchangeable. The motor is series wound, and very light, including all the parts. It is 16-h.p. capacity, single reduction, slow speed (400 revolutions) and no sparking at the brushes. The switch at the motorman's hand is universal, and is used to start, stop and reverse.

BONNEAU-DESROZIERS.

ELECTRIC TRACTION FOR RAILROADS.—The advantages of this system are: the power of the motive wheels is more regular; the gravity center may be lowered; the construction is more compact; the vertical pressure on the rails is more constant than in steam locomotives, and as a general result the speed may be largely increased. The most interesting engine planned by Bonneau and Desroziers is the high speed engine, whose drawings were exhibited. It has two motor shafts with wheels of 2.3 meters in diameter. The armature is fixed on a hollow shaft through which moves the axle of the wheels, and the connection between the wheels and the hollow shaft is made by rubber joints. There is a difference of a few centimeters between the diameters of the axle and the interior diameter of the hollow shaft, in order to allow movements of the dynamo, which is supported by heavy springs. The current may be furnished by trolley wires or large secondary cells, according to the traffic.

M. DE BOVET.

MAGNETIC TRACTION.—To pull boats in canals and rivers tugs can be used. These take their power from an iron chain put in the river. Formerly the chain operated around two large wheels on the boat. This is an inconvenience, because there is a certain length of chain always on the boat, which is carried with this one and forms an objection to putting two boats on the same section of the chain. Besides, the chain is very quickly worn out by its going around the wheels. M. De Bovét invented a magnetic wheel to give adhesion to the chain, which is put on the wheel for a quarter only of the circumference of the wheel. The magnetism being developed by a current of only four electric h.p., excellent results were obtained and a better adhesion than with any other system. He made a boat for his system with a steam engine ot 150 h.p. when acting on the screw of the boat, and of 80 to 90 h.p. when acting, at a lower speed, on the traction wheel. This boat has also a magnetic brake on the chain, made in the same way as the traction wheel. He proposes, also, in order to avoid the formation of big trains of boats in the rivers, to put on each boat a little magnetic wheel connected with a dynamo of 5 h.p. The current being taken from a trolley, each boat could move on the chain in perfect independence.

W. G. CREIGHTON.

CONDUIT AND RAIL COMBINED.—The exhibit was a combined electric conduit and street rail, forming a tube and rail together. It is a design to close a slot in the street surface and open it by a contact passing along. The slot is left open with the contact passing downward, making connection with the electrical tubes. The electrical tubes are held in their place by brackets firmly secured by being fastened to the side of the tube or rail. The contact that connects a car or motor passes through the slot in the rail; on the lower end it has two wheels or pulleys which pass through a slot in the insulated electrical tubes, thus taking the current from the inside of the tubes; this has the advantage over a wire by being cleaner and not subjected to water, dampness or dirt. The slot in the electrical tubes is on the lower side so that no dirt or water can collect. The rail conduit is not an underground conduit; it is set upon the ties above the surface and is thoroughly spiked to them, covered and surrounded by pavement. For the safety of the conduit that water, dirt or snow shall not collect, open spaces have been left in the bottom of the conduit at stated distances. A scraper is dragged along inside the conduit and the slush drops into the box or basin and the water flows into the sewer, the dirt being taken from the box and dumped. The exhibit showed a rail and conduit combined with a different contact. The downright contact that passes through the slot at the lower end, has two wheels of a different shape from those described above, and shows a connection with bare wires running from the dynamo to the car-propelling motors, the wires being placed on insulated brackets.

PULLMAN PALACE CAR COMPANY.

Passenger Car Equipment.—The Pullman Columbian Exhibition train was lighted throughout by electricity, by a system devised by the Pullman Company and adopted for its limited and other special trains. This system consists of a 3-cylinder engine, coupled direct to a dynamo and, with the necessary devices and instruments, is located in the forward end of the baggage compartment of the composite smoking and baggage car, occupying a space of 6′ 6″×3′ 3″. The steam for operating the 3-cylinder engine is supplied from the locomotive boiler, at a pressure of 60 lbs. to the square inch, the engine and dynamo running at a speed of 900 revolutions per minute. Charging wires are led from the dynamo, on the roofs of the cars, to switches and safety devices placed in the lockers of each car in the train, and from there to the lamp circuits and storage batteries. Each car is supplied with 40 16-c.p. 60-volt lamps, arranged on three separate circuits, a total of 640 candle power for each car; and 32 cells of storage batteries capable of maintaining the 40 lamps for 5 hours, should it be found necessary to shut down the engine and dynamo from any cause. A unique feature of the system is portable lamps, which can be held in the hand or placed at any desirable point, and may be used in the berths after retiring without inconveniencing any other occupants of the car. The connection between these lamps and the electric wires is made by means of a flexible cord and plug which fits in the small escutcheon or socket in the side finish of the car between the windows. When not required these lamps may be hung upon a small hook placed in the section for this purpose, or they may be removed entirely by the porter and placed in the proper receptacle, which is located in one of the lockers of the car. Fan motors of 1/8 h.p. are placed in the smoking, dining and observation cars for ventilating. All parts of the system have been designed especially for railway train lighting and are made interchangeable.

TRUCKS FOR ELECTRICAL SERVICE.

THE TAYLOR IMPROVED ELECTRIC TRUCK.

The sides of the frame of this truck are composed of two steel bars, extending throughout the entire length of the truck, firmly bolted together, and secured at each end by trussed end bars, and in the center by two heavy bars placed edgewise. The independent and rectangular frame thus made forms a rigid and substantial base for the construction of the truck. To the side bars are firmly bolted the pedestals or jaws that receive the axle boxes, which slide freely between the vertical guides of the pedestal, thus keeping the axles parallel with each other, and resisting the pressure of the brake when applied to the wheels. On each side of the bottom of the pedestal are sockets which receive the angular braces, supporting the ends and centers of the frame. Connecting the two arms of the pedestals is a strong steel stay, fitting tightly between the shoulders, and a strong brace extends from pedestal to pedestal, and with the side bars and angular braces, forming a simple and at the same time a substantial truss, distributing the weight of the car and motor over the entire frame of the truck. Above, and resting on the axle boxes between the side bars, are half elliptic springs engaging in yokes at their ends, and giving a perfectly free and elastic action. At each end of the truck frame is placed a pair of elliptic springs, securely fastened to the trussed end bar. Upon these springs but not fastened to them is placed the body bolster, extending across the car and engaging the side sills by angle irons at each end. The truck is equipped with a compound lever-brake so hung that the shoe adjusts itself to the periphery of the wheel, when the car is light or loaded. The brake shoe is entirely independent of the brake head, and a key which holds it to the brass head can readily be removed, allowing the worn out shoe to be quickly replaced by another one. The brakes are supplied with an adjustable releasing spring, and the brake rods are provided with quick acting turnbuckles, to take up the slack caused by wear, and all adjustments of the brake can be made from the side. The car body is fastened to the truck by king bolts placed in the center at each end, and passing through the bolster and end bar. These king bolts are arranged with coil springs that bear against the center of trussed end bar upon the under side, and assists the full elliptic springs in preventing the end teetering of the car body.

JOHN STEPHENSON COMPANY (LIMITED).

TACKABERRY MOTOR TRUCK.—The truck is composed of two side sills of wood, reinforced with inserted metal plates. These are joined at center by a wooden cross-bar, whose ends rest in iron sockets, the whole securely fastened to the side sills ; this wooden cross-bar is not only the center brace of the entire truck frame but always acts as the support to channel bars which receive the nose of the motor frame which rests securely upon steel spiral springs. The method of construction with wood and iron in combination is not only elastic but enables all the parts to be fitted and held securely, thus avoiding the loosening and rattling of bolts and nuts incident to iron trucks. The truck frame is supported by four dust tight journal boxes, thereby keeping journals and lubricator clean. The wheel brake supported on ends of side sills is always held in unvarying relation to the wheels. Brake is simple and powerful, requiring minimum exertion of the motorman. By the combination of levers it is possible to apply the brake with force and rapidity, while at the same time it is not necessary to have such nice adjustment as is ordinarily required. Special effort has been made to reduce the members of the truck to the minimum number, and, as a result, all parts of it are readily accessible for adjustment of motors or for working at under part of car body. The truck is complete in itself, needing only when under car body to have brake chains attached to hook of brake rod to make it ready for immediate service. The four journal boxes which support the truck frames play vertically in the jaws of pedestals secured to the side-sills of the car body. These pedestals have at their ends receptacles for rubber (super) springs, with freedom of motion in every direction, thus making flexible connection of the truck with the car body without agitation or discomfort to passengers, and promoting durability of car structure and connections.

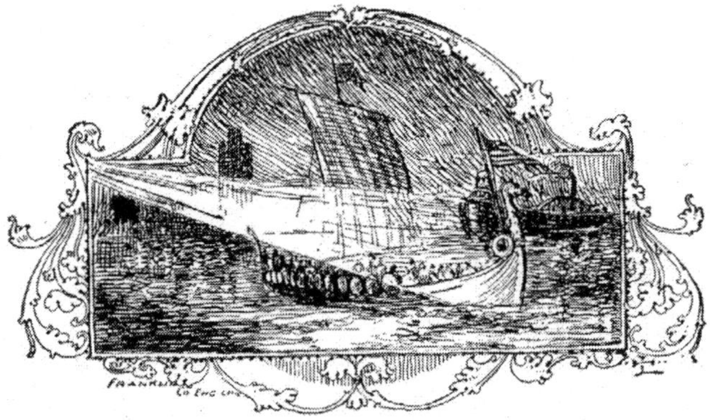

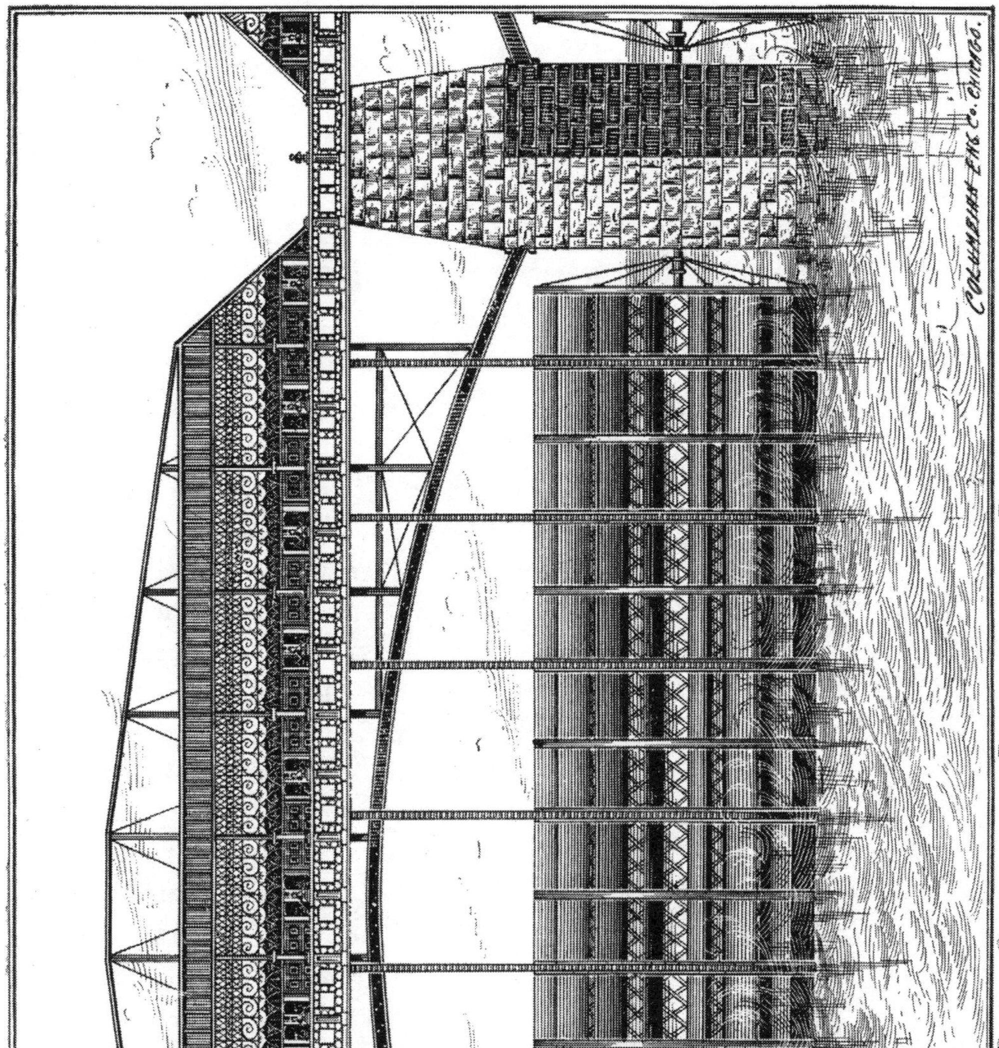

Mather Water Wheel.—Power Plant Operated by Belting to the Bridge Above.

CHAPTER XI.

RAILWAY SIGNALING AND SAFETY DEVICES.

SIGNALING for railway purposes has undergone almost a revolution since the advent of electricity. In a few instances, especially in suburban traffic, the block signal systems and safety stops have served to decrease the time of trains, but in long distance service and on all through trains, the electrical service has materially aided in developing fast time and has done more than all other influences combined to prevent accident. On nearly all the first class roads of this country the old semaphore or Napoleonic signaling system has been or is being displaced by electrical equipment, and were it not for the actual first cost of installation and the discarding of property that cost much money, all the roads would now operate electric signaling service. There are some applications of electricity that are difficult to force to the front because of the expense of installation or operation. Railway signaling devices are among the number. Any system, operated from a central office or from any central point, will do away with most of the men required by the semaphore; the instruments in this class of service are inexpensive, the road apparatus simple and not likely to get out of order, cannot be tampered with without the central station being apprised of the fact; above all is largely automatic, therefore not subject to accidents due to the carelessness, or worse, of employes. At the Exposition, not a very large, but a very comprehensive, exhibit of signaling apparatus was made, both operative, as for instance on the Intramural road, and simply presented for dead exhibition.

ROWELL POTTER SAFETY STOP COMPANY.

Positive Block Stop.—This system presents three distinctive features: The Safety stop, which consists of a method of stopping trains at points of danger, independent of the engineer or trainmen; the Duplex or Auxiliary system, by which the engineer is immediately apprised whether through any accident the signals have failed to work, and so can take means to protect his train; the positive mechanical action by which every moving train, switch, drawbridge, etc., automatically operates its own signals. The Safety stop consists of a track instrument and an engine attachment, and by contact between these two the train brakes are automatically applied. The track instrument in the danger position forms an inclined plane parallel with and above the rail, but in the safety position is

Rowell Potter Safety Stop System on the Intramural Road.

flat beside it. The engine attachment is a lever arm, free to move vertically and furnished with a friction roller, which is lifted by the track instrument when in the danger position, thereby opening a valve in the brake-pipes and applying the brakes. This vertical movement is so gradual and delicate that in actual practice and at high rates of speed the contact has not torn paper pasted upon the track instrument. The track instrument does not come in contact with the wheels of the train, but connects with the engine apparatus alone, and then only when in the danger position. Connected with the track instrument is a visual signal which indicates safety or danger, and these signals, of course, have all the advantages of any other visual systems for stopping trains. The Duplex or Auxiliary system consists in the use of two track instruments (each connected with a visual signal), one 50 or 100 feet, or any other desired distance in advance of the other, but coupled rigidly together with pipe line connections, so that one is always in the position of danger while the other is at safety. Designating these instruments in the order of approach as Nos. 1 and 2, when a train has the right of way it will find No. 1 at safety, while the auxiliary (No. 2) will be at danger. Passing No. 1, the train will automatically set it to danger, and at the same time No. 2 will go to safety, thereby permitting the train to proceed without obstruction. Should, however, the primary signal, through any accident, not go to danger, then the auxiliary will apply the train brakes, thereby notifying the engineer of the failure, so that he may take means to protect the rear of his train. The operation is such that the engineer always knows whether he is stopped by actual danger or by the signal being out of order. In the latter event he can proceed without the loss of time, and report the failure of the signal at the first station, thereby insuring prompt attention. In this system, the visual signals and track instruments are automatically set to danger by a tripping device, and locked in that position by a mechanical lock. This lock also operates to keep the trip from contact with any but the first wheels of the train, and by this means the wear to the apparatus is reduced to a minimum. When the train has passed beyond the point fixed for its protection, the visual signal and track instrument go to safety automatically. This is accomplished by an electric release, which can be worked in this system in many ways, and at such distances as has been found practical to work any electric system. It can be worked by a track circuit, or by wire circuit, with independent releasing apparatus for the wheels to operate. The company also operates a system for a rear end release, which requires the caboose of a freight train or the rear of a passenger train to pass over it before the signal is released, the other cars of the train having no effect upon it. The normal position of all track instruments connected with the system is "danger." In case of breakage or failure to operate from any cause, the track instrument and signal will go to danger, and cannot go to safety until the system is put in order. Should the electric current which is used for the release of the lock fail, the only thing that can happen is to unnecessarily stop the train; in other words, it is a physical impossibility for a train to pass a signal unless the block ahead is clear and the signal is in good working order. This fact insures the maintenance of the signal, as well as avoiding accidents. There can be no accidents under this system by the failure of signals or the failure of the engineer to see and properly interpret them, as in one case it is just as impossible to pass over the road with the signal out of order as to run without steam. In the second case, it is practically immaterial whether the engineer sees his signal or not, from the fact that if it is against him, his train will be automatically stopped should he attempt to disobey it.

In addition to the absolute block system, the company has applied this principle for stopping trains in several methods which meet the requirements of any and all possible contingencies. First, it is an absolute switch, draw-bridge and crossing protection, without adding to or deducting from the engineer's duties or responsibilities, and when his vigilance fails, it does what he should do, namely, stops the train; second, it is used as a manual block system, to protect stations and other points of danger where an operator is stationed. In this case the track instrument is located at a point sufficiently distant from the station or danger point to assure protection. When a train reaches the track instrument, it will be set automatically, so that a following train will be stopped and cannot pass. After the track instrument is set, it is automatically locked, and cannot be released until the train is out of danger; the operator can then pull the track instrument to safety. A visual signal is attached to, and moves with, the track instrument, and, if the engineer fails to obey the warning, the track instrument stops the train. The only mistake an operator can make is to neglect to pull the track instrument to safety, and thereby cause delay in movement of trains. The system was in operation on the Intramural railroad at the Exposition, where it was open to inspection. The conditions on the Intramural railroad were the severest that could be imposed on any signalling system; but the system met them all with a success which is believed not to have been possible with any system previously invented. Some idea of the severity of these conditions may be deduced from the statement that each signal operated 640 times a day, or at intervals of only one and one-half minutes. Frequently only a few seconds elapsed between the time when a signal was released to safety and tripped to danger again. The action has to be practically instantaneous, as otherwise serious delay would be entailed along the entire line.

D. D. TATE.

Block Signal System.—The Tate system is entirely an electro-mechanical device, and the elementary principles involved throughout the entire system, are reliance in its operation, cheapness in its construction and maintenance, and it is exclusively automatic in all its workings. While it is simple in design and operation, it covers all points that have been striven for in the signal service. The interlocking instrument constitutes the main feature in the system. It is so constructed as to have entire and distinct control over each and every apparatus used in the system, and will operate any other make of electric signals of either open or closed circuit design. There is no application of the system but what the interlocking instrument is always applied; even though signals and track instruments are dispensed with, it alone performs the working of the entire system; i.e., in the cases where it is applied to a high speed electric road, it operates not on signals, but directly on sections of the trolley wire. Thus, after a car enters a block, power is shut off from any other car that would attempt to enter the same block, and it matters not whether the line be double or single tracked. The target or signal used in the system is known as the "Banjo" pattern. The name originated from the similarity in exterior form of the signal case and the banjo. On the interior of the signal case is an electric device to operate a shield that is swung between the glass covering on front and rear of case. At night the rays of light form a lantern placed on rear of case, pass through the case and shield and give the tone of light in accordance with the position of the shield. In the day time, without the lantern, the position of the shield is as discernible to the engineer as it is by night with it. The signal is unaffected by lightning, the duration of current being of such short period and the armature having to go through such a great distance, that before it is reached the current has subsided. The track instrument is another very essential feature in all systems. On the eastern roads the track instrument is supplanted by sections of insulated rails. The passage of the trucks over the insulated section closes the circuit. This crude form possesses innumerable disadvantages and is only permissible on stone ballast roads, and where the blocks are short enough to permit of a very low electromotive force. The western roads are as a rule single track and dirt surfaced, about the only ones that are ballasted in the mountains, and then there is too much mineral in the ground to operate any rail circuit device. The track instrument used in combination with the Tate system is a time apparatus, but when used with the interlocking instrument it produces a deadlock. The instrument in itself is encased in a cast iron casing that prevents malicious persons from tampering with it and it consists of a differential poised vessel with a U-tube connecting the extreme ends; mercury is placed in the vessel. By the passage of a train over a lever the vessel is made to oscillate, continuing during the passage of the entire train, after which time the flowing of the mercury through the U-tube severs electric connection that has been held intact in a key in the instrument. The instrument can be placed under water without any serious results and it can be placed in road crossings or platforms without becoming an obstacle. The switch instrument is nothing more than a key to make and hold intact or sever electric connection, according to the position of the switch. The automatic gates are calculated never to

Siemen's Railway Historical Motor.

close a person or vehicle in on the crossings. There is a gong used and after the sounding of it a person would have ample time to get off the tracks before the gates close, but in case they failed and the gates closed them on the track, by driving against the gate, it will turn as an ordinary gate, and as soon as the driver has passed out, a spring returns the gate to its position of barring the street. After a train passes through a block all signals are left at danger. By a simple device in the interlocking instrument, signals can be displayed by the train in moving either direction or only one direction, if desired. On a single track it aids trains in making their meeting points, i.e., if one train is delayed a few minutes, it holds the opposite running train at the meeting point until delayed one arrives. Train orders are dispensed with to the extent of special and extra trains, and ordinarily they can be run as section of a regular train. One train dispatcher can handle three or four times the business, on account of having comparatively no orders to give. A great many night telegraph offices and all those kept open for nothing but train orders, can be closed; nothing but commercial work is needed. The serious disasters that occur at street and railroad crossings are prevented by the use of the automatic gate and bell. Should anything occur to the wiring or circuits all signals stand at danger.

SIEMENS & HALSKE, BERLIN.

BLOCK SIGNALING SYSTEM.—This exhibit contained apparatus of every possible description which can be used in connection with signaling, interlocking and operation of railway switches. The exhibit contained samples of apparatus adapted to German railways, duplicates of which have for years been in continuous service in that country, and the same appliances shown as constructed for use on American roads. The apparatus was all connected as in service, so its operation might be readily illustrated and explained. One of the important parts of the exhibit was an electro-manual system of block signaling shown in two forms, one of which was the German application and the other the American. In the German form as applied to a double track road, an instrument is placed in each signal tower. These instruments are all connected by a single line wire with a ground circuit and operate practically as follows: Assume that a line contains two terminal stations, A and B, and three towers, Nos. 1, 2 and 3. The line is all clear and all signals in the clear position. A train, designated as No. 1, starts from station A in the direction of tower No. 1. As soon as the train leaves the station the operator presses the upper right hand knob on his instrument and turns rapidly the upper crank. This rings a bell in tower No. 1 and notifies the operator at that point that a train is on the line. Each operator immediately sends the same signal to the tower next in advance and in this way the entire line is notified that a train is coming. When the bell rings a small disc drops out beneath the box on top of the machine, thus giving a visual as well as audible signal to the operator. As soon as train No. 1 leaves station A the operator sets his signal to danger, in which position it is automatically locked. When the train passes tower No. 1, the operator at that point, by the use of the large handle on the lower right hand side of the instrument, sets his signal to danger. He then "plunges" his instrument and turns the upper crank rapidly and in so doing unlocks the signal in station A, which changes the color of the eye in the instrument at that station and thus notifies the operator that he may send another train. Operator No. 2 cannot give a clear signal to train No. 2 until his instrument has been unlocked by operator No. 3, and operator No. 3 cannot unlock the instrument until the train has passed his tower and cleared it a safe distance. This is accomplished by the use of a track instrument placed a train length in advance of each tower. The introduction of this instrument also prevents the operator from unlocking the preceding instrument should a train stop in front of the tower, and thus avoid the danger of a rear end collision from a train following under the supposition that the block is clear. It will be seen from this description that it is absolutely impossible for the tower men to throw their signals to a position which will allow two trains in one block at the same time, and only a total disregard of signals can cause a wreck. The construction of the instrument is interesting. There is a double coil magnet, a magneto with an armature and brushes, the office of which is to generate all the electric current used in connection with the system. This is an alternating current and takes 21 alternations or pulsations to operate the instrument. A current of from 30 to 50 volts can be generated by this machine. There is also a shutter, by means of which the operator is informed whether his instrument is unlocked; and an escapement which is actuated by the magnet in the operation of locking or unlocking the instrument. There is a drum through which the signal is moved. After a train has passed a station the operator first sets his signal at danger and then presses down the knob. This forces down the lever and forms an electric contact with the bar. The next operation is to turn rapidly the handle of the magneto. This creates a current through lever contact bar and the magnet, (and through a similar current in the preceding instrument) which operates the escapement and allows the weight (which is loose on rod) to throw the disc into its reverse position, displaying a red signal in the instrument. When the knob is pushed down, the collar on rod allows the levers to be pushed inward by the action of springs provided for that purpose. When the knob is released the rod and the lever again assume the original position. The tendency of the rod is to move in the same direction, but is prevented from moving by the lever which is held by the lug. As long as the rod is retained the lever remains in front of the off-set of the drum. It will therefore be seen that while the parts are in this position it is impossible to move the signal, as the only means for so doing is through the movement of the drum. It is impossible to again push down the knob for the reason that the lever is under the lug. It is impossible for the operator at the tower in which this instrument is located to unlock the preceding instrument for the reason that to do so he must have an electric current through the bar and the lever, and this he cannot obtain until the lever is removed beneath the lug. The operator in the tower in advance has an electric current through the bar lever and magnet of the instrument, therefore by the use of his instrument he can send a current through the magnet and operate escapement upon which the shutter takes the position shown, and in so doing releases the lever, which in turn releases collar, which, with levers, take their normal position. The drum is now released, and the signal can be set. Setting the signal removes this lock and the instrument may then be "plunged," but it cannot be "plunged" until the signal has been cleared and returned to the danger position. The blocking instrument operates with a caution signal at each block. These signals are controlled by the same instrument, and the caution signal is always displayed on one tower preceding the danger signal. A very ingenious attachment to this system is the connection of the electric circuit to each rail of the track, so that as long as any portion of the train remains in a block the current, instead of passing through the instrument, will pass through the rail and make it impossible for an operator to unlock the signal in his rear. This overcomes one of the great objections to manual block systems, as with many of them the operators must in all cases watch for the tail lights of each passing train before they can know whether or not it has broken in two. Some serious accidents have resulted from negligence on the part of employes regarding this order, and it will at once be seen that with this system neglect of this kind will have no serious results. An important feature in connection with this system is the introduction of the track instrument. This is a great safeguard in case of employes falling asleep or assuming that a train had passed when such is not the case. This track instrument may be placed at any desired distance in advance of each signal tower, in accordance with the length of the trains run, and will undoubtedly prevent many tail end collisions. One of the strongest claims of the company for this system is the manner in which electricity is used. The alternating current eliminates many objectionable features, as with this current it is impossible for any foreign electric current or a bolt of lightning to de-range the apparatus, for the reason that such a current would have to pass through the same wire at least twenty-one times before the mechanism would be operated. Again, each instrument has its own independent electric current, which is manufactured by the operator, consequently if one instrument gets out of order it has no effect whatever on the rest of the system, and under such circumstances one block can be cut out, and the system operated as before.

THE INTRAMURAL RAILROAD.

One of the hardest questions, if not the most difficult of all that came before the management of the Fair for settlement, was how to transport visitors from one part of the Exposition grounds to another. The grounds were nearly a mile and three-quarters long in their greatest length, and three-quarters of a mile across the south end. It was, of course, impossible that visitors should be asked to walk or find their own means of travel from one place to another, and the topography of the grounds made it almost equally impossible to operate a belt road; it was of course impracticable to operate a surface road where such crowds were expected, and all of the difficulties of the situation taken together were excuse enough for the length of time that it took to evolve a settlement. Almost all the known methods of transportation were discussed and each in turn was rejected as impossible or inexpedient. Finally an elevated structure was decided upon for the reason of its facility of operation in a largely crowded area, and, because of the fact that a lighter structure would be admissible, electricity was decided upon as the motive power. That the decision was a wise one, and that most excellent judgment in a most trying case was displayed, need not be iterated, as the road, during the Exposition period, spoke for itself to those who had the pleasure of a ride about the grounds upon it, with the panorama-like view on all sides.

For the rest it may be stated without hesitation that electrical interests of all kinds owe a debt to the operators of the road for their splendid efforts in the direction of popularizing more than ever the use of electricity in railway service. The Intramural road, though operating under a concession at the Fair, made no money, for the very good reason, however, that the first cost of even the cheapest of overhead structures is comparatively great, and the very short term of the Fair only made it possible to figure what might have been realized had the enterprise been a permanent investment of capital like that in the Manhattan road of New York, or the Liverpool road. In order that a comparison may be drawn between the two kinds of service, there are carried at the end of this report some tables that will be self-explanatory.

CONSTRUCTION AND OPERATION OF THE ROAD.—Upon July 1, 1892, a contract was entered into with the Western Dummy Railroad Company and work was begun upon the same. This contract was not made, however, until after the plans for the location of all the principal buildings and for the landscape features of the Exposition had been finally decided upon, and the greater portion of that work was either completed or under way. There was then very little choice as to an available location for a line of railway, and no possible route by which the heart of the Exposition could be reached, where the principal attractions would be and where the crowds would be most likely to congregate. It was due to this fact largely that the expectations of the projectors of the enterprise as to its financial outcome were not fully realized. It was, however, from an engineering and practical point of view, an unqualified success, which demonstrates the fact that an elevated railway, with a traffic equal to that of any city, could be operated successfully with electricity as a motive power, and it has undoubtedly advanced the application of electricity in that particular direction very materially. The actual location of the road was begun early in May, 1892, and the contract for its construction was made in July. On August 3d ground was broken and work started on foundations. The very severe winter preceding the opening of the Exposition caused so much delay in the work that construction was not completed until late in April, 1893—in time, however, to operate trains on the opening day. After it was definitely determined that a belt line was impracticable it was decided to use loop terminals, the north end to be located as near as possible to the north end of the main Exposition building—the Manufactures and Liberal Arts. It was not, however, until long after construction was commenced that a definite location, satisfactory to the contractors, could be secured. It was finally built across the north inlet with an entrance to the station platforms from each side, the nearest point to the main building being about 800 feet. The south loop was changed so as to extend to a point between the Agricultural building and the Krupp building, and by the side of the south pond.

By the terms of the contract the entire structure and building were to be removed from the park after the close of the Exposition. It therefore became advisable to design a structure that would be economical in construction and from which the greatest amount of salvage could be realized. The design adopted was double track, standard gauge, with twelve feet between centers of tracks, except on curves of short radius, where the distance was thirteen feet to allow for clearance of cars. The average length of span was 25 feet. The track did not differ materially from that of any standard elevated railway; rails were 60 pound steel,

Westinghouse Direct Connected Set.—Machinery Hall.

with pine cross ties 6 inch by 8 inch and 8 feet long. There were four longitudinal guard timbers to each track, bolted securely to the ties. The stringers carrying the track were steel "I" beams 15 inches deep and weighing 60 pounds per foot. One "I" beam was placed under each rail, making four to each span, except in spans greater than 25 feet, where the beams were doubled and bolted together. The beams under each track were braced laterally by steel angles riveted to lugs fastened to the webs of the "I" beams. The ends of the beams rested on bearing plates, three-quarter inch thick, and were held in place by clips riveted to the plates. The ties were secured to the upper flanges of the "I" beams by hook bolts, the hook having a bearing on the under side of the flange and the bolt running up through the tie and fastened with nut and washer. The caps and posts supporting the track were yellow pine, dressed and chamfered on the edges. The posts rested on a sill tying the two posts together with a cross sill under each post. Braces were put on the four sides of each post, bolted to post and sills. The tops of these braces were generally even with the surface of the ground. The posts were twelve feet between centers, thus bringing one post under the center of each track. The foundations were concrete masonry, seven feet square and twelve feet thick under each post, the sills resting on the top of this concrete. The entire length of single track or the length of one round trip was 6 22–100 miles, with 25 per cent of curves, the shortest radius used being 100 feet. As far as possible the elevation of grade was lowered between stations in order to utilize the effect of gravity in the acceleration and retardation of the speed of trains. It was impossible to carry out this idea at all points. The elevation required to pass over the roof of the Transportation Annex, Terminal tracks and in the rear of Machinery Hall made it necessary to abandon this arrangement on that part of the road, where the height above the surface averaged 26 feet. The maximum gradient was 1½ per cent and the minimum head room required under the "L" beams was 12 feet. The equipment of the road consisted of 15 trains, of four cars each, the forward car being the motor car. They were all the same length, *i.e.*, 47 feet over all. Each car seated 84 passengers, giving each train a seating of 336; this number was at times, however, doubled. When loaded the motor car weighed 30 tons and the trailers 22 tons. The motor car equipment consisted of four motors, one on each axle, weighing 3,465 pounds each, or a total of 13,860 pounds. The air pump for brakes, controlling apparatus and resistance weighed 3,200, and the car bodies with trucks weighed 27,600 pounds. The total weight of 30 tons per car gives a load per wheel of 3¾ tons. With this wheel load a train of eight cars carrying over 700 passengers has been pulled over the road and around the loops easily by one motor car. The motors were designed specially for the service by the General Electric Co., and were of the T-H 4-pole iron-clad type; the armatures were gramme ring with bar winding, with three turns to the coil and cross-connected so as to require but two brushes. Each motor was rated at 133 h.p., giving a total of 533 h.p. for each motor car with a speed capacity of 35 miles per hour. The average speed attained in operation did not exceed ten miles per hour. This was owing to the excessive amount of curvature and the number of stations. The highest speed attained between stations as shown by speed indicators was 30 miles per hour, for a short time. In actual operation an average of 42 h.p. per train was required, or a total of about 500 h.p. for twelve trains, which was the average equipment in operation. At times, however, when traffic was heaviest, two additional trains were put on, making in all fourteen trains. This was the number in operation October 9th, Chicago Day, when the road carried a total of 125,476 passengers. When trains were heavily loaded and a number of them happened to start at once, a maximum of 1,400 h.p. was reached at times.

The power house consisted of a boiler room 140 feet long and 60 feet wide, in which was located a bank of Babcock and Wilson boilers, ten in number, and rated at 300 h.p. each. The engine and dynamo room was 140 feet long and 87 feet wide, having the floor raised ten feet above that of the boiler room. It contained the following equipment: beginning at the east end was a 22x44x48 inch Reynolds tandem compound Corliss engine coupled to a 500 kilowatt multipolar generator. Next was a Hammond-Williams vertical engine, cross compound, 22x44x36, driving a 750 kilowatt generator, capable of working up to 1200 h.p. In the center of the floor was the great cross compound Reynolds Corliss engine coupled to a 1500 kilowatt generator. West of this was a 13x23x22 McIntosh & Seymour tandem compound engine coupled to a 200 kilowatt multipolar generator, and at the west end of the building was located a 20x38x48 tandem compound Greene engine belted to a 500 kilowatt multipolar generator, this being the only belted machine in the house. With this equipment there were three units, either of which was capable of supplying power sufficient to operate the road.

The number of trains in service was generally 12, running with an interval of 3½ minutes. The round trip, 6 $\frac{22}{100}$ miles, was made in 42 minutes, the running time being 19 minutes from loop to loop, leaving two minutes relay at each end of the road. There were, including the loops, eleven stations requiring twenty-one stops during each trip. The system adopted for the transmission of power has been called "the third rail trolley system," from the fact that a third rail of the same section and material as that used in the track was used for a conductor.

The third rail was located at the side of the track, about 20 inches from it and 13 inches above it. The only insulation used was wooden blocks, dipped in tar and supported by dowels, extending into the block and into the tie under it. The weight was carried by light cast iron spools surrounding the dowels. This arrangement gave a dry surface on the under side of the block, which proved to be efficient, the leakage being at no time apparent. The trolley consisted of a cast iron shoe, of which there were four to each car, one on each side of each truck. Those on the inside, next to the center line between the tracks, took the current from the trolley rail. The outside shoes were used only in going over switches and in operating the electric releases for the block signal system. The trolley consisted of a cast iron frame bolted to a stout iron plank which was fastened to the truck. The frame had two arms with pins which loosely fitted two iron links, to which the shoe was attached. The shoe was 12 inches long and slid the top of the trolley rail, depending on its weight alone for contact; a flexible connection was made between the shoe and motor. The north end of the line was fed through a second rail placed at the side of the trolley rail and carried northward from the power house about two-thirds of the length of the line. The controlling apparatus on the motor cars was located at the front end of a compartment set aside for the purpose, and was what is known as the series parallel controller. Under the requirements of the contract with the Exposition Company an efficient block signal system was required. An arrangement was made with the Rowell-Potter Safety Stop Company to install their system, an account of which will be found under the heading, "Railway Signalling Apparatus." With twelve trains in operation, making a total of 196 round trips daily, there were no delays or accidents, aside from a few of an unimportant nature and such as were liable to happen on any railroad. There was but one personal injury of a serious nature, and that was due to the carelessness of the injured party and his neglect to heed the repeated warnings given by both passengers and employes. The management of the road regarded this record with a great deal of pride and satisfaction, for when it is considered that the crowds that patronized the road were constantly changing from day to day, and that a very small percentage had any knowledge of their surroundings and were entirely unaccustomed to this mode of travel, had no idea of where they wanted to go, or how to reach any given point in the Fair grounds, it was to be expected that in their confusion and excitement some one would be injured almost daily. There were many who did not realize that there was any more danger in trying to get on or off a train that was moving than would be the case with a horse car, and it required constant care and watchfulness to prevent numerous accidents from this cause. The inexperience of visitors from the rural districts was a constant menace to their safety as well as a strain on the energies of the officials of the road. This was particularly the case during the last three months of the Exposition, when both trains and platforms were crowded. When the road first opened the fare charged was ten cents to ride from one loop to the other, or from any intermediate station to either loop. This was frequently considered an excessive amount and many passengers who took trains at intermediate stations expressed dissatisfaction more or less forcibly at being obliged to leave their trains at the end of the road to buy tickets for a return trip. To meet this objection collectors were placed at the terminal stations to collect fares from those who wished to remain on their trains. This arrangement was satisfactory to a certain extent, but there were many who objected to paying additional fare to return, the majority not being satisfied unless they could make a round trip and get off at the station at which they got on. While it was the object of the management to make the road popular, yet there was no way by which passengers could be restricted to one trip, and it was necessary to either collect fares from all passengers at a given point, or remove the restriction entirely and allow them to ride as long as it suited them. Realizing the difficulty of receding from the latter plan should it once be adopted, there was great hesitation about giving it a trial. The fact, however, that trains were not filled, and the desire to attract business, finally decided the matter, the collectors were taken off and passengers remained on the trains as long as they wished to. There were some who rode for hours, looking anxiously for the end of the road, and were not conscious of the fact that they were going over the same ground many times, even after they had counted four or five Ferris Wheels. The percentage of such riders, however, was not as large as had been anticipated, and the more crowded the trains became the more the percentage decreased. Results seemed to indicate that had the plan been adopted sooner, it would have been better for the revenue of the road. As shown by the statement subjoined, the total number of passengers carried during the six months, from May 1st to November 1st, was 5,803,895. The total number of days of operation was 170, giving an average of 34,143 per day, or twenty-seven per cent. of the total paid admissions to the Exposition. The following fuel table gives some information as to the traffic and fuel, which may be of some interest. The fuel used was crude oil, supplied from the Exposition Company's tanks at the rate of 72 cents per bbl. of 42 gallons. The results shown would have been still better had the plant been installed permanently and for permanent operation. The organization was also more extensive in some respects than would have been necessary under any other conditions than those existing at an exposition of this character, the entire road and its equipment being practically an exhibit. After the completion of the road and the installation of its equipment it was sold to the Columbian Intramural Railroad Company, which operated it during the Exposition.

A few trains were run after November 1st, but the attendance at the Exposition was so small that there was not sufficient business to warrant further service, so on November 7th it was withdrawn, forces were disbanded, and the work of preparing to take down the structure was begun.

AVERAGE CURRENT PER TRAIN, COLUMBIAN INTRAMURAL RAILWAY.

DATE.	K. W. HOURS.	NO. OF TRAINS.	AV. K. W. PER TRAIN.	HOURS OF RECORD.
August 1,				
" 2,	1752	12	24.4	6
" 3,	2068	12	28.8	6
" 4,	2000	12	27.5	6
" 5,	2100	12	29.2	6
" 6,	Sunday			
" 7,	2356	12	28.5	6
" 8,	2472	12	29.5	6
" 9,	2284	12	27.2	6
" 10,	2316	12	27.7	6
" 11,	2234	12	26.6	6
" 12,	2252	12	26.8	6
" 13,	Sunday			
" 14,	2436	12	29.0	7
" 15,	2370	12	28.2	7
" 16,	1902	9	30.0	7
" 17,	2388	12	28.4	7
" 18,	2431	12	29.0	7
" 19,	2352	12	28.0	7
" 20,	Sunday			
" 21,	2312	12	27.7	7
" 22,	2440	12	29.0	7
" 23,	2656	13	29.2	7
" 24,	----	--	----	--
" 25,	2510	12	29.0	7
" 26,	----	--	----	--
" 27,	Sunday			
" 28,	3424	12	40.7	7
" 29,	2132	9	34.0	7
" 30,	2690	12	32.0	7
" 31,	----	--	----	--
Sept. 1,	2792	12	33.1	7
" 2,	2528	12	30.3	7
" 3,	Sunday			
" 4,	2668	12	31.7	7
" 5,	2648	12	31.6	7
" 6,	2928	12	35.0	7
" 7,	2968	12¾	34.3	7
" 8,	2782	12	33.1	7
" 9,	3130	13	34.4	7
" 10,	Sunday			
" 11,	2748	12	32.7	7
" 12,	2681	12	31.9	7

South End Manufactures Building.—Night Scene.

AVERAGE CURRENT PER TRAIN, COLUMBIAN INTRAMURAL RAILWAY.—(CONTINUED).

DATE.	K. W. HOURS.	NO. OF TRAINS.	AV. K. W. PER TRAIN.	HOURS OF RECORD.
Sept. 13, - - - -	2854	12	34.0	7
" 14, - - - -	3031	12	36.0	7
" 15, - - - -	2764	12	33.0	7
" 16, - - - -	2348	12	28.0	7
" 17, - - - -	Sunday			
" 18 - - - -	3396	12	40.4	7
" 19, - - - -	2752	12	32.7	7
" 20, - - - -	3432	13	37.7	7
" 21, - - - -	3380	13	37.0	7
" 22, - - - -	3018	13	33.1	7
" 23, - - - -	----	--	----	--
" 24, - - - -	Sunday			
" 25, - - - -	----	--	----	--
" 26, - - - -	----	--	----	--
" 27, - - - -	----	--	----	--
" 28, - - - -	3028	13	33.3	7
" 29, - - - -	3370	13	37.0	7
" 30, - - - -	1936	8	34.1	7

Averages total of August, - - - - 671.4
" " September, - - - - 744.4

1415.8 in 45 Readings.

Average of August and September, then give 31.4 k.w. per train, which is equivalent to 42.0 h.p.

NOTE—Certain readings, evidently errors in reading, have been omitted.

C. H. MACLOSKIE, E.E.

COLUMBIAN INTRAMURAL RAILWAY COMPANY.

STATEMENT OF PERCENTAGE OF PASSENGERS CARRIED TO PAID ADMISSION TO FAIR.

	PAID ADMISSIONS.	PASSENGERS CARRIED.	PERCENTAGE OF PAID ADMISSIONS.
May, - - - -	1,050,037	151,550	14.43
June, - - - - -	2,675,113	784,756	29.40
July, - - - -	2,760,263	854,107	30.88
August, - - - - -	3,524,493	992,958	28.17
September, - - - -	4,650,871	1,447,882	31.13
October, - - - -	6,819,364	1,572,642	23.05
Total, - - - -	21,480,141	5,803,895	27.01

MAINTENANCE OF EQUIPMENT.

Week Ending.	Repairs Car Bodies.	Repairs Trucks.	Repairs Brakes.	Repairs Electrical Equipment.	Repairs Steam Plant.	Repairs Electrical Plant.	Repairs, Tools and Machinery.	Total.
May 5,	--------	--------	--------	--------	--------	--------	--------	--------
" 12,	--------	--------	--------	--------	--------	--------	--------	--------
" 19,	--------	--------	--------	--------	--------	--------	--------	--------
" 24,	--------	--------	--------	--------	--------	--------	--------	--------
" 31,	--------	--------	$ 20 00	--------	--------	--------	--------	$ 20 00
June 7,	$ 2 60	$ 22 09	23 19	$ 43 01	$ 10 00	--------	--------	90 89
" 14,	1 95	27 36	37 96	193 86	10 00	--------	--------	271 13
" 21,	85 10	15 98	44 78	292 45	3 70	----	--------	442 01
" 28,		35 80	17 46	265 53	3 50	--------	$ 2 94	325 23
July 5,	121 50	105 79	37 77	209 69	19 30	--------	--------	494 55
" 12,	25 80	87 45	41 48	135 15	26 53	$ 9 10	--------	325 51
" 19,	21 00	48 40	13 50	179 80	16 00	16 00	--------	294 70
" 26,	22 33	24 79	22 28	166 83	10 58	1 00	--------	247 81
Aug. 2,	30 45	26 72	20 78	111 77	--------	--------	57	190 29
" 9,	12 03	16 05	9 23	88 97	10 09	--------	2 70	139 07
" 16,	7 72	20 72	9 63	159 52	8 50	--------	2 49	208 58
" 23,	7 16	34 83	9 23	167 33	1 00	--------	--------	219 55
" 30,	12 43	45 27	9 23	164 95	1 06	--------	2 12	235 06
Sept. 6,	9 92	36 57	9 23	172 51	3 49	2 01	--------	233 73
" 13,	12 43	32 71	9 22	173 42	2 01	2 50	--------	232 29
" 20,	8 67	42 12	9 22	161 21	--------	--------	--------	221 22
" 27,	8 67	39 36	9 23	166 25	--------	--------	--------	223 51
Oct. 4,	7 80	28 92	9 23	202 74	1 74	--------	--------	250 43
" 11,	17 35	33 20	10 09	232 09	86	4 30	--------	297 89
" 18,	12 39	31 72	14 23	259 11	2 48	2 50	--------	322 43
" 25,	12 89	14 21	31 74	251 46	3 24	1 50	--------	315 04
Nov. 3,	17 35	22 09	22 08	249 63	4 28	--------	--------	315 43
Totals,	$457 54	$792 15	$440 79	$4,047 28	$128 86	$38 91	$10 82	$5,916 35

ESTIMATED COMPARISON OF WEIGHTS.

	ONE CAR TRAIN		TWO CAR TRAIN.		THREE CAR TRAIN.		FOUR CAR TRAIN.		FIVE CAR TRAIN.		SIX CAR TRAIN.	
	Electric.	Steam.	Electric.	Steam.	Electric.	Steam.	Electric.	Steam.	Electric.	Steam.	Electric.	Steam.
Weight of motor or locomotive in tons, -	8.8	30	8.8	30	8.8	30	8.8	30	8.8	30	8.8	30
Weight on drivers—ton, - - - - - -	26.7	20	26.7	20	26.7	20	26.7	20	26.7	20	26.7	20
Number of passengers seats in train, - -	48	48	96	96	144	144	192	192	240	240	288	288
Weight of motors or locomotive per passenger, in pounds, - - - - - - -	366	1250	183	625	122	416	91	312	73	250	61	209
Weight of full train, all seats occupied, - -	26.7	47.9	44.6	65.8	62.5	83.7	80.4	101.6	98.3	119.5	116.2	137.4
Percentage of weight of motor or locomotive to train, - - - - - - - - -	33	62	19	44	14	37	10	29	8	25	7	21
Weight of full train per passenger in pounds,	1113	1997	928	1370	868	1162	837	1058	818	997	808	954

VARIOUS ESTIMATES.

	MAY.	JUNE.	JULY.	AUGUST.	SEPT.	OCTOBER.	TOTAL.	AVERAGE.	LOWEST.
Pay Rolls, operation, - - -	$7150.95	$14673.52	$19259.00	$19904.80	$14466.71	$14774.26	$85229.24	$14204.87	$7150.95
Cost per train mile, all items, - -	1.69	.81	.70	.63	.67	.71	5.21	.87	.63
Passengers per train mile, - -	23.75	28.31	24.46	30.97	44.29	51.32	203.10	33.85	23.75
Oil per train mile in gals., - -	8.99	2.88	3.44	2.49	3.56	4.36	25.72	42.86	2.49
Cost of Oil, - - - - - -	$991.88	$1379.89	$2079.33	$1379.17	$2011.66	$2310.79	$10152.72	$1692.12	$991.88
Cost per train mile (of fuel), - -	15.54	4.97	5.95	4.30	6.15	7.54	44.45	7.408	4.30
Operation of P. H. per train mile,	15.46	5.61	6.15	4.21	4.08	4.20	39.71	6.618	4.08
Wages of Motor-men per train mile,	8.42	6.74	6.38	6.09	6.09	6.53	40.25	6.708	6.09
Oil per passenger, - - - -	.38	.10	.14	.08	.08	.08	.86	.172	.08

DELAYS TO TRAINS DUE TO DERANGEMENT OF MACHINERY.

REPORT FOR AUGUST, 1893.

DATE.	MOTOR CAR.	CAUSE.	TIME.	DELAY.
August 2,	1	Hot fields in pump motor	6 P.M.	2 minutes.
" 2,	1	" " " "	6:48 P.M.	2 "
" 9,	64	Lead wire broken	10:05 P.M.	5 "
" 11,	all	Short circuit	8:40 P.M.	1 to 3 "
" 12,	all	Low voltage (failure of water supply by storm forcing sand in pipes supplying condenser)	10:25 P.M.	20 "
" 14,	53	Lead wire broken	---------	8 "
" 14,	5	" " "	7:56 P.M.	5 "
" 14,	13	Controller finger caught	8:45 P.M.	10 "
" 14,	all	Short circuit	5:15 P.M.	2 to 3 "
" 15,	25	Controller finger caught	6:10 P.M.	7 "
" 15,	1	Lead wire broken	5:02 P.M.	5 "
" 17,	17	Lead wire broken	1:30 P.M.	8 "
" 18,	13	Hot box	11:35 A.M.	3 "
" 18,	53	Lead wire broken	10:06 A.M.	8 "
" 19,	17	" " "	3:54 P.M.	7 "
" 19,	64	" " "	2:06 P.M.	10 "
" 21,	65	" " "	8:24 A.M.	10 "
" 21,	all	Short circuit	5:42 A.M.	1 "
" 22,	all	Short circuit	1:25 P.M.	1 "
" 23,		Unknown	12:54 P.M.	1 "
" 24,	13	Hot box	3:04 P.M.	4 "

System shut down 28 minutes in all, 20 of which was steam trouble. Other stoppages of individual trains as above, represent simply delay in arriving at destination, and not actual time stopped at one point.

MAINTENANCE OF WAY AND WORKS.

WEEK ENDING.	REP. ROADWAY AND TRACK.	REP. BLOCK SIGNALS.	REP. AND RENEWALS TROLLEY RAILS.	REP. & RENEW. BUILDINGS.	TOTAL.
May 5,	------	-----	$ 60 00	-----	$ 60 00
" 12,	------	-----	85 00	-----	85 00
" 19,	------	-----	85 00	-----	85 00
" 24,	------	-----	80 00	-----	80 00
" 31,	------	-----	80 00	-----	80 00
June 7,	$ 138 53	-----	4 50	$ 22 70	165 73
" 14,	160 82	-----	25 25	6 05	192 12
" 21,	238 22	-----	9 21	40 32	287 75
" 28,	252 77	-----	------	41 07	293 84
July 5,	274 83	-----	33 12	46 64	354 59
" 12,	254 84	-----	43 47	75 71	374 02
" 19,	228 05	-----	29 70	51 68	309 43
" 26,	174 15	-----	9 00	21 87	205 02
Aug. 2,	216 95	-----	6 75	24 96	248 66
" 9,	225 24	$ 11 63	6 50	71 57	315 14
" 16,	206 10	27 44	7 00	39 43	279 97
" 23,	116 76	21 58	75	17 19	156 28
" 30,	144 15	22 44	3 25	13 77	183 61
Sept. 6,	223 20	20 82	2 00	26 47	272 49
" 13,	221 83	20 33	3 26	29 67	279 09
" 20,	183 79	35 66	5 00	36 55	261 00
" 27,	149 42	38 44	1 25	53 65	242 76

MAINTENANCE OF WAY AND WORKS.—CONTINUED.

WEEK ENDING.	REP. ROADWAY AND TRACK.	REP. BLOCK SIGNALS.	REP. AND RENEWALS TROLLEY RAILS.	REP. & RENEW BUILDINGS.	TOTAL.
Oct. 4,	$115 98	$14 07	$4 00	$71 65	$205 70
" 11,	103 32	16 83	75	71 74	192 64
" 18,	146 56	14 44	1 25	35 48	197 73
" 25,	149 95	14 33	-----	34 11	198 39
Nov. 3,	132 74	13 96	-----	25 52	172 22
Totals,	$4058 20	$272 17	$586 01	$857 80	$5774 18

TRANSPORTATION EXPENSES.

WEEK ENDING.	TRAIN SERVICE.	STATION SERVICE.	CAR EXPENSES.	OPERATION POWER HOUSE.	MISCELLANEOUS EX.	TOTAL.
May 5, - -	$173 31	$550 12	$3 30	$150 74	---------	$877 47
" 12, - -	171 42	647 82	11 55	172 49	---------	1003 28
" 19, - -	326 66	806 03	11 55	180 80	---------	1325 04
" 24, - -	264 56	582 25	8 25	135 19	---------	990 25
" 31, - -	613 73	1049 02	27 55	266 26	---------	1956 56
June 7, - -	1246 50	736 92	98 07	243 59	---------	2325 08
" 14, - -	858 69	1158 79	138 59	328 67	---------	2484 74
" 21, - -	987 71	1213 55	35 45	383 76	---------	2620 47
" 28, - -	1344 71	1275 09	131 61	404 93	$20 09	3176 43
July 5, - -	1358 93	1290 25	175 20	458 61	35 10	3318 09
" 12, - -	1201 58	1197 32	156 58	545 22	---------	3190 70
" 19, - -	1289 38	1163 34	148 75	517 01	---------	2760 52
" 26, - -	1156 21	1020 60	123 80	459 91	---------	2760 52
Aug. 2, - -	1108 26	997 53	125 72	461 50	---------	2693 01
" 9, - -	1109 99	980 33	71 62	344 05	50	2506 49
" 16, - -	1062 06	969 44	71 57	321 35	1 50	2425 92
" 23, - -	1084 25	969 84	66 77	303 65	---------	2424 51
" 30, - -	1084 10	991 96	75 25	294 89	---------	2446 20
Sept. 6, - -	1088 57	960 29	75 08	333 85	---------	2457 79
" 13, - -	1101 35	891 87	75 41	329 33	---------	2397 96
" 20, - -	1117 23	893 67	76 23	327 35	---------	2414 48
" 27, - -	1184 78	836 36	81 52	328 37	---------	2493 96
Oct. 4, - -	1118 54	1836 36	81 52	328 37	---------	2364 79
" 11, - -	1229 16	959 68	89 87	330 35	---------	2609 06
" 18, - -	1053 47	907 70	72 73	331 40	---------	2365 51
" 25, - -	1191 96	948 70	72 32	331 40	---------	2544 38
Nov. 3, - -	942 88	852 62	64 26	342 81	---------	2202 57
	$26559 99	$25740 39	$2174 37	$8961 80	$57 19	$63493 74

Following are the total earnings, expenses and net of the Intramural for the five months per train mile :

DATE.	EARNINGS	EXPENSES.	NET.
June, - - - - - - - - - - -	$2 82	$0 81	$2 01
July, - - - - - - - - - - -	2 44	70	1 74
August, - - - - - - - - - -	3 08	63	2 45
September, - - - - - - - -	4 40	67	3 73
October, - - - - - - - - -	5 12	71	4 41
Average, - - - - - - - - -	$3 56	$0 70	$2 86

Cost per passenger, .021 cents.

Western Union Exhibit Pavilion.

CHAPTER XII.

TELEGRAPHY.

TELEGRAPHY in commerce is one of the oldest branches of electrical science. This fact emphasizes the very newness of electricity in the sciences, arts and industries. When one reflects that fifty years ago, Senators of the United States sneered at and ridiculed the idea of the instantaneous transmission of intelligence, and demanded of Morse, when he sought an appropriation of $30,000 to demonstrate his invention, that he should send their soiled linen to their homes by the same agency, one may almost appreciate the grand strides made in all branches of electricity in the last half-century. In telegraphy itself we have ceased to wonder at its influence—indeed, we have ceased to feel it almost. A network of wires are strung about the earth, which has shrunk to an almost insignificant size so far as the relations between the different nations and countries are concerned, and the states of the American Union are not more intimately connected with each other, in the matter of interchange of intelligence, than the United States with the antipodes. This fact has been brought rudely home to us recently on account of the political disturbances in Hawaii, in the Pacific ocean. These islands are about the only parts of the business world not connected with all other points by telegraph. In consequence, the happenings in that country are more than a week old before we know them, and information of what is going on in the rest of the world is withheld from those islands until the ocean steamer bears it to them.

No new principles in telegraphy have been presented within a decade, if we except improvements in the mechanical construction of apparatus, and excepting also the development of the multiple and quadruplex transmitters of intelligence, by means of which the commercial possibilities of telegraph service are greatly improved.

The story of the telegraph told by exhibits at the Columbian Exposition was one of the most instructive in the whole Fair, and the two great American companies, the Western Union Telegraph Company and the Commercial Cable Company, and the foreign companies, gave demonstrations of the birth, growth and development of that branch of the science that called forth the unstinted admiration of all visitors who had the fortune to see them.

THE WESTERN UNION TELEGRAPH COMPANY.

The exhibit of this mammoth corporation was necessarily retrospective and historical rather than actually operative. The section of space occupied by it was divided into two departments, telegraphy proper, and cable work, and at either end stood a marble bust, the one of Prof. Morse and the other of Cyrus W. Field, typifying the two branches of the work. In the telegraph department was the original receiver made by Prof. Morse and exhibited at the University of New York in 1836. It is a very primitive looking affair, consisting of a wooden frame at right angles to which is held an ordinary horse shoe electro-magnet, actuating a second frame suspended from the first and carrying the tracing pencil. The mechanism for drawing the paper strip upon which the message is traced is simply an old clock train and weight. In the same case was shown the latest form of sound receiver for purposes of comparison. The development of the modern instrument from the crude experimental one is most striking, while the principle involved is the same. Here were also shown photographs of the original message "What hath God wrought!" A set of instruments for showing the working of the quadruplex instruments occupied a table, the far end of the line being worked by means of an automatic sender at the exhibit end. These several exhibits were intended to show at a glance the advance from the early days of telegraphy to the latest developments of the science.

Two early messages were shown, framed, the one an original bearing the signature of Daniel Webster and the date 1851, while the other was an operator's transcript of a message received from a certain Mr. Hathaway in 1850. Some specimens of telegraph poles were also presented, showing the destructive action of air, water, insects and birds in different climates and were very interesting from a constructive and historical point of view. In the cable section were cases containing models of cable laying and repairing vessels, most prominent among which, of course, was that of the "Great Eastern." Then came the "Mirror," the repair vessel of the Eastern Telegraph Company, with whose lines the Western Union Telegraph Company connect in England for points in the East, and also for points in Africa and Australia. Next was shown the "Relay," belonging to the Central & South American Telegraph Company, with which the Western Union lines alone connect at Galveston, Texas. Sections of the original Atlantic cable hung upon the walls and in a large case was shown the method of making and insulating a modern cable, each step being given, from the crude rubber to the finished armored cable ready for deep sea service. Leaning against the case containing the model of the "Great Eastern" was a fac simile of the original grapnel with which on September 2d, 1866, Capt. Anderson recovered the broken cable of 1865, from a depth of 1950 fathoms in mid-ocean. Previous to this feat the greatest depth from which a cable had been raised was 500 fathoms. A profile chart of the ocean's bed, showing the route and positions of the ten existing cable lines, formed an interesting feature of the display, as did also two large telegraph charts of the entire world. The latest development of the well-known syphon recorder was shown in operation in the center of the space, while near by was the old mirror system, in which a ray of light was reflected from a small mirror carried on a small magnet, influenced by the sending current, dots and dashes being indicated by the movement of the ray to one side or the other of the zero point of a scale. By this method one man was required to read the signals while another transcribed them, and, of course, no automatic record was possible.

THE COMMERCIAL CABLE COMPANY.

The exhibit of the Commercial Cable Company illustrated especially the manner of operation of their Atlantic system. For this purpose there were installed an artificial Atlantic cable, properly duplexed, and a full set of automatic sending apparatus and receiving instruments, whereby constant transmission and reception are carried on in both directions at the same time. One extremity represented Europe, the other America.

Artificial Cable.—The artificial cable exhibited was the invention of Dr. Muirhead, of London. It simulated perfectly, with all its peculiarities, a cable 2,380 knots long. A continuous strip of tin-foil represented the copper core, the insulation being obtained by the use of paraffined paper, while the tin-foil in plain sheets took the place of sheathing and sea water. The apparatus was put up in boxes of suitable size, on the outside of which were two sets of terminals, one set being connected at intervals with the tin-foil strip, the other with the plain sheets. Their use was to make connection from one box to another and to vary the electrical conditions in any manner that might be found necessary. As previously stated, the artificial cable exhibited was duplexed. In doing this the Wheatstone bridge method was used. The current from the sending apparatus in dividing was made to act across two condensers of Dr. Muirhead's make and of twenty microfarads capacity (variable for adjustment) each. To the opposite sides of these condensers were connected the artificial cable which represented the cable proper, and a second artificial cable to complete the balance. The receiving instrument was placed across the points of equal potential and remained neutral so long as the equality was maintained. As this arrangement was necessary at both ends of the cable, three complete sets of "Artificial" line had to be set up.

RECEIVING INSTRUMENTS — The Cuttriss siphon recorder. The receiving apparatus exhibited was identical with that in daily use on the Commercial system. It was the Thomson siphon recorder almost entirely altered and improved by Mr. Chas. Cuttriss, the company's electrician. Two of these instruments were shown. A light rectangular coil was pivoted in a powerful magnetic field generated by a large, laminated steel magnet. The coil was wound from the finest copper wire that could be drawn. On the top of the coil were two minute ears of soft iron and the lines of force were deviated in their vicinity by a simple arrangement so as to lift the coil and almost suspend it in the air. The same magnetic arrangement also served to adjust it. A fine glass siphon, drawing ink from a reservoir, was attached to a vertical suspension, slightly twisted, by torsion, to keep the siphon slightly off the zero. The siphon was connected to the coil by a fine fibre. The coil in moving drew or released the siphon, which gave a magnified record of the coil's movements upon a tape drawn beneath it, but which it did not touch.

THE VIBRATOR.—Were the siphon to trail upon the paper, the sensitiveness of the instrument would be totally destroyed. The ink is fed to the tape by means of a vibrator, of which two were exhibited, one in connection with each receiving instrument. On the nose of the siphon is a small particle of soft iron and immediately beneath an electro-magnet, connected with the vibrator. The vibrator which interrupts an electric circuit passed through itself and the electro-magnet immediately below the siphon, on the same principle which operates an electric bell, possesses the property of having the rate at which it makes and breaks the circuit varied. This is done by attaching to its armature a glass tube, which is practically lengthened or shortened by forcing into, or drawing from it, a column of mercury. When the period of vibration in the tube is the same as the natural period peculiar to the siphon in use at the time the latter will vibrate vertically in unison and deliver the ink in an apparently continuous line upon the tape.

MOTORS.—The paper upon which the record is made requires to be drawn at a perfectly uniform speed. For this purpose the company used a special motor, of which two were exhibited. By means of a peculiar governor and heavy fly-wheel they can be run at absolutely the same speed when placed in a circuit with five gravity cells, or in an electric light circuit of 110 volts with one sixteen candle power lamp in series.

AUTOMATIC SENDERS.—Sending on cables has hitherto been done by hand. For this purpose a double key was used, one lever of which sent a positive current and the other a negative, thereby causing a movement of the coil in the receiving instrument either to the right or left. There were two of these keys set up in the exhibit in such a manner that they could be manipulated directly by hand or by automatic machines with which they were in electrical connection. Two styles of automatic senders were exhibited. The messages are first prepared by being punched on tape. The punched ribbon in passing through the automatic transmitter sets in motion two small levers corresponding to the two levers of the double key. In one of the patterns shown the tape is drawn by clock-work. The second style is novel in design. The feed wheel which draws the tape is intermittent in its motion and is operated by a small motor running equally well in the electric light circuit or by a small chemical battery. An important feature is the possibility of regulating by means of a cam on the shaft of the motor the length of the electrical impulses without varying the total speed.

PERFORATORS.—Two kinds of punching machines were shown. One for the Wilmot transmitter, which is very similar to the ordinary Wheatstone, the other specially adapted for the Cuttriss transmitter. The latter machine merely makes indentations, the little blades doing the work remaining sharp indefinitely. In the older machines, where the piece is actually blocked out, the perforators require constant attention. The center row of perforations, which is necessary for the perfect feed of the paper, is made by a specially designed instrument, which makes 2,500 perforations per minute. A plain roll of paper is put in the machine which requires no further attention, because when the roll is completely perforated the machine stops itself automatically.

The company had also on exhibition a fine model of the repair ship, the "Mackay-Bennett." She was represented in the act of picking up an injured cable. Specimens of all the different types of cable used throughout the Commercial's system were shown, in all sizes, from the small deep-sea section to the large, heavy shore-end, protected by powerful articulated rings of steel. In addition, there was a model of a Mushroom anchor used for securing cable buoys; also a model of a grapnel such as is used for picking up cable. Specimens of cable injured by ice or anchors were exhibited, and several minute "faults," the location and removal of which are a monument to the skill of the company's engineer-

Commercial Cable Company's Pavilion.

ing staff. All visitors to the Commercial Cable Co.'s booth were given specimen cablegrams actually sent and received over the fac simile Atlantic cable.

THE E. S. GREELEY & COMPANY.

This firm exhibited, among other things, a full line of Morse and District telegraph apparatus. The Victor telegraph key was shown in connection with other instruments. This key is the only telegraph key that has no trunnions, the fulcrum being a knife edge. This prevents side motion and consequent poor contact and also eliminates the so-called jamming experienced in some telegraph keys. A short, stout spring near the fulcrum holds the lever in place. This key is also modified to meet the requirements of cable transmissions and railway signal work. The Victor relays and sounders also embody the principle of the Victor key, viz: discarding troublesome trunnion screws and using points as fulcrums. The relay has a searcher or plunger by which the adjustment of the spring is quickly determined and the adjusting screw then secures the spring. By virtue of the mechanical construction, a stout spring is used, thus doing away with the troublesome and short-lived ordinary relay spring. Various modifications of the Greeley ink writing, self stopping and starting registers, were connected with Victor keys and district call boxes. These registers are entirely automatic, starting and stopping automatically with the opening or closing of the circuit. They can be used on either open or closed circuit, or on both. They also work readily as a main line instrument, as is the case when used as a fire alarm register. The instruments are made with one, two and four pens. The four-pen instrument is the first of its kind built, and four signals of various lengths of duration are readily received at one time. These instruments were connected to a single Jack switchboard. This switchboard was of the standard Western Union pattern, each jack to accommodate four loops and the board built for twenty-five lines. In the center of a large show-case was exhibited what is destined to become historical. that is, the gold Victor key used by President Cleveland to open the Exposition. This key was connected in series with an automatic engine starter and Nash & Eddy water supply regulator, which instruments respectively started the 2000-h.p. Allis engine and 5,000,000 gallon Worthington pump. The current was supplied by the Exeter dry battery. The key is elaborately mounted on a three-step dais covered with blue and yellow plush, representing the American blue and the Spanish yellow. The figures "1492" and "1893" in silver letters are on opposite sides of the base.

J. H. BUNNELL & COMPANY.

This firm made an exhibit of telegraph apparatus in connection with the U. S. Naval Observatory in the Marine section of the Exposition. The exhibit consisted of :

POLE-CHANGER.—Quadruplex form, arranged with adjustable contact springs, so placed that their operation is in plain view of the operator for better convenience and accuracy of adjustment.

POSTAL RESONATOR, with 1892 local sounders—being similar in

MAIN LINE SOUNDING RELAY.

form to the resonators used on continental lines, and containing the 1892 one-cell local sounder.

BUNNELL RESONATOR.—Resonating box (on swinging arm) in which sounder is placed and connected, the object being, first, to amplify the sound ; second, to bring it as nearly as desired to the ear of the operator, —especially adapted for receiving operators using the typewriter, and for noisy railway stations, or for all situations where the sound of the instrument is subject to interference from outside noises.

MAIN LINE SOUNDER RELAYS.—One hundred and fifty ohms, with key on base. An entirely new form (1893) of main line instrument, which combines the sound-giving qualities of the best local sounders with the large fine wire magnets and special adjustments of a relay, and which, in practice, gives more than three times the volume of sound given by the very best box relays, or any other form of sounding relays, for short or long main lines. They are unequalled for wrecking instruments, testing sets, field telegraphs, switchboard sets, temporary offices, and all places where it is desirable to dispense with the use of local batteries. These instruments are made with full size relay magnets of 150 ohms resistance, having both magnet and armature adjustments, and fully suited for doing, in a greatly improved manner, the work of regular main line box relays. They are highly sensitive to the weakest currents, and will give a clear working sound on circuits where, without them, a relay with local sounder and battery would be absolutely necessary.

AUTOMATIC REPEATER FOR OPEN CIRCUIT.

NEW MODEL OPEN CIRCUIT COMBINATION SET.—In this apparatus is used a transmitter operated by the key and actuated by the same local battery as that by which the relay actuates the receiving sounder. By this means the operator is enabled to hear his own transmission, and the transmission of the signals is not subject to the variable peculiarities of the hand on the key, as is the case when the transmission is direct by the key. In this way the work of the operator is better because he can distinctly hear it, and the liability to imperfect contacts is much less.

STEEL LEVER KEYS WITH LEGS.—Usual form as at present adopted in the United States, the especial feature being the very light and at the same time very strong steel lever in one piece.

LEGLESS STEEL LEVER KEYS.—Being in all respects like the usual steel lever keys except that the "legs" are omitted and the connections made at binding posts so as to permit the key to be screwed down to desk or table from the top.

MAIN LINE RELAYS, one hundred and fifty ohms each.—Usual form of standard relay as used on all American telegraphs, except as to the armature, which in this case is stamped out in one piece, giving needed rigidity with much lighter weight of metal than required if otherwise put together.

NO. 2 POLAR RELAYS, twenty ohms each.—Special form polar relay, particularly for use requiring convenient manipulation of the magnetic adjustments and also for powerful polarization.

POCKET RELAY IN CASE.—Special form pocket relay devised with the aim of containing within the smallest limits all the working parts of as nearly full size and power as possible, with special view to the military telegraph service, and for service in temporary stations on railways, in accidents, etc.

AUTOMATIC REPEATER.—Bunnell patent open circuit. This is a special system of repeaters for open circuits, so arranged that the two sets of instruments constituting the complete repeater may be used at will, either for repeating from one line to another or as two separate instruments, each one operating independently its own individual line.

CIRCUIT TIME REPEATING SOUNDERS.—Repeating sounder arranged with four sets of contact springs which are simultaneously actuated by the sounder lever for the purpose of transmitting on four lines at the same moment, the signals used in sending time. The same form is, with a greater or lesser number of contact springs, adaptable for from one to eight circuits, and is used at the Naval Observatory, Washington, for the transmission of time signals over the wires of the Western Union Telegraph Company, covering the greater part of the United States.

SPRING POINT REPEATING SOUNDERS.—Ordinary repeating sounders with spring contacts for repeating in direct Morse circuits. In addition there were in operation in the exhibit, giant sounders, circuit preserving transmitters, single transmitters, local sounders, etc.

WESTERN ELECTRIC COMPANY.

TELEGRAPH APPARATUS.—Exhibited in the show cases of the Egyptian Temple were telegraph instruments of the latest design, comprising keys, sounders, relays, repeaters, quadruplex and duplex apparatus, switchboard and cut-outs. Alongside of the apparatus made at the present day were shown some of the types of instruments made fifteen or twenty years ago. A telegraph switchboard, such as is seen in the larger offices of the Western Union Telegraph Company, was exhibited in the space along with telephone switchboards, and a complete set of the latest improved quadruplex apparatus set up and in working order, the other end of the circuit being in the Western Union Telegraph Company's exhibit in the gallery.

BAUDOT'S QUADRUPLEX TELEGRAPH.

Apparatus allowing the simultaneous transmission on the same wire of four messages, these being sent all in the same direction, or some in a certain direction and some in the opposite direction. Worked with keys, transmitters and a revolving switch, the keys being connected one after the other with the line wire through the switch ; a similar device acts at the other end of the line, a revolving switch synchronized with the former one connecting the line wire at the proper time with electro-magnets and the electro-magnets working as local relays for the printing instruments. The transmission of the various letters, results of the various combinations made at once with the five sending keys (the first depressed alone or at the same time as the third); the printed strip of paper at the receiving end is afterwards pasted on the sheet of paper sent to the customer.

MUNIER'S TELEGRAPH.—Set of apparatus devised for quadruple telegraph transmission on the same line and combined for using the Hughes printing telegraph apparatus.

M. E. MERCADIER.

MULTIPLEX TELEGRAPH SYSTEM.—The multiplex telegraph was placed in the center of the exhibit of the French Administration of Posts and Telegraphs and occupied two tables arranged symmetrically On each table there was placed a complete terminal station with twelve "legs." The two terminal stations were connected by means of an artificial line made up of a combination of resistances and a variable capacity so as to imitate the conditions of an external telegraph or cable circuit. The exhibit showed a receiving and a transmitting side of a system designed for the transmission of twelve messages simultaneously ; also in detail the manner in which the signals are received and transmitted without interfering with each other. Electrically actuated tuning forks are constantly vibrating and sending the intermittent currents from the batteries into the primary wires of the induction coils to which they are connected. When the key of a station, for instance, is depressed, the induced currents produced by the induction coil bring into action the primary circuits of the two induction coils. The secondary induced currents produced by these two coils, acting differentially on the tele-microphonic relay, do not affect the latter. In this way transmission passes to line without affecting the receiving relay. When received, the secondary currents sent by the corresponding transmitter at the other end of the line pass from the line or ground into the secondary circuit of the coil, and to the line coil of the tele-microphonic relay ; the diaphragm of the latter acts upon the microphone contacts, and the induced currents produced in the induction coil

The Gray Telautograph Exhibit.

of the relay pass into the mono-telephonic electro-magnets and vibrate their plates respectively. Each of the plates of these monotelephone receivers is tuned to respond to only one fundamental note, which corresponds to the number of vibrations sent out by the corresponding tuning fork at the other end of the line. The theory of the monotelephone has been the subject of extended inquiry by M. Mercadier, and its practical application in this special system of multiplex telegraphy has been greatly facilitated by the results of his researches. With the arrangement as shown at Chicago, any combination of numbers, so far as transmission and reception are concerned, can be effected; thus twelve messages can be sent simultaneously in either direction, while any less number can, of course, be transmitted either way and in any proportion of twelve. It is also evident that the same wire can be used for Morse or Wheatstone transmission, while the wire is simultaneously transmitting the rapid alternations required by the Mercadier system, as those do not interfere with the current of the former.

DE BRANVILLE ET ANIZAN.

Artificial Lines.—The question is to build a line having same resistance and same capacity as any overground or underground line. The inventors made two types of boxes, one being an underground line of 50 kilometers long, the other one an overground line of 500 kilometers long. The boxes are symmetrically disposed; just in the middle of each resistance coil is inserted one plot which can be connected through the plots to the condensers. The surface (or capacity) of each condenser is in proper proportion with the connected resistance; that is, the resistance of a certain length of overground line, the capacity is the capacity of the same length of same line. The working of the box is very simple. Take out the pin between the two plots, you have a line of resistance, and without any capacity. Put the same pin between the two and you introduce the capacity in that line. Proper contacts are provided for allowing double or single resistance, as is desired. Practically the boxes are 60x32.50 centimeters, and the plots are in the center of the boxes. The overground line box is divided into sections of 5-10-15-20 and 50 kilometers. This box contains two series of resistances, the first one made of bronze coils, the second one of iron coils, in order to facilitate the study of a telegraphic or telephonic line, or of any combination of them. These boxes may be used as artificial lines, or to take independent measurements of resistance or of capacities. Their particular feature is the disposition of the condenser just in the middle of the corresponding resistance. Another box showed the use of these boxes for the study of a double telephonic line and the measure of its electrostatic capacity.

THE GRAY TELAUTOGRAPH.

THE TELAUTOGRAPH, as its name—a compound of three Greek words: tele, at a distance; autos, one's self; grapho, to write—implies, enables one to transmit instantaneously a fac-simile of his handwriting to a distance. It consists of a transmitter and a receiver associated for use by each subscriber. The mechanism of these instruments is extremely simple, and so arranged that electrical impulses sent over the connecting wire put the receiving pen in synchronous movement with the transmitting pen. The impulses are transmitted by direct and positive methods, which are therefore independent of, and uninfluenced by, atmospheric conditions.

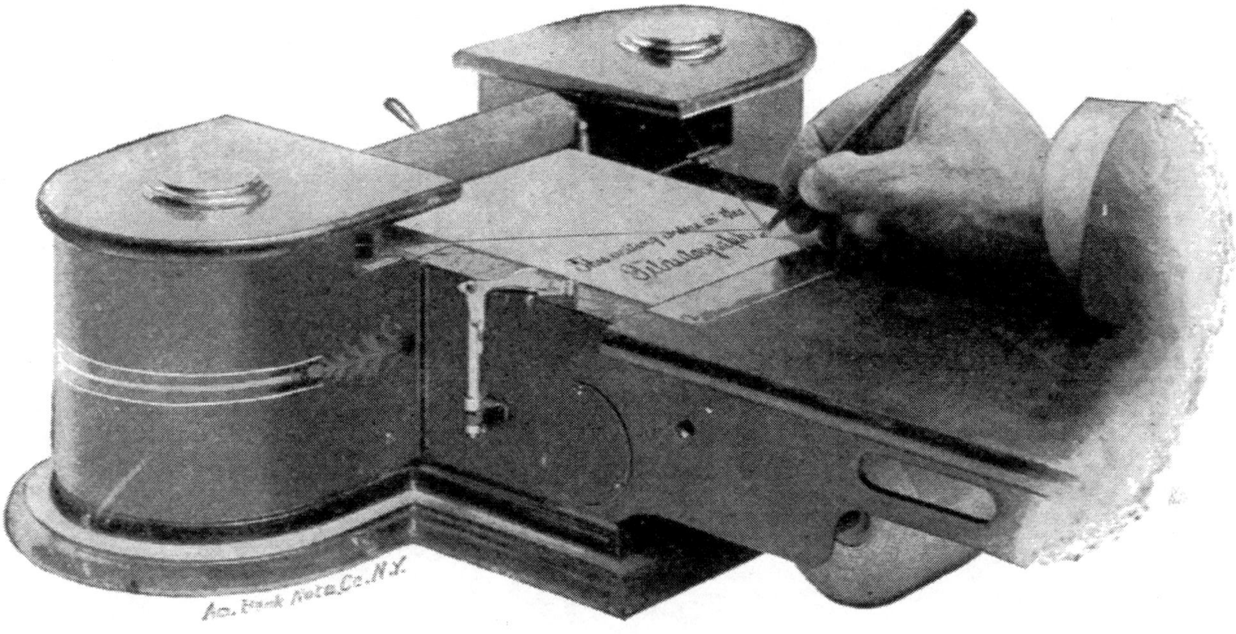

GRAY TELAUTOGRAPH.—TRANSMITTER.

THE TRANSMITTER.—An ordinary lead pencil is used, near the point of which two silk cords are fastened at right-angles to each other. These cords connect with the instrument, and, following the motions of the pencil, regulate the current impulses which control the receiving pen at the distant station. The writing is done on ordinary paper, five inches wide, conveniently arranged on a roll attached to the machine. A lever at the left is so moved by the hand as to shift the paper forward mechanically at the transmitter, and electrically at the receiver. The receiving pen is a capillary glass tube placed at the junction of two aluminum arms. This glass pen is supplied with ink which flows from a reservoir, through a small rubber tube placed in one of these arms. The electrical impulses, coming over the wire, move the pen of the receiver simultaneously with the movements of the pencil in the hand of the sender. As the pen passes over the paper, an ink tracing is left, which is always a fac-simile of the sender's motions, whether in the formation of letters, words, figures, signs or sketches. With the telautograph a business man may conduct his ordinary correspondence and transact matters of importance, without the medium of an operator. From his office he can send instructions to his factory—close by or many miles distant—having the same delivered in his own handwriting. In like manner he can receive at his office, daily or hourly, reports from his factory. The superintendent of a large estab-

GRAY TELAUTOGRAPH.—RECEIVER.

lishment may use it for quick and accurate communication between his office and the different departments. A banker may sign a check in one city without moving from his office in another. A broker, dealing by wire, can give quotations and execute orders to buy or sell securities without danger of dispute. Important negotiations may be carried on between two parties, widely separated, and a memorandum of agreement drawn up and signed by each without the parties coming together. The other systems of electrical communication—the telephone, telegraph and printer—are not adapted to the above and similar uses, since they do not transmit autographic writing, and consequently afford no means of identifying and binding the corrrespondent at either end of the line. A physician may wire his prescription to the druggist, using the arbitrary code of the profession, in the confidence that no mistake will be made in the transmission. A reporter, writing up a fire or railroad accident, can send to his paper a sketch of his subject taken on the spot.

The telephone, because of frequent mistakes arising from the similarity of articulate sounds as heard through it, is but little used in transmitting messages to and from the telegraph offices. By means of the telautograph this service may be quickly and correctly performed without the intervention of the messenger boy. Railroad companies can use these instruments on their own lines without the services of telegraph operators. Much of the telegraphic business of a railroad company is done by its freight department; and with the telautograph car numbers, rates and tracers can be accurately sent. And so, a train dispatcher can have train orders delivered to conductors and engineers in his own handwriting, thus minimizing the danger of errors. It will prove invaluable for quick and correct reports of races, base ball games, and other sports. In cities and towns the telautograph may be operated on the exchange or central station plan, in much the same manner as the telephone is now worked. At the Exposition the telautograph exhibit was one of the most interesting and instructive in the Electrical building. Prof. Gray made the exhibit in a handsome pavilion in ivory and silver finish, carved with allegorical electrical figures. There were half a dozen instruments in constant operation before visitors, and as many operators in attendance to answer questions and demonstrate the uses of the machine.

TELEGRAPH SERVICE FOR THE COLUMBIAN EXPOSITION.

There were two systems of telegraphic communication in the Exposition grounds, one installed by the Western Union Telegraph Company, and the other by the Postal Telegraph & Cable Company. At the time of the dedication of the grounds and buildings, in October, 1892, these two companies ran temporary cables from Sixty-second street entrance, on poles through the galleries of buildings and in the main subway, to a point under the grand stand, situated in the Manufactures building, where, in a small space reserved for them, they each installed a force of operators and sent over the country the details of the various ceremonies to the news depots of the world. But these lines were all removed soon after the close of the dedicatory exercises, both companies leaving but one office each with an operator in the Service building, for the transmission of messages in the ordinary course of business. Permanent quarters of both companies were located on the ground floor of the southeast pavilion of the Administration building, and as it was necessary for them to enter the grounds at Sixty-third street, some means by which they could reach their permanent office, and also communicate with the underground system, was necessary. For this purpose a double duct line was laid by the Exposition from the south side of the Transportation Annex to the office in the Administration building, and another double duct line from their office to Machinery Hall. This work consumed 4,016 feet of duct, sixteen manhole boxes and covers, and cost, for labor of installing the same, $132.00, the companies dividing the expense equally between them. Through these ducts the cables were laid, and connections made through Machinery Hall to the main electric subway. Through the Transportation Annex the cables were laid in a trough built along the south and west walls for the fire alarm circuits, it being ample to accommodate all three cables. Inside the Annex connections were made with the Intramural elevated railway structure, so that the west side of the grounds could be easily reached, the main subway giving access to the east side of the grounds. Telegraph stations were scattered about among the main buildings and reached by wires tapped from cables run on the elevated track and in the main subway. Messages were sent from these scattered stations and received in the main offices of the companies in Administration building, and from there sent to their several destinations.

INDEPENDENT SIGNAL SERVICE.

When the power plant in Machinery Hall was started, oil was adopted as fuel, and owing to the fact that the supply pipe might break at any time and flood the furnaces, and a disastrous fire might ensue, some quick signalling device was necessary to stop the supply of oil should such an accident occur. It was necessary that such a system should always be ready when needed, and therefore so constructed that it would automatically notify those in charge should the line from any cause become disabled. An adaptation of the common telegraph circuit was devised so that when the circuit was closed and current flowing in it, relays placed at the operators' stations would keep open a local vibrating gong circuit. But should the main circuit be broken from any cause whatever, the relays would instantly close the bell circuit, indicating by the loud ringing of the gong that the line was in trouble, or else that a signal was to be transmitted. To further avoid the danger of a breakdown, a double line wire was installed, one run underground, buried about two feet, the other overhead on the elevated railway structure, from the boiler-room to the oil pumping station. These two wires were used as one conduotor, so that if either one should break down from any cause, the other could be used. The earth was used as a return conductor. Along the whole length of the boiler-room closed contact switch buttons were placed in series on this main line, one push button

Administration Building, Electric Fountain and Search Light Effect on Machinery Hall Towers ard MacMonnies Fountain.

for each set of boilers. At the office of the engineer in charge of the boiler-room, in the cabin where the valves of the pump line were placed, and in the oil pump station itself, a relay was placed in series on this main line with a local vibrating gong circuit, the connections of which were so arranged that as long as the main circuit was closed the local circuit would be open and the bell silent, but if, from any cause whatsoever, the main circuit should be open even for an instant, the relay would close the local circuit and the gong would ring loudly; therefore, in case of any accident a fireman could step to this push button, and, by pressing upon it, open the main circuit a given number of times, sending in an alarm of accident. The operator at the valve station instantly closes the oil valves and shuts off the supply of oil, and the engineer at the oil pump house reverses his pumps and cleans the oil out of the pipe line, effectually removing any large amount of the oil from the scene of danger. This system gave very good satisfaction, and was tested twice every twenty-four hours, at twelve o'clock noon and twelve o'clock midnight, so as to be sure that it was in perfect working condition. As oil is liable, unless carefully adjusted at the burners, to emit a dense smoke in burning, some inspection system became necessary so that the firemen in charge of boilers could be notified when the furnaces were not in perfect working condition. On account of the length of the boiler-room it was not feasible to do this by messenger, so the inspector of smoke was stationed in the cabin where the oil valves were situated, whence he could keep a close watch over the chimneys and note whenever they were emitting smoke, besides being always on hand to close the valves in the oil line in case of accident. A set of ordinary push buttons were placed on his desk, one for each set of boilers, and an annunciator with a large vibrating gong was placed at each boiler. These gongs and annunciators were connected in regular order to the push buttons so that whenever a chimney smoked an undue amount the inspector could, by pressing a button at hand, notify any fireman to attend to his fires. This system was operated by ten dry cells and served well the purpose intended. The lines were carried in mouldings along the raised platform in front of the boilers and were concealed from view and protected from mechanical injury.

CHAPTER XIII.

TELEPHONY.

THERE be three things," said Bacon, "which make a nation great: a fertile soil, busy workshops, and easy conveyance of men and goods from place to place.

To these three things, a fourth and a fifth have now been added: instantaneous transmission of intelligence by means of the telegraph, and instantaneous transmission of spoken words between people widely separated.

At the Centennial Exposition in Philadelphia, in 1876, was given the first general public exhibition of the telephone by its inventor, Alexander Graham Bell.

To-day, seventeen years later, more than half a million instruments are in daily use in the United States alone, six hundred million talks by telephone are held every year, and the human voice is carried over a distance of twelve hundred miles without loss of sound or syllable.

THE AMERICAN BELL TELEPHONE COMPANY.

At the World's Fair in Chicago, The American Bell Telephone Company exhibited local and long distance telephone service in full operation, and also a complete collection of telephones, maps, statistics, and apparatus, illustrating in detail the beginning, the progess and the present state of development of this great industry.

THE BUILDING.—For this purpose the company constructed a building, one hundred and three feet in length by sixty-seven in width, within the Electricity building, directly opposite the main entrance which opened upon the Plaza. The structure was of classic design and a beautiful specimen of architectural skill. The front was a reproduction of that of the Erectheum at Athens. Approaching the interior by a broad flight of steps through a portico of Ionic columns, visitors entered an open court or atrium, surrounded on all sides by peristyles with domed corner rooms. At the rear of this open court was a domed temple large enough to accommodate one hundred persons. The rotunda was twenty-four feet in diameter and surrounded by an ornamental entablature, the drum of the dome, and by the dome. The exit from the building was through a peristyle and portico, similar to the entrance.

THE SWITCHBOARD.—The telephone once invented, for two to talk together over a single wire was a simple matter. But it at once became obvious that this was only a beginning, and that, to fulfill the possibilities of the invention, means must be found to enable large numbers of subscribers to interchange communications at will. Such an apparatus was soon devised; it was called a switchboard, and the invention of the switch-

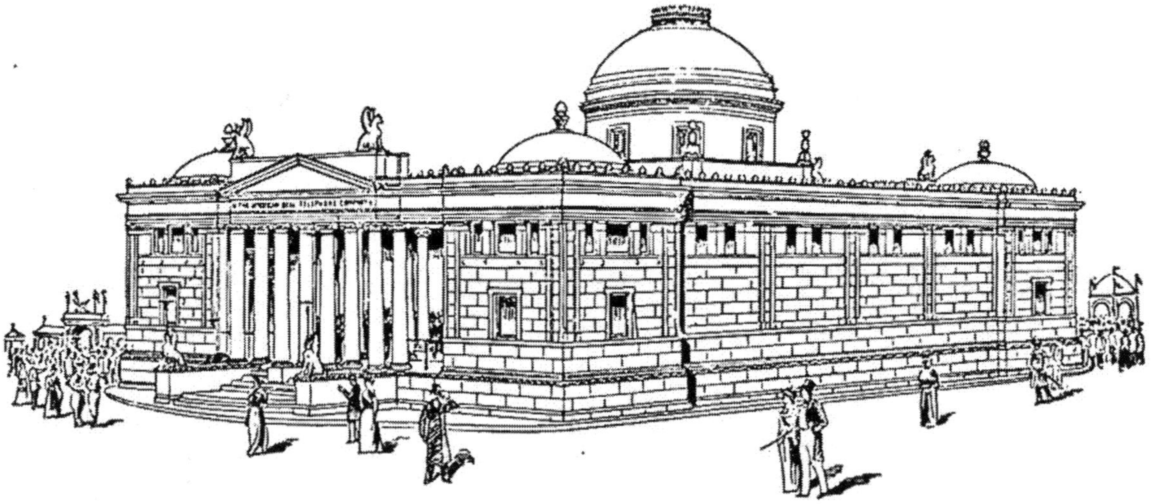

THE AMERICAN BELL TELEPHONE COMPANY'S EXHIBIT PAVILION.

board gave rise to the introduction of that most useful institution of modern times—the telephone exchange. The first switchboards were crude in design and limited in capacity. To-day, one of the switchboards in New York—that in the Cortlandt street office—is fitted for six thousand lines, and is so arranged that each operator attending to the calls of fifty subscribers can, in a moment of time and without leaving her place, connect the wire of any one of her allotted subscribers with any other one of the whole number of subscribers connected to the switchboard. The switchboard which did the work for the World's Fair was in the west peristyle of the atrium. It constituted for the time being one of the many branch offices of the Chicago Telephone Company. To this switchboard came the wires of over 350 telephone stations in the World's Fair, and during the time it was in operation it was the busiest telephone exchange in the world. Ample space and opportunity were afforded the public for witnessing the work, and attendants were at hand to give any desired information. At regular intervals a plain and concise description of the board and of its method of operation was given. The greatest interest was shown by the public in the novel spectacle of a telephone exchange in ordinary, every-day operation. With the rapid growth of the telephone business the problem of dealing with the enormous number of wires used became one of extreme difficulty. The highest technical and engineering skill has been brought to bear upon it. Aerial lines are now carried on large poles, in many cases reaching sixty feet in height, while in the principal cities the wires are, to a great extent, placed underground. In carrying out this work the telephone companies had no previous experience to guide them, and were obliged to expend large sums in experimenting before they succeeded in producing a type of cable adapted to the peculiar characteristics of the telephone current. The underground wires are placed in heavy lead-covered cables, drawn into conduits specially devised for the purpose. Within the World's Fair grounds all the wires were underground. At the southerly end of the west peristyle was an illuminated manhole, covered with wire grating, through which were to be seen a number of these cables, each containing fifty pairs of wires, entering through underground conduits, and terminating in what are known as cable heads From these the wires were led upward and again gathered into cables to connect with the switchboard. All of this part of the exhibit was an exact illustration of a modern telephone exchange, and demonstrated clearly the elaborate methods necessary in the arrangement of a large number of wires. The cables from the air-tight "heads" of the underground cables proper passed to a distributing frame by means of which connections can be readily made between different lines, then to a similar arrangement called the intermediate distributing frame, which allows of altering at any time the number of subscribers to each operator, and then to the switchboard itself. A telephone switchboard is built up in sections; in this particular board the first was the trunk line section. To this section came calls from other exchanges in the city of Chicago. The remaining sections were the subscribers' section proper. Here all calls from the World's Fair subscribers were received. The lines of the American Telephone and Telegraph Company—the long-distance company—also came to this switchboard, and here the necessary connections were made in response to calls to and from Boston, New York, Philadelphia, Washington, and other points reached by the long-distance lines. On the wall above and beyond this switchboard were three large photographs, showing the left, center, and right sides of the great switchboard at Cortlandt street, New York, already referred to, the largest of the nine exchanges in that city. This switchboard is two hundred and sixty-four feet in length and is of horseshoe form. Other photographs on this wall showed the switchboards in the main offices in Boston, Milwaukee and San Francisco. Upon the wall near the manhole were numerous drawings and photographs illustrating methods of underground construction.

CABLES.—A large collection of samples of underground cable, designed especially for telephone working, was an important part of the exhibit. These samples showed the various changes in construction and methods of insulation which have resulted from experience with many forms of cable and from prolonged study of the problem of underground telephony, a study involving the highest electrical knowledge. In the latest and most improved form shown, the wires are insulated from one another by wrappings of paper, and are laid up in pairs, each pair being separately twisted. These twisted pairs are laid up together in the form of a cable and are covered with an armoring of lead which must be absolutely air-tight along its whole length. It is found when the pairs are properly twisted and laid up that there is no induction or "cross-talk" between different circuits.

LONG-DISTANCE SERVICE.—At the rear of the Temple, on the easterly side, was a long-distance booth, an object which, within the last few years, has become very familiar in all the principal cities of the Union. How welcome it is to the busy man can only be judged by experience. The introduction of sleeping and dining cars but a few years ago saved time and strength to the active business man. It may be possible to calculate this saving, but the economy of time and the added efficiency in the transaction of business which the telephone gives can hardly be estimated. The manufacturer and merchant regulate the production and sale of their goods; large business enterprises, having establishments in different cities, practically unite them into one by inter-communication which is immediate, satisfactory, and confidential; the client consults his lawyer; the patient his physician; the sensitive ear of the leader selects a musician for the orchestra; the mayor of a city, absent from his post, keeps informed of its affairs; the politician keeps in touch with what is going on at the nation's capital, and the man who has sickness in his family, yet is forced to be away, talks with his wife and children, and knows on the minute exactly how things are at home. And yet, in each case, the distance between them may be more than a thousand miles. To test the ease with which one can talk over this distance, an opportunity was given for visitors to speak with an operator in one of the offices of the American Telephone and Telegraph Company, in New York, from the long-distance booth, every afternoon between three and five. This privilege was very largely availed of by visitors and it was evident that to most of them it was a decidedly novel experience. Near the booth, upon the

Block in the Power Plant.—Machinery Hall.

wall, was a map showing the route of the 1,200 mile line from Chicago to Boston, the longest telephone line in the world. There also were a number of photographs showing the construction of these lines as they actually appear upon the highway; and samples of the heavy, hard drawn copper wire used, were also shown. The lines which connect Chicago with Boston, via New York, are of copper wire of extra size. It is about one-sixth of an inch in diameter, and weighs 435 pounds to the mile. Hence each circuit contains 1,044,000 pounds of copper. A circuit of this length of the wire ordinarily used between cities less distant from each other, running up to distances of 500 miles, would weigh about 415,000 pounds, or less than one-half the weight of the circuits between Chicago and the East. The longer the line the heavier the wire required. The American Telephone and Telegraph Co. has in operation over 55,000 miles of copper wire. The lines now built and the principal local lines connecting with them were shown on two large maps, and many of those interested were surprised at the extent of the territory now supplied with long distance telephone communication.

BELL RADIOPHONE.

TELEPHONES.—In the east peristyle, in three double show-cases, were shown a large number of telephones, illustrating the development of the electric telephone from the earliest forms of the instrument up to the various standard forms now in use. Much interest naturally attaches to the earlier instruments, which are of great historical as well as scientific importance. There was also what remains of Bell's telephone of 1875. A complete reproduction, showing what it was, was placed by its side. Here, also, were Prof. Bell's Centennial instruments—the identical telephones which were used at the Centennial Exposition in 1876, when Prof. Bell, in the presence of Sir Wm. Thomson, Prof. Henry, Dom Pedro, and others, gave the first general public exhibition of the telephone. On the walls in this part of the building were instruments adapted to a great variety of uses. Standing in the corner was a diver's suit, in the helmet of which a telephone was fixed. It was connected by a rubber tube with the northerly end of the peristyle, where, upon a table, were the boat instruments used in connection with those in the helmet. The tube was the ordinary diver's tube used to supply the diver with air, and through it passed two wires, which rendered conversation possible between the diver, at the bottom, and his assistants in the boat above. Here, also, were several forms of telephones used on shipboard; also a telephone for use in mines. This last differs from the others only in special strength of construction, and in devices to keep out moisture. In the atrium were telephone sets designed for use by the army, for connecting the different parts of the field—for the navy, to connect ships and boats with one another and with the shore.

Speaking Tube Telephones.—There was exhibited a special switchboard, shown in actual operation, adapted for what is known as "speaking tube" service. The American Bell Company has paid particular attention to the use of the telephone in large buildings, such as hotels, office buildings and factories, in place of ordinary speaking tubes, and has devised a complete system of instruments, wiring and switchboards for this purpose, which has been adopted by many large concerns throughout the country.

Automatic Exchange System.—In the north and east peristyle were three telephone sets, so connected that from either of them any one of the others could be called at will without calling upon "Central" to make the desired connection. A number of small towns have been equipped with automatic exchanges of this kind, as many as forty-five subscribers being connected in some cases. For purposes of exhibition, only three subscribers' stations were installed, the principle of operation being the same for any number. In the north peristyle stood another sound-proof booth, which had a peculiar and fascinating interest for all scientists. It was used for the exhibition of Prof. Bell's radiophone, an instrument which transmits speech by means of a beam of light. The operation of the radiophone was shown by means of a transmitting instrument placed in a gallery one hundred feet distant from the receiver. The transmitter is a circular mirror about four inches in diameter, on which are condensed, by means of a suitable lens, the rays of light. The source of light may be the sun, or any powerful light, such as an electric arc lamp. From the mirror the beam of light, passing through the lens, is reflected to the receiver. This consists of a parabolic reflector, which focusses the beam upon a small glass tube, containing carbon in the form of lamp-black. To this glass bulb a rubber tube is connected, from which branch two smaller tubes to convey the sound to the ears. The person at the transmitter talks into a mouth-piece at the back of the mirror: this causes the mirror to vibrate, and those vibrations affect the beam of light in such a way that it causes similar vibrations in the carbon of the receiver, upon which it is focussed, and this, in turn, causes corresponding vibrations in the air in the bulb, reproducing the sounds delivered in front of the transmitter.

An extremely interesting example of the difficulties sometimes met with was illustrated by three large photographs showing the method of line construction in crossing mountains. These views were taken at different points along the Colorado Telephone Company's line through the Mosquito Pass. To maintain this line against wind and storm-

PROF. BELL IN NEW YORK TALKING TO CHICAGO.

especially in winter, it is necessary to use one hundred and fifty-nine poles to the mile, making the interval between them only a little over thirty feet. Where the line crosses the crest of the Pass it has been found necessary to bury it, as it is impossible to construct an overhead line that will stand up against the storms in winter. At the north end of the temple were large sectional diagrams, four in number, showing the principal instruments invented by Prof. Bell, in his solution of the secret of the electrical transmission of speech. These are the drawings shown in his patent of 1876, one of the Centennial instruments, the box telephone, having a horseshoe magnet, and the hand telephone in ordinary use. At the southerly end of the temple were four drawings of typical transmitters, illustrating in a striking manner the evolution of the powerful instruments in use to-day. These were the Berliner microphone, the Edison carbon transmitter, the Blake transmitter, and the long-distance or "solid back" transmitter. On the east side of the temple were a large number of photographs of buildings in the principal American cities, all of which have been specially constructed by the various telephone companies to comply with the requirements of modern telephone engineering. These buildings are all fireproof structures, and are designed primarily to contain the equipment of telephone central offices. Opposite these photographs, upon the west wall, was a large oil painting of Prof. Alexander Graham Bell, the inventor of the telephone.

Loud Speaking Telephones.—A very popular demonstration of the properties of specially constructed telephones was given. Within the temple, standing upon a pedestal, was an exceedingly delicate transmitter. This instrument is so sensitive that it took up, and conveyed to the listener at the other end of the wire, words spoken in an ordinary tone of voice in any part of the temple. The listener, who, for convenience in this experiment, was in a near room at an ordinary telephone, could answer in a low tone, and his words were reproduced by the loud-speaking receivers which hung in the temple immediately over the transmitter mentioned. The voice was so magnified that it filled the entire temple, and what was said could be distinctly heard by every one present. Another very popular part of the exhibit was the transmission of music from different parts of the Fair. Upon either side of the temple were placed a number of small telephones. These were connected with the German Village in the Plaisance and with other points, whence music was transmitted at stated times. It could be distinctly heard by holding these small telephones to the ear.

Business Statistics.—Upon the walls of the east peristyle were large diagrams giving statistics illustrating the growth and development of the telephone business in the United States. These related to miles of wire in use, miles of wire underground, the number of telephones in use, the number of subscribers, the number of connections throughout the United States per annum, the number of exchanges, and the number of employes in exchange service, for each year since the establishment of commercial telephony. They showed how rapidly the telephone, which had no commercial use fifteen years ago, has come into prominence as a factor in human affairs. For instance, in 1881 there were 47,880 telephone subscribers in the United States; at the end of 1892 these had grown to 232,140. In 1884 123,000 subscribers exchanged 215,000,000 conversations, or an average of less than six per day per subscriber. In 1892 there were 232,140 subscribers who exchanged over 600,000,000 conversations, an average of nearly nine conversations per day per subscriber. In eleven years, then, the number of subscribers increased over 500 per cent, and the use of the telephone, *by each individual subscriber*, increased over 50 per cent.

WESTERN ELECTRIC COMPANY.

TELEPHONE SWITCHBOARDS.—The almost endless variety of telephone appliances now in daily use was a surprise to the average visitor of the World's Fair. Never before has the public had such a chance to examine the appliances in use for this work. The telephone receiver has been a mystery to him, and the transmitter and magnetic call bell have been literally locked up so that he could not examine them; while, as far as switching apparatus is concerned, he has never formed a definite idea of what takes place in that mysterious and much-abused place called "Central." In one corner of the space allotted to the Western Electric Company was shown almost every instrument or piece of apparatus used in telephone work. Near the entrance to the "Scenic Theater" were the magneto call bells and allied apparatus, and at this point an examination of this very comprehensive portion of the exhibit should begin. No attempt was made to conceal the apparatus or even to put it out of reach; the many types of magneto bells, transmitters and telephones were within reach and could be examined and handled by all. Indeed, the intention seemed to be that the bells should be handled, as the lids were not locked and were arranged so as to lie open. All the familiar types of magnetos were displayed. Telephone men easily recognized the "No. 1," "No. 2," and "No. 3" Chicago magnetos, the "bridging" magneto, with its high wound ringers and ingenious arrangement for bridging the generator in on the line; a bell with two sets of ringers for private line use; the old "Gilliland" magneto, and many special forms of magnetos. There was one magneto designed on the "nickel-in-the-slot" principle, for self-collecting toll work; also an instrument designed for either alternating or direct current. The Antwerp branch of the Western Electric Company sent a very fine contribution to this portion of the exhibit, and a comparative examination of the domestic and foreign made apparatus was interesting. In one case in particular where two complete telephone sets—one of American, the other of European make—lay side by side on the counter, the difference could be noticed. The Antwerp factory seems

Telephone Switchboards. — Western Electric Company.

to have preferred to punch and form the cog and pinion wheels, while the American shops cast these parts. The finish on the foreign made instruments is better than on the domestic, which may possibly be explained by the fact of the latter being exactly such apparatus as is shipped every day; no extra finish was put on. The Antwerp instruments were in most cases provided with telephones and transmitters, and in some cases a noticeable departure from American custom, two telephone receivers. It is worth while to notice, in passing, the small receivers used abroad. They are about two-thirds the size of our well known instruments, and are of the "bi-polar" type. Even smaller than these are the "pony" receivers used where there are two on a set. These latter can be entirely covered by the hand and are so light as to be extremely convenient. The foreign exhibit included some desk sets in which the calling and signal apparatus were arranged on a neat box, with a Hunnings type of transmitter and the usual two receivers. Such a set enables one to place the entire set on his desk, and is movable to a certain extent. The "Ader" transmitter seems to be somewhat in use abroad, as several samples were displayed, much to the curiosity of the American telephone user. Not satisfied with having shown the telephone, transmitter and magneto bell complete, the Western Electric Company with great care dissected each of these instruments and mounted the parts on boards or glass plates in a show case, to educate the public into the mysteries of instrument construction. The 330 complete parts of the magneto bell, the 88 parts of the Blake transmitter and the 28 parts of the Bell telephone, were all displayed, although to the unskilled to assemble these parts would no doubt prove a Chinese puzzle. The Western Electric Company manufactures for the American Bell Telephone Company the telephones and transmitters used throughout the United States, and a very complete assortment of these instruments was displayed in a show case near the apparatus just described. The familiar Blake transmitter, the ordinary Bell receiver, the Hunnings transmitter, the new "solid back" transmitter, the long-distance receiver, and the various forms of switchboard transmitters and receivers, were among the articles displayed in this case. The small transmitter and receiver, together with the compact stand and call bell for use on short lines (speaking tube system), were worthy of a careful examination. Many forms of extension bells and special switches for use in connection with them, together with generators and their parts, went to make up a nicely arranged show case. Leaving the strictly telephone portion of the exhibit and coming to the switchboards, there were cases filled with the many forms of "drops," "springjacks," "plugs," "condensers," "retardation coils," "listening and ringing keys," and the many parts of complicated apparatus used in making up a switchboard. It must be confessed that to the average visitor, even if he be fairly well posted in electrical matters, but not a practical telephone man, these parts were more confusing than instructive. They served to show, however, the many intricate details of the work. The earliest kinds of switching apparatus shown date back to about 1883, an ancient date in this recently developed business. Previous to that time there were switchboards, but the models have not been preserved. The 1883 model of "standard" switchboard was an exact reproduction of the boards in use at that time; the spring-jacks were large, of cast brass; the annunciators were also heavy and the keys and plugs were of what now seem ancient designs. By the side of this board stood a "standard" switchboard as made in 1893, and used in many small exchanges throughout the country. The difference between the two styles is less than might be expected, and goes to show that in the working of small exchanges quite a degree of perfection was reached at an early date. There were small switchboards of more elaborate design for use on metallic circuits, as well as on strictly "long-distance" lines. In these last the care to secure good insulation was noticed, one line requiring two springjacks, one for each circuit. Numerous small "toll" boards, and boards for special work, together with an elaborate desk and switchboard for use in a building having an exchange of its own, completed the smaller boards and brought the visitor to the large "multiple" switchboards. In arranging the exhibit of multiple switchboards the Western Electric Company tried to show the evolution that has taken place since 1884, the time when the multiple principle was first applied to this work. This company invented and perfected this form of apparatus and has manufactured it for the telephone companies of this and other countries, and took much pride in arranging an historical exhibit. The models shown were full size and exact duplicates in every respect, except perhaps the kind of wood used in some cases, of the switchboards made at various times. The oldest form, 1884, had large springjacks and annunciators, while the cords came down from overhead. It is needless to say that the operator at this board worked in a standing position, as when seated but a small portion of the frame could be reached. There are exchanges of this type still in use; indeed, while this exhibit was being prepared, an exact duplicate of the switchboard on exhibition was sent to Austro-Hungary to add to the exchange at Buda-Pesth. That this style did not meet with entire satisfaction is shown by the next model, of very little later date but radically different. The need for reducing the size was exemplified. The springjacks and annunciators were much smaller than in the first example; the cords came up from beneath, and, although the annunciators were above the springjacks, there was an opportunity for the operator to sit while at work. This is the form of board now in use at Nashville, Tenn., and its working has been most satisfactory. The next model was that known as the "Antwerp" style, for the reason that such a frame was sent to Antwerp for use there. The use of local (answering) springjacks was here noticed, and a different arrangement of the keys and cords made, the cords being on a shelf above the annunciators, with the keys below. This pattern was a favorite one for several years, and exchanges of this style are found in many cities; as for instance, San Francisco, Omaha and Cincinnati. That this style meets with favor still is shown by the fact that a switchboard of this type is now being made by the Western Electric Company for Para, Brazil. In all the switchboards on exhibition, with one exception, the two-cord system was used, i. e., one cord is used for making connection with the calling party, and a second cord is used for connecting with the party called. The exception was the model "single cord" board, being similar to boards now in use in Cleveland, Portland Oregon, as well as in many European cities. In this system there is one cord for each subscriber's line, making the connecting process a short and simple one, as but one cord has to be handled to connect the desired parties. The chief disadvantage of this system was in the fact that for each line there were, necessarily, a cord as well as a somewhat complex key or switch, increasing to some degree the liability for trouble from bad connections. The model was of interest, however, as showing the tendency at one time in switchboard construction. With the introduction of metallic circuits a new plan of switchboard construction became necessary. The single wire system for keys and switches had to be changed for double wires. The frames of the boards also became somewhat changed, as may be seen in the two models of about 1888 and 1889 for Buffalo, N. Y., and 38th street, New York City. The keys on those are quite complex and the use of tubular drop annunciators is made. Double conductor cords and plugs are also used. Two models are yet to be noticed before the "Model Exchange" is reached—one of them for a small exchange such as are much in use throughout New England, and the other a large sized board like some shipped to Liverpool, Eng., in 1891. These were also on the metallic circuit plan. In the exhibit of the "Model Exchange" the entire course of a telephone line was shown from the point where it enters the exchange through an underground cable to the switchboard itself. One hundred pairs of underground cables came up through the floor and terminated in standard form of iron box terminals, from which Okonite cables led to the distributing frame. Here "jumper" wires connected the lines to the strong current protectors, from which cables led to the intermediate distributing board, thence to the multiple springjacks and the answering jacks and drops. One and one-half sections of multiple switchboard of the latest form were connected up as in actual practice. There were 300 of the new self-restoring annunciators, and the board was of the full height and one-eighteenth the length needed for an exchange of 5400 subscribers. In the face of the board, in a space of 69⅜ by 26 inches, the entire 5400 lines and 240 trunk lines, a total of 5640 lines, is reached. As each line through the switchboard requires three conductors, the mass of cable at the back of a sectional area of 26 by 6 inches contains 16,920 separate conductors. When it is understood that each of these must be entirely separate and thoroughly insulated from the others, the great care needed to erect and maintain such apparatus becomes apparent. The front of the board was designed with especial care to make the operator's work as easy as possible. The connecting cords and keys are on a shelf within easy reach, and while the annunciators are above the springjacks; they are arranged on a terrace-like frame so the upper row can be distinctly seen. Three operators have to work within a space 69⅜ inches long, and each operator in this particular board would attend to 100 subscribers. A feature of this new board is the form of annunciator used. It has a double magnet coil, one operating in response to the call, throwing open a shutter, the other operating from a battery current when a plug is inserted or a switch thrown, restoring the shutter. By its use the operator need never touch the annunciator, as the process of answering the call automatically restores the annunciator to its original position. The keys or switches are arranged on a sort of trap door, so that they can be inspected and repaired with greater ease than in the case of the old switchboards, where the only way to reach these parts was from beneath. A working switchboard of this type, of almost the same details, was in the exhibit of the American Bell Telephone Company's booth. The Western Electric Company designed and installed that plant for the use of the telephone service of the World's Fair. It was operated by the Chicago Telephone Company. Switchboards of this type have lately been sent to Zurich, Switzerland, to London, England, as well as to Cincinnati, Ohio, and Detroit, Michigan. One very noticeable feature of all the switchboards on exhibition was the great care given to the finish of the woodwork. The finish was not a highly polished one, but a dead hard finish, well fitted to stand the hard usage of the exchange room. Mahogany seems to be the favorite wood for this purpose, making an exchange room present a handsome appearance. The frame itself of the modern boards is of iron, giving strength and allowing the sizes to be reduced. This allows the length to be reduced from the old standard of 76 inches to 69⅜ inches for a full size three-operator section—a considerable reduction where every half-inch saved in the distance necessary to reach is of importance. In addition to all the features noted there was a display of telephone and connecting cords and plugs, also lightning arresters and strong current protectors of various kinds. Several patterns of distributing boards and exchange furniture, as inspector's desks or chief operator's desks, were also displayed. Many diagrams of the circuits used on the various switchboards were displayed, by the help of which the workings could be understood, and the evolution of the switchboard from its simplest form to the present complex apparatus followed. The entire display of telephone appliances was worthy of more than a passing glance even from the average visitor, while to the telephone man the careful study of the various pieces of apparatus could not but be of profit.

THE STROWGER AUTOMATIC TELEPHONE EXCHANGE.

In this exhibit it was intended to represent an automatically working central station of a telephone system, and connected therewith were nine telephone stations outside of said area, and one telephone within, making ten in all. They were so connected up that communication could be held between any two of the ten 'phones. The space was fitted up as a central telephone office. Upon a frame work of shelves were one hundred switches, but only ten were connected up, each one with its own particular telephone as regarded manipulation, but together for telephonic service. The object of these switches was to accomplish by mechanical means the work done by the operators at the central office—in other words to dispense with the services of the telephone girl—and is described as follows: The Strowger automatic telephone exchange is a combination of old principles to accomplish a new result hitherto deemed by telephone experts as among the impossibilities. As its name designates, its purpose is to obviate the necessity of an operator at the central office; to place the subscribers in communication with each other, each subscriber performing this work himself at his own telephone. The Strowger switch is governed mainly by the underlying principles of the old Morse telegraph; that is, magnets, armatures, levers and electrical energy, to which is added an old principle of pawls, ratchet wheels and shafts, together with a contact arm. The magnet, with its attendant armature, lever and pawl, are so constructed as to be considered as only one element, and as such they will be treated. The Strowger system is decimally operated, therefore is intimately related to the Arabic notation of numbers; that is to say, there is a units movement, a tens movement, a hundreds movement, and so on. A circular hard rubber disk is used to hold the contact points of the terminal wires of the various telephones connected with the system. These contact points are arranged in circles, one hundred in each circle, and as many circles as there are hundreds of patrons. In the

Bell Telephone.—Central Station.

center of these circles and at right-angles with the disk is a perpendicular shaft to which are attached a pair of "units" ratchet wheels, a pair of "tens" ratchet wheels, a "hundreds" and a "thousands" arm, if so many are needed. One of each pair of ratchet wheels is for the purpose of locking and stopping the motion when the pawl has reached its ultimate; in other words, to overcome the momentum. This device is such that, which ever movement is used, the ratchet or rack is firmly clasped and held so that the contact point of the arm is above the contact point of the disk. When the circuit is broken the contact point of the arm drops on the contact point of the disk, completing the telephone circuit. On the telephone shelf in the patron's office is a keyboard having a release key, a units key, a tens key, etc., connected with its corresponding magnet of the central switch. It will be seen that the shaft and contact arm have a rotary motion, and if more than one hundred telephones are in the system the contact arm has also a longitudinal motion. The rotary motion is for the units and tens, and the longitudinal motion of the arm is for the hundreds and thousands. The units ratchet wheels have one hundred ratchet teeth to correspond in number with the one hundred contact points on the disk, so that for each closing of the units key the contact point of the arm moves from point to point on the disk; for each closing of the tens key, the contact point of the arm jumps nine contact points at a time, touching every tenth contact point; and for each closing of the hundreds key, the contact arm and point are moved outwards from row to row, and so on. A patron wishing to converse with any other patron through his telephone, say with 432, he will firmly press his hundreds key four times, his tens key three times, and his units key twice. His telephone is then connected with telephone No. 432, for the contact point of his switch arm at the central is on the four hundred and thirty-second contact point on the disk. The patron calling then rings his bell, which also rings the bell of 432, but no other. This signals the party wanted to come to his 'phone. After conversation is ended, the earphones are hung up and the calling patron presses his release key for about one second, allowing his contact arm at the central to resume its normal position. An automatic release, accomplished by the hanging up of the earphone, is attached to telephones when desired. To overcome the friction which would arise from moving the contact point of the arm from point to point on the disk, a device is used to lift the shaft and arm from off the contact point of the disk, which ever motion, longitudinal or rotary, may be given to the shaft or arm. This is done by a peculiar method of wiring, by having a lift magnet between the other magnets and the battery. The return motion of the shaft and arm is accomplished by a weight and cord, or by a spring and fuse. This return motion is necessary in order to have the contact arm at the starting point or normal position before commencing to call another patron, therefore, it is necessary that when conversation is ended the release key should be pressed. The automatic releasing device is now put on all telephones, as stated above, which renders it unnecessary to think about pressing the release key. At the central office there are on shelves as many switches as there are telephones, one to each telephone, so that each patron controls his own switch and no other, therefore no confusion can arise. The aforementioned wire terminals are legged or attached to corresponding numbered bank wires. Thus, wire terminal No. 1 of every disk is attached to bank wire No. 1, wire terminal No. 2 to bank wire No. 2, and so on. The normal wire terminal (the wire terminal upon which the contact arm rests at starting—in other words, the initial point) of each switch is attached to the bank wire numbered the same as the switch; for example, the normal wire of No. 8 is attached to bank wire of No. 8. It is through this normal wire the patron is called up. If a patron so desires he can have his normal wire attached to the frame of the switch. This will enable his telephone to be called up whether it is in use or not (this might be desirable for physicians), or the normal can be attached to another 'phone, to be used as a receiving telephone—that is, a transmitting telephone, to be used whenever another patron is to be called, and the receiving telephone is used whenever another calls. There are various methods of wiring the bank wires, and also the manipulating wires. Thus, there may be a single wire system, a two-wire system, a three-wire system, etc. Also, there may be more than one of each. Each has its advantages, likewise its objections. The one to which preference is given is termed a two-wire and twin battery system. This furnishes a secret service and at the same time a metallic circuit. This system of wiring is preferable when secret service is desired, and in towns where high tension currents are used, as electric lights and electric cars. The Strowger system is applicable to trunk line service, with corresponding cheapness. Only one operator is required for a number of towns, which operator is needed to keep an account of the toll.

THE ELGIN TELEPHONE COMPANY.

The exhibit of this firm was an acoustic telephone system, a model of the installation being on exhibition. The instrument is so constructed that the call, receiver and transmitter are all in one. It has no electrical connection, hence is for use strictly on private lines and limited to short distances. The instruments are made wholly of metal and are self-adjusting, requiring practically no wall space to rest upon. The line for use by this system must be strung so that the movements of the diaphragm will vibrate the whole length of the line. Hence the line cannot be laid in moulding or on a fixed support, but must be suspended or set on light springs.

GENERAL ELECTRIC COMPANY (LIMITED).

This exhibit contained telephone apparatus for central station uses as well as for isolated plants or stations, as large offices, factories, etc., where they are intended to act as substitutes for speaking tubes. The company has a thoroughly worked out system and in consequence thereof has supplied, and still is supplying, the British Government, General Post Office, railway companies, and other prominent corporations with these devices. The receiving instrument employs usually a horseshoe magnet instead of the more familiar bar magnet, the two poles of which are brought close together, and around each of which is wound the exciting coil. The instrument itself, instead of being wholly of hard rubber, is of metal and covered with a sheet of gutta percha. The cup is wholly of brass, and the seat upon which the diaphragm rests, holds a thin zinc ring loosely laid on and the diaphragm is brought into contact with this by a screw on the metal face plate. There were shown several complete sets of very small instruments whose receiver is known as "watch receiver." This name is significant in expressing the favorably small dimensions, which are slightly larger than an ordinary large watch. In this instrument the magnet has also a horseshoe shape, occupying the inner circumference of the box, the two poles being turned in and then upward, so as to be central to the metal diaphragm.

The transmitters are the well known Hunning cone transmitters, which provide every instrument with a receiving cone in front of the diaphragm, and the latter is protected by a gauge to prevent its being injured by either tapping with the finger or pencil. In this transmitter granulated carbon is used between two upright diaphragms.

MECHANICAL ADJUNCTS TO TELEPHONE SERVICE.

O. L. WULLWEBER.

TELEPHONE STAND AND ARM REST.—This apparatus consisted of an adjustable stand attached to the telephone, by means of which the telephone user could have the use of both hands while using the instrument. There is a holder for the receiver that remains fixed at the user's ear, when once placed, and a little shelf for taking notes. The exhibitor claimed for his apparatus as follows: The apparatus does away with the tiresome and often painful holding up of the receiver to the ear. It gives one the use of both hands for twisting and handling papers, books, etc. It is equally comfortable for tall or small persons. It enables the operator to write orders, etc., direct into his order book, thus saving time usually consumed by copying from slips. It enables the operator to write on book or paper at a proper distance from his eyes. It enables the operator to sit down comfortably at his table or desk and converse or write while sitting without being obliged to hold the receiver to his ear by hand. It enables the operator to converse in a low whisper, by means of the extension mouthpiece, as the sound waves are carried from the mouthpiece through the elastic tube, directly to the center of the transmitter. It secures a perfect understanding by the person spoken to, as the extension mouthpiece is so arranged that when the ear of the operator is at the receiver, his mouth is always directly opposite the mouthpiece; the slightest sounds are thus accurately transmitted. It secures, as far as possible, the correct understanding by the operator of the words spoken by the person with whom he may converse, for, as the muscles of the neck, unlike those of the raised arm, do not become tired or painful, the operator can maintain, for an indefinite time, a close connection of his ear with the receiver, thus excluding all other sounds from his ear. By the conveniently arranged extension note or memorandum holder, the operator is released from the trouble of holding up notes or memoranda, which he may desire to transmit. By means of the segment, the apparatus can be placed in any desired position and can also be so arranged that it can be used for the right ear as well as for the left.

WILLIAM N. MARCUS.

AUTOMATIC CONNECTION FOR BUSINESS HOUSES.—This exhibit was an invention, and model for a mechanism providing means by which one telephone in a building may be used by persons in various rooms or offices without leaving the room. The object of the invention is that one telephone may be made to communicate with several rooms in the same building, so that communications may be received and sent from a room in which no telephone is located. Speaking tubes are placed throughout a building, running from a central office in the building. A circuit wire is placed in connection with each tube, and in the guest room if in a hotel, or office, is a telephone receiver, which, when not in use, hangs from a peg. If this apparatus is used in a hotel and a call comes to the office telephone for a guest, the person in the office calls through the speaking tube to the guest's room that some one desires to speak to him over the telephone. The guest has only to take the receiver from the peg in his room and carry on a conversation the same as if he were directly at the telephone; proper connection has been previously made in the office. A guest may also send a telephone call by requesting, through the speaking-tube to the office below, to be connected, and, when the connection has been made, carry on a conversation as already described. In large office buildings, as well as in hotels, this invention may be used advantageously, and, where employed, one telephone can be used by all the various offices, thus avoiding the necessity for every office having a separate telephone. There was also in the exhibit auxiliary mouthpieces, and supports for telephone receivers.

M. E. MERCADIER.

Bi-telephone.—The Mercadier bi-telephone is a head telephone device, allowing, therefore, to write with the right hand. The left hand can be employed to work a commutator, the connections of which with the microphone and bi-telephone allow to bring down, nearly to one-half, the resistance and the self-induction in the two telephone stations.

Microphone.—This microphone is formed by means of carbon rods. The microphonic contacts occur only through the superior part and all the microphones move without difference of phase. A special disposition allows to change at will, mechanically, the points of contact. Another disposition which may be applied to all microphones, permits their use for either long or short distances.

TELEPHONE SERVICE OF THE COLUMBIAN EXPOSITION.

The need of a telephonic service for rapid and accurate communication was felt early in the beginning of the work at the park. At first, the necessary instruments were attached to the Oakland Exchange of the Chicago Telephone Company. There was also a private line, supplied with long-distance American Bell instruments, connecting the first Service building at 62nd street to the main offices of the company in the Rookery building in the city, but as the office forces from the city were removed to the park an increase of instruments was required in order to establish communication with the city, and a contract was made with the Chicago Telephone Company for the establishment of a separate exchange upon the grounds, for the express service of the various offices, and with trunk lines connected to the various other exchanges of the Telephone Company. It was also found necessary to have a telephone system by which communication could be carried on within the limits of the grounds, entirely separate from the Chicago Telephone Company's system, and so, on the 20th of April, 1892, a contract was drawn up between the Chicago Telephone Company and the Exposition Company, and duly signed by the representatives of both. By the terms

Bell Telephone Exhibit.—South Front.

of this contract, the Exposition Company was allowed the use of 300 telephones and transmitters for its private use within the limits of the grounds, and also in connection with its fire and police alarm system. In consideration of this, the Chicago Telephone Company was given full and exclusive right and authority to construct, maintain, and operate an exchange in the grounds under certain conditions and to rent long-distance telephone instruments connected with this exchange to parties at the rate of $125.00 for a one year's, or $100.00 for six month's, service. They were also allowed space in the various main buildings for telephone pay stations and district messenger offices, and were allowed space free of charge in the old Service building for offices and the requisite requirements of service, said space, exclusive of the store-room of apparatus, wire and material, not to exceed 900 square feet of floor space. In November, 1892, the World's Fair Long-Distance Telephone Exchange was established, temporarily in the old Service building. It was but a small board, adapted for but a few instruments. This exchange, with some enlargements for new instruments, remained in service until a new and permanent switchboard was installed in the American Bell Telephone Company's exhibit building, in the south center of the Electricity building. The wires for this temporary exchange were run as best they could be with the means at hand. No. 2 double and twisted Okonite conductors were used, no cable work at all being attempted. They were run over buildings, on poles and in any spare conduit that had been laid for the fire alarm and police telephone service.

Early in the spring of 1892, cable work was started for the permanent exchange, and was pushed rapidly. These cables were run in conduits laid by the Exposition for possible emergencies on the fire and police signal system, in the main electric subway, and on the Intramural railway structure, the cables being supported on the ends of the sleepers of the elevated track by broad zinc straps.

The permanent cables were Western Electric paper telephone cables, and were of three different sizes, 100 pair, 50 pair and 25 pair, the larger sizes being laid near the exchange. The wires within them were twisted in pairs to avoid induction and were insulated by paraffine paper. For mechanical protection they were covered with a thick coating of lead, metallic circuits and long-distance telephones only being used. Iron terminal boxes were located at various points in the main buildings, under floors of the buildings and also on the Intramural track in several instances. The locations were so arranged that a new subscriber in any portion of the grounds could be easily and quickly reached, without running any very great amount of extra wire. There were over five hundred miles of wire in these cables as they were laid. All the cables terminated at the central exchange in the Electricity building which was known as the World's Fair Exchange of the Chicago Telephone Company. Entering Electricity building mostly through the subway, they were run into a large brick manhole which was built under the floor of the building after the most approved designs. They were terminated in a rack in this manhole containing iron cable terminal boxes, in which each wire of the cables was numbered, so that any particular conductor in any cable could be easily found. From there the wires were run in smaller cables to the main distributing board, where they were all spread out so as to be easily accessible. From this last board they were taken to what is called the intermediate distributing board, and from thence to the main switchboard itself. This switchboard occupied the whole west peristyle of the American Bell Telephone Company's building. The description of the service switchboard will be found elsewhere. The board gave the best of service, conversations being carried on over it to New York City, Boston and other eastern points as easily and as distinctly as to points within the city, and the service to the city was exceptionally good.

LOCAL TELEPHONE EXCHANGE FOR THE SERVICE OF THE EXPOSITION.

A private telephone system within the limits of the grounds for the private use of the Exposition employes was an early requirement. One hundred and fifty telephones and transmitters were set aside for use in police boxes and the rest were used as required about the grounds. The first instruments were placed at the various gates of the park, seven being on gates, and a private line run between the temporary power house and the office of the electrical engineering department in the Service building. The telephones on the gates were connected in two circuits, north and south, the instruments on each circuit being in series, the two circuits terminating in the fire alarm office at the Service building, and they were used to notify the men at the gates of the location of any fire, so they could inform outside fire operators entering the grounds which roads to follow in order to reach the location. Early in August, 1892, three cables were installed in the Service building, a 24 pair, a 12 pair and a 5 pair cable, and telephones were installed in the various offices of the buildings. These cables were installed and connected to a small telephone switchboard, located in the fire alarm office, and connections were made and calls answered from there. Telephones were also placed at various points about the grounds during the fall and winter of 1892, as they were required, and temporary lines were run to connect them in service, using a double and twisted police wire. To economize material as much as possible, the instruments were connected in series on the same line, and although these lines were only temporary affairs and were constantly in trouble from grounds, more or less, they gave very good service under the circumstances. Early in August, 1892, plans were made for a permanent system of telephones and a map of the grounds was laid out showing routes and lengths and sizes of the cables. The plan called for a total of 701,500 feet of double conductor in the cables. Specifications were sent out and a contract made with the Safety Insulated Wire & Cable Company of New York City, for about 750,000 feet of double conductor, on a rental basis of 1.4c. per foot. The telephone cables arrived early in March, 1893, and they were installed as per the terms of the contract as rapidly as possible. The switchboard for this local exchange was situated in the fire alarm office at the Service building, and presided over by one operator who worked the board and answered calls. All ringing on the board was done by a small magneto machine, mechanical power being obtained by a water-motor. The board was wired with 100 drops and jacks sufficient for 100 subscribers, but there was room enough on the board for the installation of double that capacity if it had been necessary. There were 89 subscribers' telephones connected directly to this board. They comprised all guard and ambulance stations, and the various offices of the departments of the Exposition. An addition to the local telephone system was the connection with the secret service department. This department had a small 45 drop board, located in the office of the chief of the department and connected with a system of telephones distributed over the grounds in various places for the express use of the secret service department and the department of admissions. These locations were at entrances and exits and the various sub-offices of the secret service department, beside which there were various instruments located in out-of-the-way corners, at points where it was likely they would be useful. This latter system was connected into the local exchange cable system, but the special conductors for this service did not run to the regular local board, but were connected by a 50 pair Jumper cable to the private 45 drop exchange in the secret service headquarters. There was a trunk line connecting the two local exchanges so it was possible to reach the secret service switchboard from any local telephone on the grounds that went to the central office. Besides all these local exchange 'phones, there were several private telephone lines connecting two or more isolated points independent of the central office.

CHAPTER XIV.

FIRE AND POLICE APPARATUS.

FIRE and police telegraph service is almost one with the history of the Morse telegraph.

Almost immediately after the first successes in telegraphy, it was pressed into service for fire alarm work and police service.

Prof. Moses G. Farmer, just recently deceased, and Dr. Wm. F. Channing, of Boston, devised a system which in 1851 was accepted officially for the city of Boston. One year later the system was put into actual service; since that time improvements have been made by various inventors, until now the service has reached a degree of perfection and usefulness not exceeded in any other branch of electricity. At the Columbian Exposition the apparatus exhibited was in every way typical of the latest developments. The exposition itself had in operation the latest improvements in this class of service, and never before has so important a territory as the Exposition grounds been so well patrolled for fire protection, or so fully covered for police service. Under the proper head at the close of this subject, this service will be treated somewhat in detail. At present, the various classes of apparatus will be discussed under firm names and headings.

THE GAMEWELL FIRE ALARM TELEGRAPH COMPANY.

The Gamewell Company, the pioneer concern in this business, made a comprehensive exhibit, covering all apparatus used for fire and police service. The exhibit was in actual operation in the Electrical building, and this same service gave the fire service for the Exposition.

THE AMERICAN FIRE ALARM AND POLICE TELEGRAPH.—The fire alarm telegraph consists of a central or battery station, the wire circuits which connect the central station with the street signal boxes and the alarm apparatus consists of electro-mechanical bell-strikers in church or other towers, electro-mechanical gong-strikers located in engine houses, public buildings and residences of fire department officials, and indicators, showing in plain figures the number of the signal-box from which an alarm originates.

THE CENTRAL OFFICE.—The central office, whether worked by operators or automatically, must have batteries sufficient to supply the necessary electro-motive force, and galvanometers for measuring the same, with the switch-board connections necessary to determine the location and character of hindrances to the reliable working of the system. Lightning arresters are indispensable, and where there is more than one circuit, a manual or automatic repeater is necessary.

THE MANUAL CENTRAL OFFICE.—In this system there are signal circuits and alarm circuits—the signal wires connecting the signal boxes into one or more signal circuits as may be desired, and the alarm wire connecting the bells, gongs and indicators into any required number of alarm circuits. As originally introduced, all alarms from the signal boxes were received at the central office only on call-bells, and also recorded by self-starting Morse registers on strips of paper. An operator in constant attendance re-transmitted these alarms, by setting the hands of a manual repeater to the proper number and putting the repeater in motion, which instantly and simultaneously sounded all the alarm bells and gongs on the alarm circuits. The operator, by Morse keys, also transmitted the alarm to a small bell in every signal box, and the same could be distinctly heard and understood by any person listening. The more modern method is to combine the several registers into a self-starting multiple pen register, one pen for each signal circuit. This instrument, by making all the records on one strip of paper, enables the operator, with greater certainity and less delay, to re-transmit the alarms by means of the manual repeater, consisting of one or more dials. This manual repeater operates as follows: Suppose an alarm has been received from signal box 154. The operator sets the index of the first dial to the left at figure 1, the second index at figure 5, and the third index at figure 4. He then releases the detent of the clock-work and the repeater, once started, transmits the alarm over all the alarm circuits, striking these numbers upon all the alarm-bells and gongs, and repeating it as many times as may be desired. By this it will be seen that the manual central office system requires the presence of an operator at all times. This fact—together with the possibility of error on the part of the operator in re-transmitting alarms—suggested the most important improvement which has been made in fire telegraphs since the original patents of Channing and Farmer. These improvements are comprised in what is called the automatic system, which is simple, yet perfect, and its introduction and opera-

Fire Alarm Telegraph Switchboard at the Columbian Exposition.

tion involve little expense compared with the benefits which it confers. By its use there is no possible delay in the transmission of alarms, the fire department being reached instantly by the party giving an alarm, without any intervention whatever. The automatic repeater is essential to the general use of the fire telegraph, and of the systems now in use, more than two-thirds are automatic as distinguished from the original central office plan. This principle has become so universally recognized that the so-called "Joker" system, invented by the Chief Electrician of Chicago, has been adopted by nearly all of the larger cities which originally introduced and are still using, with some modifications, the original central office plant.

THE AUTOMATIC CENTRAL OFFICE.—This office, in its simplest form, is furnished with a battery, lightning arrester, switchboards and galvanometer, located in an engine house, city hall, police station, or other public building, connected by telegraph wire with as many street signal boxes and gong strikers as the size of the village or town may demand. Bell strikers and indicators may also be embraced in the system. It may, if desired, have call bells and registers similar to the manual central office, but its essential peculiarity consists in the automatic repeater, used wherever the apparatus comprising the system is sufficient to warrant its distribution over two or more lines of circuits. This repeater was referred to by Sir William Thomson as the finest and most ingenious piece of telegraphic mechanism ever exhibited. It is so arranged that a signal on any one circuit is instantly repeated on all the others; but if the signal is caused by a break of the wire, the repeater, after sounding one blow on the other circuits, throws the one disabled out of service, leaving the others intact. As soon as the broken circuit is repaired, the repeater automatically takes it into service. The use of this instrument enables the operator to connect the signal boxes, bell strikers, gong strikers and indicators, indiscriminately and directly with each other, on two or more circuits through the central office, all the circuits acting and re-acting on each other without intervention, at the central office.

In case a battery becomes too weak to work efficiently, or an intentional or accidental interruption occurs to any part of the wire, in an instant notice is given by one blow upon all the alarm bells and gongs, calling attention to its temporarily disabled condition, thus, not only keeping watch over the city, but actually watching itself and guaranteeing reliability every moment. Both the manual and automatic systems require an arrangement for changing, testing and combining the various circuits into which a fire telegraph may be divided.

STANDARD GALVANOMETERS.—These instruments are adjusted to a standard fixed after the most careful tests, and are perfectly uniform and reliable. They indicate at all times the electro-motive force of the batteries, and the exact electrical condition of the circuits by the deflection of a magnetic needle, as plainly as the gauge on a steam boiler indicates the pressure of steam by its index. In the larger cities a rheostat combined with the galvanometer is found to be invaluable to those in charge of telegraphs for testing circuits, apparatus, etc., and in locating breaks and ground escapes.

The Battery.—The form of battery known as the gravity, or sulphate of copper battery, is the best adapted to the requirements of a fire telegraph, it being so simple that a fireman of ordinary intelligence can take charge of it. It is easily fed and readily cleaned.

Construction of Circuits.—Much of the apparatus must be placed out of doors, exposed to wind and rain, and the constantly recurring changes of variable climate, hence the finer parts must be substantially made and carefully protected. Portions of it may stand unused even for months, but when needed it must respond quickly and correctly. Metallic circuits are always demanded, never using the earth as any part of the circuit except temporarily in case of a break in the wire. In such cases this system of building affords facilities for the immediate restoration of the electric circuit through the earth. The lines should be built with the very best quality of galvanized iron or hard-drawn copper wire, weighing about three hundred and twenty-five pounds to the mile for iron, and one hundred and seventy for copper, every joint soldered, thoroughly insulated, and secured to poles not less than twenty feet above the ground. Where the circuits leave the central office, and enter bell-towers, engine houses, or street signal boxes, iron wire should be well soldered to thoroughly insulated copper wire leading to the apparatus. All street signal boxes are of cast iron, cottage shaped, and contain clock-work with spring or weight motors, so arranged as to open and close an electric circuit a definite number of times at certain intervals, indicating, by the number of blows upon the alarm bells, and the intervals between, the exact location or number of the signal mechanism in action.

Improved Non-Interference Signal Box.—The clock-work is enclosed in a dust-tight, cast iron, round box, which is placed in one corner of a square iron case having a hinged door and lock, with key under the exclusive control of such city officers as may be authorized or required to give special signals. This case contains, in addition to the round box, a lightning arrester, also a telegraph key and call bell for engineers or police signals; thus every signal station may be used as a telegraph office for any city purposes. A small starting lever, spread into a thumb piece or hook on the outside, passes through the door of the case and projects through a closely fitting aperture into the round box, where its end rests immediately under the detent lever of the signal mechanism. Lifting the detent lever, by pressing the starting lever lightly downward once, sets the signal mechanism, and through it the whole system of fire telegraph, in operation. The outside boxes are locked with combination locks, the keys of which are placed in the hands of the police and one or more responsible property holders in the immediate vicinity of each box. It will thus be seen that the mechanism is secured inside of the inner of three distinct iron boxes, and that opening the outer door reveals nothing but a square iron case with a thumb-piece projecting from its door. So perfect is this mechanism that it is utterly impossible for any other than the correct signals to be transmitted, and when once started it can in no way be interfered with, not even by the pulling of another box, until it has performed its entire work.

THE SECTOR SIGNAL BOX.—The Sector non-interference signal box, as exhibited, is the style of box best adapted to sections of the country where excessive and long-continued cold prevails. These Sector boxes may be made to run with either weight or spring. The distinguishing feature between them and the regular non-interference boxes is that these have a much longer pull, and are wound by the pull at each alarm of fire; while the regular boxes are wound up like a clock, and need but the slightest touch of a finger to set them in action.

MAGNET SHUNT.—This is a simple arrangement for cutting the entire box out of circuit when the outer door is closed, making an additional protection against damage by lightning, and also saving battery.

PATENT TEST SWITCH.—This is an ingenious double switch for testing the electrical, as well as the mechanical, condition of the boxes at any time. When used, the operation of the break wheels, instead of opening and closing the main circuit correspondingly with the number of the box, only shunts the bell magnet in the box in and out, thus tapping the number on the little bell and efficiently testing the electrical condition of the break wheel and line, as well as the mechanism of the box, without giving any blows upon the tower bells or gongs.

LIGHTNING ARRESTER AND KEY COMBINED.—This is a very simple and compact arrangement for discharging atmospheric electricity, and grounding the circuit on either side of the apparatus, or cutting it out entirely by the use of the plug normally placed in the middle of the ground plate, combined with a simple break circuit key, which, for greater safety, may be always cut out by the plug, except when wanted. By this same firm were also exhibited electro-mechanical gongs and bells in a great variety of forms.

THE POLICE TELEPHONE AND SIGNAL COMPANY.

The combination of the telephone and telegraph, as an auxiliary to the police, was first introduced in Chicago in 1880. Placed at the outset in the most turbulent district of the city, it so speedily increased the efficiency of the "force," by enabling it to concentrate promptly at any needed point, that within a few months the district was as easily cared for and protected as the average districts of the city. Patrolmen soon learned that in cases of necessity they, or some one for them, could literally, with the rapidity of lightning, summon assistance from the nearest station, and that they could reckon with certainty on a response, and the criminal and riotous discovered that there was little chance for them where electricity was utilized so successfully in aid of law and its agents. The practical working of the experimental system in this one district in Chicago was so satisfactory that its extension into other districts was generally demanded. The demand was met by the city government; and at this time Chicago has a very complete system of police patrol service, with about 1,000 street stations, and many hundred private boxes from which instant communication can be had at all hours of the day or night with all stations of every precinct.

Description of Street Stations and Boxes.—Radiating from each police station are telegraph lines or circuits connecting with street stations or boxes of various sizes and shapes.

The Booth Street Station.—It is octagonal in shape, two feet four inches in diameter, and about eight feet in height. These stations are quite extensively used, and, wherever sufficient room can be found for them, are for many reasons preferable to wall or lamp post boxes. The street station also guarantees to the officers, while within it, perfect immunity from disturbance while communicating with his superior through the telephone, and insures for his communication a secrecy which is sometimes important. Keys to the street stations and boxes are given to the patrolmen of the district. Locks are uniform throughout the city, and one key will open any station. Keys are also given to responsible citizens with which they can operate the signal boxes inside of these booths or street stations, without opening any door. In this connection the recent improvements of the double door box, which can be operated both from the inside and outside of the booth, are being universally adopted. All boxes furnished by this company can be similarly operated by the citizens' key. Citizens' keys when used are trapped; these keys are numbered and the name of the holder kept on record. This establishes responsibility and prevents the giving of needless alarms, for when a key has been used it can be returned to its owner only by the proper officer, who holds in his possession the master or release key. Patrolmen need carry but one key, the locks being so constructed that the keys furnished patrolmen will not only open any box, but may be used as the citizens' key without being trapped. The citizens' key turns in a call for help only, but from the inner box patrolmen may transmit a number of different calls, such as "report," "telephone," "fire," etc. All boxes are furnished with telephones and transmitters by which patrolmen may communicate with their headquarters. Wall boxes are designed to be placed in depots, warehouses, factories, etc., and on poles or walls of buildings where it is unnecessary or impracticable to place the sentry street station, or as a matter of economy, and may be interspersed between the more expensive and elaborately equipped street stations.

Individual Calls.—A certain number of calls in every signal box can be, if desired, reserved for the use of individual patrolmen to indicate their personal presence at the box; but the safest and probably the only method of identifying a patrolman at the box is by the sound of his voice while using the telephone.

Automatic Signals.—The advantage of automatic signals over those given vocally cannot be over-estimated. In cases of excitement, experience has abundantly proven that wrong impressions are often unwittingly conveyed by the speaker or received by the listener, while with an automatic printed message there can be no variation or mistake. The mechanical result is invariably the same. Suppose trouble of any kind occurs, demanding the presence of a police force in any part of the district covered by this system. The signal is instantly communicated to the district headquarters, and the reserve immediately hurries to the place whence the signal has come. Where horses are used, but a very short space of time elapses after the call before the aid required is at hand ready for any service. Patrolmen, as they go their rounds, may be required to report at the several stations on their beats as frequently as it may be deemed advisable. This will go a long way toward securing faithfulness of the force.

Visual Signals.—In connection with the system of police patrol telegraph, a combination of visual signals is attached to the top of the booth stations or lamp posts. These signals combine the use of semaphores by day and flash-lights by night, together with the continuous ringing of a large bell. An important feature of this part of the system is that the user is not limited to the use of one or two of these signals upon one line or circuit, but can operate as many visual signals as there are signal boxes on one circuit. In almost every city there are points at which these visual signals could be placed and operated to great advantage, for by their use the patrolman who comes within sight or sound of

Fire and Police Central Station.—Columbian Exposition.

them can be summoned to the box or booth for conversation, or to receive orders from a central station.

GENERAL ALARM.—The bells of the fire alarm telegraph may be utilized, where this system is in use by the police department, to summon, in cases of emergency, the entire force to rally at the nearest street station, and then directions may be given them by telephones, so that, when desired, the facts in detail may be known to every policeman in the city.

THE TELEPHONE.—While it is true that emergency calls, or calls for help, may be transmitted by a multiple signal box by any one, no matter how unskilled or excited, it is also true that the importance of the telephone in connection with a system of police can not be over-estimated. In fact, it is indispensable as an auxiliary, and in its place accomplishes what would otherwise be impossible. If the expense connected with their use is considered too great, they may be dispensed with in perhaps one-half of the street stations.

PATROL WAGONS. — The patrol wagons generally used with this system are models of convenience and adaptability for the work. They are furnished with alarm gong, and under the seats, which are run lengthwise on each side of the box, are compartments for hand-cuffs, clubs, come-alongs, blankets, canvas stretcher, ropes, a medicine chest and other articles necessary and convenient. One of the force accompanying the wagon is an expert, trained in the necessary expedients for resuscitating a case of suspended animation, stopping a flow of blood from a wound, and other temporary appliances for saving life and alleviating misery.

The stretcher is an ingenious arrangement. When not in use it is rolled up on poles, and placed in one of the compartments under the seat. When required for use the stretcher is suspended on hooks between the seats of the wagon in such a manner that the jolting of the wagon is not noticeable. The stretcher can be lifted in and out of the wagon without disturbing the patient, and for obstreperous persons there is a ring in the floor of the wagon, to which the belligerent party can be tied down and secured.

PRIVATE PROTECTION.—A small signal box is specially constructed for private residences, banks, hotels, or business offices, to be connected directly with the system. When a signal box is placed in a private residence, a key of the house is left at the station, under seal. In case the occupants of the house have occasion to call for assistance of the police at any time, they can do so by simply pulling the lever attached to the box, and they can also indicate the nature of the want by using any of the different signals; that is, they can indicate burglars, drunken servant, fire, etc. When a call is made, the occupants of the house can, if desired, remain quietly in their bed-rooms, while the police answering the call take the key of the house from its place, and having reached the house, step quickly in at the front door to the utter surprise of the thieves or burglars, who find themselves trapped. The owner of the alarm box can, each night if he wishes, before retiring, give what is called the "test of the line" signal; and the answering ring on the bell shows that the line is in working order. Another important feature of the signal apparatus is that after transmission of the call a small bell attached to the instrument is made to give one blow; this assures the party calling for help that his call has been heard, and is sure to be answered.

Advantages of the System.—The experience of the various police departments in the working of this telephone and signal system has been such that new uses and advantages have been and are being constantly developed, but the following may be mentioned as among the principal and most important points of value. It increases the certainty of punishment for crime, and must therefore exercise the most powerful influence for its prevention; it gives to every citizen, however remote from headquarters, the means of instantly summoning police assistance; it enables the authorities to humanely care for the unfortunate victims of accident of sudden illness; it affords the opportunity to convey to the station, in a properly arranged wagon, the uproarious drunkard or street brawler, without the lamentable exposures so common under the old style of arrests. The city is the custodian of the morals of its citizens, and owes to all ages and both sexes proper immunity from indecent exposures, blasphemy and impure language, which are too often the accompaniment of such arrests. It is an excellent auxiliary to a fire alarm system, and possesses the additional advantage of enabling the possessor of a private box to call the fire department directly to his door without leaving the premises. It enables the force, whatever it may be, to work with the greatest efficiency, and to accomplish infinitely more in the preservation of peace and order than could be possible without this valuable aid. By making a comparatively small force extremely efficient, and therefore a large force unnecessary, it saves a large annual expenditure for the maintenance of the department, which saving will in most cases pay for the construction of the system in a year or two, as well as its running expenses. By promptly furnishing the means of dispersing crowds, checking disturbances or riots, arresting disorderly persons, it infinitely decreases the chance of serious outbreaks, which often end in murder, robbery or arson. It enables the patrolman to remain on his beat; if he arrests a party or wishes assistance, he has only to go to the nearest box and make his wants known, and in a short time he is relieved of his prisoner, and assistance is at hand. It makes every key-holder, to a considerable extent, a policeman, for he carries with him the power to summon the police to any point wherever he may see that their services are required. It increases the power and dignity of a police officer, and to a very great degree will lessen the chances and necessity of personal encounters with the roughs who frequently combine for his injury or to divert him from his duty, while evil-doers seize the opportunity to disturb the peace and invade the proper rights of citizens. To provide against such exigencies by largely increasing the number of policemen is obviously much less economical than to quicken the working of the police system by putting every patrolman within reach of instant communication with station headquarters, at the same time giving every orderly citizen, in case of need, the means of calling upon the same authorities with least delay.

E. S. GREELEY & COMPANY.

Mounted in a case of this exhibit was a burglar alarm instrument with clock cut-off. This clock may be set in the same manner as an ordinary alarm clock, but will not only ring a bell at the set time, but will, in addition, cut the main circuit of the burglar alarm system, allowing the doors and windows to be opened without ringing the burglar alarm. Another burglar alarm instrument was shown having an electric light attachment. When a drop falls an electric light is lit so that the indications could be sure should the room be dark. This company also exhibited a police and fire alarm apparatus. The boxes have a positive non-interfering movement. The main circuit runs through a magnet, which holds an armature having a lever attached. This lever has a pin in it which catches in a slot of the main lever or pull of the box. The movement of the box cannot be operated unless the pin is in the slot, consequently when another box is sending in an alarm, the current only flows momentarily through the non-interfering magnet, not allowing the pin to catch in the notch. This effectually prevents interference. The non-interfering magnet is short-circuited by the closing of the outside door of the box, thus saving the battery. Testing keys, lightning arrester and bell were also mounted in the box.

W. F. GARDNER.

THERMOSTAT.—The exhibit was essentially a thermometer, having an open top and a bulb coiled into a flat spiral. Electric connection is made through the column of mercury by fine platinum wire inserted at the top and another fused into the spiral. A condenser, or spark arrester, is placed in the circuit between the binding posts of the thermostat. It has an automatic valve, a spring keeping it open for a flow of gas, also an electro-magnet, working by a lever having attached to its armature the stem of the automatic valve. When in use for testing chronometers at a fixed temperature, the platinum wire is adjusted so that the end in the tube indicates the temperature at which the room is to be kept, and the gas jets are lighted. Suppose the temperature of the room to be below that required, the mercury column will not be in contact with the platinum wire, and the circuit will be open. In this condition the valve will be open, gas flowing, and burners burning, which will continue until the temperature rises and closes the circuit by contact between the mercury and platinum. Then the armature is attracted and closes the valve. This continues until the temperature falls enough to break the circuit, when the action is repeated.

THE ELECTRIC HEAT ALARM COMPANY.

This exhibit represented a four-story factory showing a complete system of thermostat for fire alarm purposes, wired up and connected with the fire department; also journal alarm and hot grain alarm, to be used in any place where extreme heat is dangerous; also testing and recording devices.

THERMOSTAT.—The thermostat really constitutes the fundamental basis of both fire and journal alarms, and is constructed in the following manner: The base is made of vulca beston, into which is inserted a glass tube firmly imbedded in the plug. The lower part of the thermostat is a Bessemer steel cup forced on to the fibre plug under great pressure, which is filled with chemically pure mercury. On one side is a steel screw permanently fixed in the mercury for one contact, and the other point is an adjustable screw, set to any particular degree of temperature at which it is desired the mercury shall make contact; the contact between the positive screw and the negative mercury is made by the rising of the latter as in a thermometer. Every part of the thermostat is so compactly and accurately fitted together that the mercury is practically hermetically sealed in the steel cup. This feature enables the thermostat to be placed on journal bearings in steamships or other dirty and wet machinery; also in corn bins, grain elevators or wherever there is liability of overheating. The contact of this thermostat is made outside of all atmospheric influence.

TESTING AND RECORDING.—The testing and recording devices cannot be tampered with or rendered useless, as the cylinder is moved by clock-work This clock-work would have to be loosened on the permanent screw and turned back in order to disturb the mechanism. To get at the cylinder would necessitate tampering with the lock in the door of the case, and then manipulating the cylinder, which would show at once.

Every inch of wire can be tested at any time by simply turning a key in the recording case. The days of the week being on the top lines of the cylinder, one only has to look in the column of weeks to see what days of the week the test was made, as a full year's record is accurately kept for the whole year on the cylinder.

HOTEL FIRE ALARM.—An entirely new device exploited by this company is the hotel alarm, which has been perfected to an extent that makes it an extremely valuable protection for hotels and other large buildings.

CHARLES E. LEE.

TEMPERATURE REGULATOR.—This device belongs to that class of apparatus which controls the temperature of the entire house or one of its rooms, and a damper on the heating appliance which, being electrically connected with the thermostat and controlled by its movement, will open and close automatically according to the rise and fall of the temperature. The thermostat is made in the usual manner, of two metals of different expansibility, preferably zinc and iron (ordinary sheet tin) on account of ease of working in process of manufacture, and because it produces a thermostatic bar that is very permanent in its adjustment. The strip is mounted on a metal base covered with a metallic case carrying a thermometer, while at the top is an indicator which may be moved in either direction on a scale, and determines the degree at which the room is to be kept. The damper consists of a cast iron frame having a ring of suitable size to connect with the pipe of the furnace or the heating flues supplying the room, and at one side of this, a case containing the mechanism for moving the damper. There is a shaft running through this case, and through the damper ring, on which the damper-fan is fastened. There is also (in the movement case) an armature fastened to this same shaft. This armature is cone shaped in section and passes freely under the poles of a magnet hung from the top of the box. When the damper is at rest, in, for instance, its open position, the armature is below and at one side of the poles, but the passage of current through the magnet draws the armature towards itself, but, just at the central point, the circuit is broken by a suitable commutator, and the armature, together with its shaft and the damper, continues by its own momentum over into its closed position. The advantage of this form of regulating device claimed over others is in the fact that it is purely electrical and does not depend on the help of clock-work, compressed air or other outside power, and it can be installed by the ordinary electrician or wireman.

Agricultural and Machinery Halls by Night.—From Electricity Building.

JOHN F. HURD.

Coat Thief and Pickpocket Detector.—This mechanism will sound an alarm when a garment is removed from a coat hook or a pocket meddled with. The exhibit contained four styles of coat hooks, one which only sounds an alarm when pocket is picked; another will sound a continual alarm when pocket is picked or garment removed; a third, temporary alarm when garment is removed, and continual alarm when pocket is picked. The other, temporary alarm only when the garment is removed. The hooks work automatically, the alarm being sounded by removing any garment. It also contained two coats with two styles of circuit-making and breaking devices in garments, one of which was in the pocket, being a small switch which closes the circuit when the hand is placed in the pocket, by forcing a small insulator from between the two wires of said circuit. The insulator is attached to a cord which is extended across interior of pocket; the switch and cord being concealed by a muffle or flap in the pocket. The other circuit-making and breaking device is placed inside of lining of garment, which works on a pivot, and is attached to bottom of pocket by cord or tape, so arranged that any pressure on bottom of pocket will turn the switches, thereby closing the circuit. Inasmuch as the circuit is made and broken in the garment, the pickpocket and coat-thief device could be carried around in the garments by means of a pocket battery and bell or buzzer, using the same style of circuit-making and breaking devices for pockets which are in garments when hanging on the hooks, and the alarm for removal of garment may be sounded by said pocket appliance by removing garment from hook, the circuit being formed by relieving weight from a spring switch in the collar of the garment, so that an alarm may be sounded by picking a pocket when the owner is wearing the garment or has it hanging on a coat hook; also by removal of garment. The alarm sounds either in the garment or by a bell or buzzer attached to hooks. The owner may remove his garment or use his pocket by turning a switch, which will prevent an alarm.

REPORT OF THE FIRE AND POLICE PATROL SERVICE FOR THE WORLD'S COLUMBIAN EXPOSITION.

When I. R. Jenkins was appointed chief of the fire and police telegraph and telephone department of the Exposition, on Jan. 4, 1892, the Exposition had already contracted with the Gamewell Fire Alarm Telegraph Co., of New York, for one hundred and fifty fire alarm boxes and the necessary instruments for a central office, also contracted for a police patrol system of one hundred and fifty street boxes and a central station outfit. There were a number of buildings nearing completion and very much in need of fire protection. Lines for fire and police were at once laid out, also telephone service, interlacing lines where they were practicable for greater protection in case of an accident to the wires. Boxes and instruments were located in buildings and throughout the grounds, covering the entire park and Midway Plaisance, leaving no part unprotected. As subways and conduits were not completed, thirty-two miles of aerial lines were erected on the electric light poles. It is bad practice to operate fire and police wires on the same poles with electric light or power wires, yet in this instance there was no alternative but to use them temporarily. As soon as subways and conduits could be prepared, and suitable wire procured and placed in conduits, the aerial lines were abandoned, connections with the underground system were made, boxes removed from poles and placed on buildings where practicable. At the north end of the grounds, where the state and foreign buildings were located, Midway Plaisance and a few other points, boxes were placed on wooden posts of a pretty design.

Underground Wires.—In subways, conduits and under buildings the wire used was No. 14 B. & S. copper wire, rubber-covered. One hundred and twenty miles of this wire were used for fire alarm and police service; forty miles being single and forty miles double and twisted, divided into twenty-two circuits—six circuits, for fire alarm street boxes, on which there were one hundred and forty-nine boxes. There were six circuits for police patrol boxes on which there were one hundred and forty-four street boxes with telephone connections. Forty-three private boxes under buildings for watch and fire service were placed as follows: Under Manufactures building, 8; under Machinery Hall, 14; under Agricultural building, 15; under Electricity building, 6; total, 43.

Six alarm lines, running to engine houses and guard quarters, were placed, on which were operated forty-one gongs for sounding alarms of fire and emergency calls for guards in cases of robbery, riot, etc. There were also four lines to engine houses, two for joker lines for registering alarms and signaling calls, and two for telephone service. A metallic line was run from the central office to the Englewood fire alarm office, giving direct communication with the city fire department, to which all alarms of fires were transmitted. The central office also received through the same source all alarms in the city of Chicago.

Local Telephone Exchange.—A two-section switchboard, with a capacity of one hundred lines was placed in the office, on which ninety-four telephones were operated in the various offices throughout the grounds. It was the improved standard switchboard made by the Western

Electric Company of Chicago, and used in the principal cities of this country and abroad. Seven hundred thousand feet of copper wire, doubled, were used in establishing this service. The fire alarm, police patrol and local telephone exchange were all operated from an office located in the northwest corner of the court in the old Service building. Fire alarm instruments consisted of ten nickel-plated relays, mounted on hard rubber bases, six of which were connected with the signal wires extending throughout the grounds and buildings, on which boxes were placed for sending in alarms of fire. Two relays were connected with the city fire alarm office to transmit alarms of fire to the city; these were also used in receiving alarms of fire from the city. Two relays were connected with lines going to the various engine houses, and were used for sending and receiving signals throughout the grounds. Nine standard galvanometers were used on these lines for testing and showing the conditions of same. Sixteen lightning arresters were placed, through which all wires passed leading to instruments, also two registers for receiving and recording alarms, six relays or sounders, six keys for testing alarm lines, one ten number annunciator to indicate the line on which the signal was received, twenty switches, three pole changers for reversing the lines, and one four-dial manual repeater to which the six alarm lines were attached.

Mode of Sounding an Alarm of Fire.—All street boxes were protected by an iron box with a glass door. To send in an alarm of fire it was necessary to break the glass, open the door, pull the hook down and let go, which set the machinery in motion in the box; the operator in the central office, by moving a switch, threw the pole changers in circuit and placed all registers in engine houses on the grounds and throughout the city in direct communication with the box, giving the exact location of box pulled. The repeater was then set by moving the dials showing the number of box, then by throwing the lever the box was sounded on gongs in all engine houses, guard dormitories and offices. All engine houses were fitted with the "Barrett Joker" outfit, consisting of polarized relays with a free armature, keys, switches, sounders, registers and telephone connections, also gongs giving various ways of receiving alarms.

Police Patrol.—A large cabinet was used on which were set ten fuse plugs and lightning arresters through which all lines passed going to the six galvanometers and chemical registers, on which all calls were recorded. There were twenty switches and thirty keys for testing lines and answering calls, six shuttle drops, a small bell for denoting trouble on lines, also two telephones for communicating with street boxes. The boxes used in this system were the wall box with telephone. To operate this box, guards were all provided with keys with which they unlocked the box and set the pointer to telephone, pulling the lever down which records number of box. In central office the operator struck a key which sounded a bell, sending out the O. K. button in box, notifying the officer that his call was received.

PATROL SERVICE.—There were two patrol wagons, one located at the south end of the old Service building, which was on duty day and night. No. 2 was located at the railway terminal and was on duty during the day and early in the evening only. They responded to 2,929 calls.

AMBULANCES.—Three ambulances were in commission, located as follows: No. 1, south end of old Service building—on duty constantly; No. 2, located at Fire & Guard Station No. 7; No. 3, located rear Fire & Guard Station No. 3 on Midway Plaisance. The last two ambulances were on duty during the day and part of the evening only. They responded to 2,307 calls through the central office during the Fair.

TREASURY DEPARTMENT.—A system of signals was placed for the protection of the Exposition treasury department in the new Service building. Lines connecting with the central office were run on which were alarm bells by means of which the department would be notified instantly should any attempt be made to rob the treasury.

FUSES.—All lines entering the office were protected by a fuse known as the "Kansas Grasshopper," which proved of great value. The fuse acts very quickly, opening lines when charged with too heavy or foreign currents, and gave much satisfaction in checking what is commonly called a "sneak" current. Fuses were replaced without danger to the operator.

BATTERIES.—Gravity batteries were used on main lines for fire and patrol service, and two hundred and fifty-eight cells were used. For striking lines and locals, two hundred and seventy-four cells of Phoenix dry batteries were used. For telephones in police boxes, one hundred and ninety-three special Phoenix dry batteries were used which were made to fit in patrol boxes. All batteries gave entire satisfaction.

TESTING.—The fire alarm, police patrol boxes and telephones were inspected each week and were always found in good working order. Instructions were given to detachments of the guard three times a week, in fire fighting, by the late Capt. Fitzpatrick of the Chicago Fire Department, and by Capt. Jenkins in patrol duty in case of necessity to use the apparatus. In every case, with one lamentable and noted exception, fires were extinguished in their incipiency, the exception being that of the cold storage warehouse where so many brave men lost their lives. The working force under Capt. Jenkins was as follows:

Supt. of Electrical Alarm and Patrol Service	1
Capt. commanding Company No. 27	1
Operators	9
Inspectors	3
Sergeant of Patrol Wagons	1
Patrolmen	7
Total	22

Also two lady operators for local telephone exchange switchboard, and one clerk, making total force twenty-five.

Electricity Building.—Night View.—Looking North.

CHAPTER XV.

PRIMARY BATTERIES.

BATTERIES of primary type have been much restricted in their field of usefulness by dynamic electricity, yet there are a number of purposes for which this class of electrical generators seem peculiarly fitted, and notwithstanding some weighty predictions of their early decadence in commerce, there seems to be a general feeling that they will not be wholly supplanted—or, indeed, supplanted at all—until new principles are evolved which will revolutionize present methods of electrical procedure. A number of firms made displays of primary batteries, and numerous forms were shown.

Primary batteries may be divided into two classes; for use on open circuits and for use on closed circuits. The former class furnish a current of proper voltage and quantity for a short time and after a period of rest renew themselves, while the latter kind are constant during their life. Of the open circuit type there were exhibited various kinds of the sal-ammoniac battery, some with the carbon element around the zinc, and vice versa, and others of the porous cup packed with granulated carbon and manganese. A number of styles of cells, designed to give a large voltage (as high as two volts), and of great quantity, were shown—such as the Grenette type, which use sulphuric acid, bi-chromate of sodium and water as a fluid and carbon and zinc elements. These give excellent results for a short period and the same type, with a smaller voltage but of large quantity, are successfully used for electrolytic purposes. Two kinds of these cells were shown, one with zinc and carbon elements, the other with zinc and platinized silver elements. Dry cells of various kinds, consisting of a zinc containing element holding a carbon element and packed with granulated manganese and plaster of paris and chloride of silver cells for testing purposes, giving an accurate current of one volt and one ampere, were also on exhibition. Double fluid cells, with a potential of two volts and of low amperage of the Fuller type, were shown which are used with success on telephone lines. Grove cells, Munson and carbon batteries were shown by several exhibitors. Of the closed circuit type the gravity crow-foot type so much used for telegraphic purposes, the Daniel cell and the Hussey cell were shown. The Daniel cell, used much for testing on account of its reliable, even performance, is composed of a jar, porous cup with zinc and copper elements with an auxiliary reservoir which constantly feeds to the cell a sulphate of copper solution.

THE EDISON MANUFACTURING COMPANY.

THE EDISON LALANDE BATTERY.—The exhibit contained many forms and applications of the Edison Lalande battery. The elements employed in the cell are zinc, from which the negative current is obtained, and black oxide of copper (Cu. o) the positive current. The exciting liquid is simply a solution of caustic potash. The oxide of copper is obtained by the process of roasting copper turnings; the oxide is then ground into a fine powder and compressed into solid blocks, from which plates of a suitable size for the different cells are cut. These plates are suspended from the cover of the containing vessel, a porcelain jar, by means of a light framework of copper, one end of this framework carrying the binding post for the positive pole of the battery. This framework is fastened rigidly to the cover. On each side of the copper oxide element in the larger type cells (but only on one side in the smaller types) is suspended a rolled zinc plate. These zinc plates are fastened by a bolt to a knob on the cover. This prevents any movement in the relative position of the elements, and does away with the necessity of using vulcanite separators to prevent any short-circuits occurring in solution. The zincs are amalgamated, and, as in most batteries, the zinc is attacked more vigorously near the top than at the lower part of the plate; the zincs for this cell are made slightly tapering, the thick part being uppermost. The exciting liquid employed in the battery consists, in all types, of a 25 per cent solution of caustic potash in water, or, in other words, of a solution of one pound of caustic potash in three pounds of water. When the circuit is closed and the cell is put in action, the water is decomposed, the oxygen forming, with the zinc, oxide of zinc, which, in turn, combines with the potash to form an exceedingly soluble double salt of zinc and potash, which dissolves as rapidly as it is formed; the hydrogen liberated by the decomposition of the water reduces the copper oxide to metallic copper. A layer of heavy paraffin oil 3/8 inch deep is then added to keep out the air and prevent creeping. As for inspection or supervision, so fruitful a source of trouble and expense with other batteries, this cell requires absolutely none. Polarization and local action are, as we have already intimated, entirely absent, the zincs never require cleaning, the solution does not crystallize or creep, and there is no porous pot to crack or lead cap to be converted into white oxide. No fumes or other noxious chemical products are given off by the battery at any time during its existence; in fact, its presence could never be detected by the sense of smell, even in a lady's boudoir, and, as for convenience of form and dimensions, what could be better than a cylindrical porcelain jar containing the two elements suspended from the cover? In the exhibit there were models of specially prepared batteries for various purposes, among them being batteries for telegraph work, phonograph batteries, annunciator cells, special telephone models for local transmitters, and small telegraph and motor models, also batteries specially designed for fan and other motors.

E. S. GREELEY & COMPANY.

A prominent feature in the exhibit of this firm was that of primary batteries. Attention was called to the exhibit by a pyramid of primary batteries surmounted by a statue of Liberty 36 inches high, holding aloft a 16-c.p. lamp. This exhibit of batteries comprised a large number of styles of batteries, among which was the Fotte Primach battery. The battery is of an improved form and construction; to the center of the copper element is fastened a wooden post covered with asphaltum that supports the zinc element. The zinc is star shaped and has a recess on the under side into which the post fits. This post keeps the two elements always the same distance apart, thus preventing the copper working up and all short-circuits are in this way avoided. A new telephone battery was shown which is a modification of the Fuller cell, which is used so extensively in that class of work. Instead of having a bell-shaped zinc, it has a pencil at the bottom. The zinc feeds down and the life of the battery is said to be much longer. Its voltage is two, and the initial current about three, amperes. Another battery in the exhibit was the Exeter dry battery, which was used to generate the current operating the starting devices which opened the World's Fair. This cell is known for its long life and quickness with which it receives rather than its initial high current strength. The Primach battery of 30 cells operated the Greeley fire alarm system on exhibition, consisting of four non-interfering boxes, one indicator, an electro-mechanical gong and two registers.

APPLICATION OF PRIMARY BATTERY TO OPERATE A DOOR OPENER.—An interesting application of the Primach battery was shown in the form of a door opener. There was a sample double door showing how a Wildt door opener operates. One door showed the opener without a door check and the other door with a door check. One cell of battery is all that is required. The door opener will operate successfully with only one cell of battery.

THE S. S. WHITE DENTAL MANUFACTURING COMPANY.

This exhibit was properly divided into two parts: first, the Partz electric batteries and sulpho-chromic salt; second, electric dental outfits and accessories, consisting of electrical dental engines, tools, etc., and the necessary electric motors, rheostats, magnetic clutches, and foot treadle to operate them, together with some useful accessories, such as electric mouth lamps, hot air syringes, cauteries, electric mallets, etc. The Partz batteries are divided into three varieties: first, the No. 5, which is an ordinary open circuit zine carbon cell, similar to the many forms of zinc carbon sal-ammoniac batteries in the market and of about the same strength; second, the Partz motor battery No. 6, which is only a strong Bunsen or Poggendorff cell, nothing more, except that it is well designed, compact of form and has the Partz patent slotted carbons, giving very large carbon surface; and, third, the Partz acid gravity cell, of which they manufacture several different styles, all depending on the same underlying principles. It is claimed that the Partz acid gravity batteries are the best open circuit batteries now on the market. They are the only gravity chromic acid batteries in existence and in that respect are novel and peculiar, and any accidental weakening due to a short-circuit or extra hard work can be immediately overcome by adding some of their sulpho-chromic salt to the funnel in each cell. It combines the advantages of the gravity and Bunsen forms of cell. The carbon is at the bottom, the zinc near the top. The jar is filled with a solution of sulphate of magnesia, or common table salt (sodium chloride), and Partz sulpho-chromic salt is fed through the funnel. This salt dissolves and forms around the carbon a depolarizer, the superior density of which keeps it at the bottom and away from the zinc. On closed circuit it gives the nearly constant current of the Bunsen, while on open circuit it stands like a sal-ammoniac cell. E. M. F. = 1.8 to 1.9 volts; current on short circuit, 1 to 4 amperes; working current from 1 ampere down. Intended for all open circuit work, but not for battery motors. Specially adapted to long distance telephone work, railway signal work, and open circuit fire alarm work. Nothing wears out but the zinc. Cells in constant use for over five years with the original zincs still in (July, 1893).

In the exhibit two Partz acid gravity batteries, No. 3 and two No. 4, were set up May 25th, 1893, and so arranged that each pair could be connected to a set of ten 3½ in. electric bells, and ring the whole ten at once.

These cells were almost continuously in use ringing these bells, and, during their five months' use, sulpho-chromic salt was added through their funnels two or three times. At the close of the Fair each pair of batteries were connected to five bells, and allowed to ring them continuously until exhausted. Sulpho-chromic salt was then added, and the batteries again allowed to run out on the bells.

Below the tests are given :

CELL.	VOLTS.	
Partz A. G. No. 3 - -	1.93*	Time to run down on 5 bells, 20 hours.
" " " - - -	1.9 *	Time to run down on 5 bells, 20 hours.
After sulpho-chromic salt added		Time to run down on 5 bells, 24 hours.
Partz A. G. No. 4 - -	1.9 *	Time to run down on 5 bells, 15½ hours.
" " " - - -	1.8 *	Time to run down on 5 bells, 15½ hours.
After adding sulpho-chromic salt		Time to run down on 5 bells, 22 hours.

These tests could have been repeated some four or five times before the fluids would have become saturated and the battery would require recharging.

A set of thirty No. 3 Partz acid gravity cells were charged on May 25th, 1893, and connected to a McIntosh medical wall cabinet.

At the end of the Fair these thirty cells were tested and gave 54¾ volts. After adding some Partz sulpho-chromic salt to each cell the voltage rose to 59 5-6 volts. Two each of cells Nos. 3 and 4 were charged on May 25th, 1893, and then connected on short circuit and allowed to remain so. On August 14th, 1893, one No. 3 and one No. 4 of the above cells were disconnected from short circuit and tested. The e.m.f. was 0.03 volts. Some sulpho-chromic salt was then added and in a few minutes the e.m.f taken : No. 3, 1.7 volts ; No. 4, 1.6.

These cells remained on open circuit until the end of the Fair, when the e.m.f. still remained 1.7 and 1.6 volts. The remaining two cells were kept on short circuit until the end of the Exhibition, when the circuit was opened and the e.m.f. found to be 0.03 volt. On adding some sulpho-chromic salt through the funnel, the e.m.f. rose in 15 minutes to 1.6 and 1.7 volts.

PARTZ SULPHO-CHROMIC SALT.—Sulpho-chromic salt is, as its name indicates, a combination of sulphuric and chromic acids in an amorpho-crystalline state, the efficacy of which is attested by the superiority of the batteries based upon its use.

It was specially devised for use with the Partz acid gravity batteries, but when dissolved in water makes a standard electropoion fluid without the necessity of handling liquid acids, having practically the same amount of stored energy as an equal quantity of fluid made in the ordinary ways, giving with a zinc carbon element an e.m.f. of 1.9 to 2 volts, and an ampere hour capacity of about the same.

It is asserted that this substance affords the best, cleanest and most

*At beginning of final test, after ringing bells daily for five months.

MacMonnies Fountain.—Search Light Effect.

convenient method of making electropoion fluid for medical batteries, Grenet cells, Bunsen batteries, for lecture tables, experimental work, etc., and motor batteries.

J. C. VETTER & COMPANY.

The exhibit consisted of the Vetter dry cell Leclanche battery and the Vetter complete portable dry galvanic battery.

The dry cell Leclanche battery embraced all of the valuable qualities of a battery having the zinc and carbon elements, the well known polarizing agent peroxide of manganese in a porous cup, and sal-ammoniac as its excitant. For open circuit work and many special purposes this is an excellent battery: No glass, no liquids, no acid, and inexpensive, neat, light and portable, high electro-motive force, great recuperative power, long of life and does not lose power when not in use. They are made in five standard sizes. The Vetter complete portable dry galvanic battery may be conveniently and quickly operated. It is only necessary to open the cover, connect the electrodes, when, upon final closing of the circuit, the exact amount of current can be turned on or off and correctly measured in milliamperes. This instrument is also provided with an improved pole changer, and three binding posts. The binding posts are to receive the current through an adapter. Thus can be utilized direct incandescent light current of 110 volts for all galvanic treatment.

THE LECLANCHE BATTERY COMPANY.

The goods exhibited by this firm consisted of batteries. The method of arrangement, the design of the arch that spanned the entrance to the exhibit, the appropriateness of the details, as well as the taste displayed in the colors of the decorations, elicited the admiration of visitors. At the top of the arch miniature incandescent lamps spelled out the word Gonda, the well known trade mark of these batteries, adopted some 20 years ago by this company. Prominent on the face of the arch appeared the name of the company, the names of the different cells manufactured by the company, and near the bottom of the columns the words, "Strength, Endurance," qualities that these batteries are said to possess in a marked degree.

Forming three sides of a square, of which the arch formed the front, were the glass cases that contained the exhibit of batteries. The glass sides of the cases enabled the batteries to be seen from both within and without the space. The batteries shown were designed for open circuit work, and in all the improved battery connection was used. The washers used were made of non-oxidizing metal that is always free from rust and corrosion, and no local action can be set up between the connection and the carbon. Among the batteries shown was the Gonda porous cup cell, which is the original form of the Leclanche cell. It is about seven inches high and four and one-half inches square. The Axo cell shown was an improved form of battery in which a porous cup is used. The special form of cup used in this battery has a flange that fits over the top of the glass jar, making a closed cell. Thus all dust is excluded and evaporation is prevented. This cell is especially adapted for physicians' use. Among the other styles of batteries exhibited were the Gonda cell, well known to the trade, and the cylinder cell that possesses a large amount of surface and low internal resistance. The Vole cell is the latest battery introduced by this company. It is constructed for work where a quantity of current and continuous service is required. It is eight inches high and four and one-half inches square and differs greatly in construction from the other cells. The negative electrode consists of a carbon having six vertical wings, over which is stretched a bag so as to form pockets between the wings for the depolarizing compounds. From its upper end a carbon rod projects through the cover of the jar, by which it is suspended. This cover screws down on the neck of the jar, and the cell is thus rendered water-tight. A cap, of the same material as the cover, completely shields the top of the carbon, with its connection, and preserves it from injury of any kind. The zincs of the Vole cell are semi-cylindrical, tapering from the bottom upward, an arrangement which allows the cell to depolarize itself much more readily than when the negative element is surrounded along its entire length by the zinc. They are attached to the bottom of the cover by zinc screws, which pass through them and also through the zinc strip which connects them. One of these screws serves as a binding post on top of the cover. The zincs may be thrown out of connection and only one used wherever a very low internal resistance is not an object. In such case there is, of course, less consumption of zinc and less quantity of current developed by the cell when in action.

C. J. HIRLIMAN.

This exhibit included the Champion primary battery and an improved porous cup battery. The Champion battery possesses the distinctive feature of being two batteries in one, as it is used with a rod zinc for telephone and other similar work where there is a constant strain upon the battery, or with corrugated zinc for electric gas lighting, multiple bell ringing, clock circuits, induction coils and all intermittent work where large currents for short periods of time are required. An important feature of this battery is that it has duplicate facilities for attaching carbon connections, so that, in the event of the destruction of either one by electrical or mechanical cause, the other one remains.

THE IMPROVED PATENTED POROUS CUP BATTERY.—The points of merit possessed by this battery are steadiness and reliability of action and great recuperative powers. It is an improvement on the regular Disque Leclanche battery, consisting in valuable additional parts, improved construction and new qualities that are absent in the regular or original Disque form. There is reduced evaporation and absence of crystallization of the salts of the carbon inside the cup—a serious though outwardly invisible defect, which is present in all other forms of Disque battery. The battery has a glass cover to reduce evaporation and assist sealing. The vents in the porous cup are of wood instead of glass, therefore cannot be broken or clogged up. But it is chiefly in the interior construction of the cup and in the ingredients used that the advantages claimed are arrived at. While this battery is ordinarily made in the standard or popular size of the original Leclanche battery, it is also made for special purposes, such as fire alarm work, physicians' use, and in varying sizes from very small to very large.

WILLIAM BURNLEY.

This exhibit was characterized as the Burnley Cartridge battery and the peculiar merits or points of excellence for the exhibit are as follows: first, the porous cup is comparatively very small, being only 5⅜ inches in height, and in diameter permitting the use of a small containing jar and consequent economy of space; second, the porous cup is made of paper or card-board having sufficient rigidity to withstand the necessary compressions and having a prepared paraffined top to resist the action of creeping salts. The porous cup will not slump or disintegrate after long immersion in liquid. The advantages claimed are: first, a reduction in cost of cup; second, a reduction in weight of cup and consequent reduction in cost of transportation, as the weight of the cup filled and complete for market is between 14 ounces and 15 ounces, while the ordinary Leclanche cup weighs from 2½ pounds to 3 pounds; third, in this form of cup there can be no choking by the deposition of reduced salts within the pores of the cup, as is often the case in earthenware cups, but, on the contrary, the reduced salts as formed adhere to the extremely fine shreds always present on paper and, having a flimsy support, soon drop by their increased weight to the bottom of the jar; fourth, liability to breakage is eliminated, and this, coupled with the fact that a large number can be packed in a small box for shipment, and the cell has a life equal to the best forms of Leclanche cells for all classes of open circuit work. Another important feature of the battery is that the form and internal arrangement of the depolarizing compound of manganese and carbon surrounding the negative electrode is such as to utilize all the oxygen present in the manganese, and thus a smaller bulk of that material can be used, thus reducing the size of cell, while retaining its efficiency as compared with other forms of open circuit batteries.

THE MIAMISBURG ELECTRIC COMPANY.

This firm exhibited its Imperial dry battery. The peculiar merits or points of excellence claimed for this battery are as follows: first, the possibility of internal short-circuit is eliminated; second, in this battery it is almost impossible to so deplete the cell by exhaustive work that it will not do comparatively good service—that is, it will do good service so long as the zinc cup lasts, and as there is no local action on zinc, the efficiency of the cell is the test desired. The internal resistance is very small, due to the internal arrangement and compounding of material surrounding the negative element, and under this arrangement it is not possible for the battery to deplete itself by used-up depolarizing matter. It is also claimed that this battery has superior merit in that it possesses the quality of holding its initial force and value over long periods of time.

OTTO RAUDA.

Students' Plunge Battery.—It is intended for the use of students and general laboratory work. The batteries are very convenient to set in action and out of action, very portable, wearing parts can be renewed very easily, and are of general neatness of desgin.

J. H. BUNNELL & COMPANY.

Standard Dry Battery.—This improved form of dry battery is used on open circuit work. Its compact form, being smaller than any other of equal capacity, and its great strength, durability and recuperative power, are its chief merits. Being in all respects a dry battery, without liquids, it is not subject to freezing in winter, or to evaporation at any time, and is clean. Its connections are so arranged that no creeping salts whatever are at any time developed. For portable purposes, railway signals, etc., where the battery is often placed out of doors, it is especially recommended.

THE UNION ELECTRIC WORKS.

This firm made an interesting display of the Crowdus primary battery operating sewing machines, small rotary pumps, dental engines, etc. It also exhibited small, low volt motors and dental engines especially adapted for use in connection with, and operated by, the primary battery.

GENERAL ELECTRIC COMPANY (LIMITED), LONDON.

DYNAMO-ELECTRIC GAS LIGHTER.—The gas lighter, which does away with matches for lighting gas, consists of two cylinders of hard rubber, concentric with each other, the outer one forming a handle, while the second, fitting loosely in the former, is revolved by a ratchet movement actuated by thumb pressure. This inner cylinder is coated inside with tin-foil and has on its outer surface at one end a number of small brushes, corresponding to the brushes on a Voss machine, which bear upon the interior surface of the outer cylinder, which, also, is coated with tin-foil. The spark which is generated by the rotation of the inner cylinder passes between two points at the extreme end of the instrument to which the charge is led by suitable conductors—thus it will be seen that there is no consumption of material and the instruments will therefore never become exhausted.

THE SLOSS ELECTRIC GAS LIGHTER COMPANY.

GAS LIGHTER OPERATED BY CHEMICAL BATTERY.—This lighter is attached directly to the gas burner and the small chloride of silver battery may be hidden in the ceiling hood of the fixture. The result is accomplished by bringing the two poles into contact immediately at the point of egress of the gas. A spark is created, and the gas lighted. The weak point in apparatus of this character is the failure of the batteries after a short period of use.

A. C. BECQUEREL.

TWO-LIQUIDS BATTERY.—This battery, which was designed in 1829, is the first two-liquids and constant current battery ever made. It is formed essentially of a glass tube U-shaped, one of the branches of which being filled with a solution of some copper salt, the other branch filled with a mixture of water and sulphuric acid. The two solutions are separated by some kaolin powder, placed in the lower curved part of the tube, and working as a diaphragm. A zinc plate is immersed in the acidulated water, and a copper plate in the solution of copper salt.

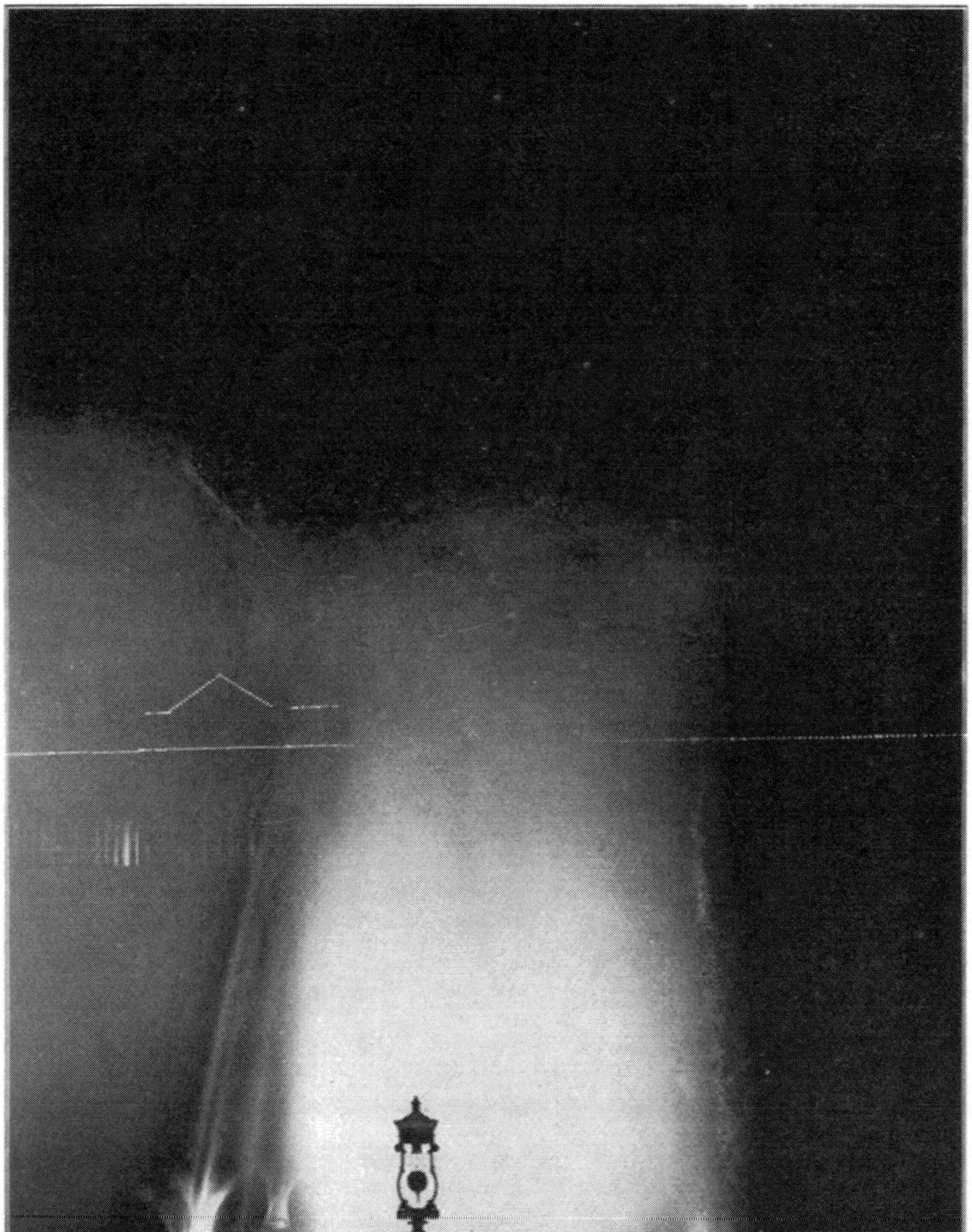

Electric Fountain Effect.—Machinery Hall in Distance.

CHAPTER XVI.

SECONDARY OR STORAGE BATTERIES.

MUCH good work was shown at the Exposition as the product of storage batteries, and the exhibit, while not large, was quite complete and the use of this class of batteries was exemplified extensively in a number of ways. In the French section the first storage battery ever made was shown as an historical exhibit, and elsewhere its evolution could be accurately followed. The storage battery depends for its principle on the electrolytic action of the electric current and consists primarily of two plates of lead, one coated with sulphate of lead in the form of a paste, the other uncoated. Both are immersed in a dilute acid solution and, the coated plate being connected to the positive pole of the dynamo and the uncoated plate to the negative pole, the current causes the transference of the sulphate of lead to the uncoated plate, it becoming hypo-sulphate of lead by absorbing oxygen from the acid solution and freeing hydrogen by the process. When the battery is used the hypo-sulphate of lead transfers to the original plate, again becoming sulphate of lead by giving up the oxygen it has absorbed. All modern storage batteries embody this principle, the difference being in their method of manufacture, zinc being in some cases substituted for lead and other electrolytes and active materials being used, and, as a rule, the plates are made in such form as to preclude the possibility of their buckling and the active material is formed on them primarily by the action of an electric current and in such a way as to insure its retention under all circumstances, as in the Plante type. In the paste type the active material is applied to the plate in the form of a paste. The Plante type meets more nearly all conditions, better resisting sudden and violent discharges such as are occasioned by accidental short-circuiting, and, owing to this construction, bear the shocks of transportation. The advantages of storage batteries were conclusively shown in their use as movers of the electric launches on the lagoons, and another use was made of them in propelling a six-seat carriage, which attracted much attention by its smooth running, it being displayed on the roadways near Electricity building. The application of storage batteries to street railway service is a problem which the ablest electricians are attempting to solve. The question is purely one of cost, the other questions having been pretty well settled. There is an underlying belief among the electrical people—that is, the practical men—that the principles of the storage battery will not admit of their use independently as an economical proposition, but that, in an auxiliary capacity to dynamic service, they will be of much use, that is, for starting purposes. Carried on the cars, and gradually charged, they can be made to serve the purposes of a good many horse power on the line. It requires some thirty horse power to start well and run grades, while only a small fraction of this power is at work while the trains are running. It is contended that the power on the line can be reduced to a very little above the amount necessary to run steadily on level, the small excess of power being used to constantly work on the storage batteries so that they will be ready for storing and pulling up grades. As a factor in reducing expenses of large lighting stations by storing away the surplus energy developed by their dynamos during the day when the load is light, it is attracting attention. The whole storage battery problem is in an unsettled shape, largely due, no doubt, to the present, past and prospective litigation over patents. One use for this service, which has been placed beyond discussion, is for operating small water craft, especially pleasure boats.

THE ELECTRIC LAUNCH AND NAVIGATION COMPANY.

This company exhibited fifty electric launches which operated for the transportation of passengers upon the lagoons of the Exposition grounds. These boats were constructed of seasoned white oak, planked with selected white cedar, copper fastened. The decks were of mahogany, calked. The finishings were of mahogany with high combings. They were fitted with brass cleats, brass stern bands and rails and stanchions for awnings—awnings of neat design covered the entire cock-pit. The hull and finishings were highly polished and varnished. The length of the boats over all was 35 feet 10¼ inches. Length at water line 31 feet 6 inches. Extreme beam 6 feet 2½ inches. Greatest draught at no load, 27 inches. Greatest draught with full load, 30 inches. Weight of hull, 2000 pounds. Displacement when completely fitted and without passengers, 5000 pounds. Each launch was fitted with storage cells of S. 17 type, manufactured by the Consolidated Electric Storage Company, which were placed beneath the flooring and under the seats, leaving the entire boat space for passengers. Ampere capacity of each cell was 150. Electromotive force of each cell, 2 volts. Each boat was fitted with a magnetic controller which gave four speeds ahead and two speeds astern. A Thomson-Houston motor having a normal horse power for a continuous run of 3¼-h.p. was directly connected to the propeller shaft and placed beneath the flooring. All bearings were self-lubricating. The normal revolution of the armature in each boat was 625, while its maximum revolution was 1100. The motor used was a series wound four-pole machine, and weighed complete 397 pounds. The propeller used had a diameter of 18 inches, with a pitch of 12 inches. Each launch had a seating capacity for thirty persons. The charging current used was 75 volts and 20 amperes per cell, and the boats required from 5 to 6 hours to recharge. These boats were in constant service from the opening of the Fair, making a daily average of 25 to 40 miles each. The average rate of speed was about 5½ miles per hour. The cost of charging for power did not exceed 60 cents per day per boat. Some of the boats made over five thousand miles with the renewal of but four cells, and the general records show renewals of less than 10 per cent. The exhibit met with the greatest success in the pleasure its use afforded visitors to the Fair, while it excited the liveliest admiration of experts, both electrical and mechanical, as giving illustration of the great things to be done by electricity as a motor for boat use. Two kinds of trip were provided for by this fleet of boats—the round trip, distance about four miles, touching all the principal buildings of the Exposition (fee, 50 cents), and the shorter trip (25 cents) for passage between one terminal point and the other, or intermediate points. Not a single accident or injury to a passenger occurred during the Fair, nor any serious damage to any of the boats. This remarkable fact was due to two causes, namely, the weight of the boat (about 5000 pounds, of which about 3000 is for the battery and apparatus), was so placed as to render it very safe and solid in the water, and the great ease with which the boat could be guided and manipulated. There was one other noteworthy feature about these launches, and that was the great economy effected in operating them. They required no licensed engineer, whose wages would be at least $3.50 per day. Any person of ordinary intelligence—any child, in fact—could operate and guide them, once the battery was charged, and a daily pay of $1.50 for such easy work would be enough. Then there is the saving in other respects—no coal, no costly repairs, no extras of any kind.

THE CHARGING STATION.—In a comparatively out-of-the-way part of the grounds, toward the south end, was the charging station for this fleet of electric launches. It was located in the east and south of the annex to the Agricultural building, below the level of that structure, and consisted of a shed-like building extending over the south pond, directly communicating with the lake itself. There were twenty-five slips in this station, providing safe and commodious anchorage for fifty boats. The station was L-shaped, the southern section forming the longer wing. In this place, between the hours of 11 p. m. and 7 a. m., all the launches underwent the process of being charged with new electric power in place of that exhausted the day before. There was a small force of men at work, employed in cleaning, overhauling, and, if needs be, repairing the boats, polishing the brass work on them and attending to similar mechanical duties. The power needed every night for this recharging of the batteries was conveyed direct from Machinery Hall by a cable, which, on reaching the station, was subdivided. Two wires were directly affixed to each battery by means of plugs that fitted into them, and, with the aid of an amperemeter, the degree of the power was regulated to the proper pitch. The power absorbed averaged about ten horse-power per boat, and as high as 2000 amperes of current at eighty-five volts were drawn in one night for fifty-two boats. The record of the launches as a whole was a most excellent one. Their chief points of excellence may be summed up as follows: Beautiful outlines, perfect safety for the passengers, swiftness and pleasure in riding on them, easy management, and large saving in their operating expenses. Besides that, however, these electric launches proved a great benefit for the Fair. Not alone did they become a great source of income to the Fair management by reason of the large sums taken in by their owners in the shape of fares, of which a percentage went to the Exposition treasury, but they permitted all those anxious to obtain a good and close view of the notable outdoor events at the Fair, such as gondola and canoe races, swimming matches, fireworks, processions, etc., to gratify their wish. The managers of the company kept a launch which they dubbed the "Press Boat," specially for the use of the press, and thus rendered valuable service to the newspapers and their writers.

THE AMERICAN STORAGE BATTERY COMPANY.

The American Storage Battery Company exhibited at the World's Fair an improved form of the Plante cell. No red lead oxides or active material is mechanically applied. The plates are formed by oxidation in an electro-chemical bath. The plates consist of a series of horizontal strips of rolled lead crimped on both sides ⅜ and ½ inch wide and of such thickness and separated by such distance as by long testing and experiments give the greatest capacity combined with the longest life. These strips are solidly united at the ends by a process which leaves the sides of the plate a solid bar of metal of special composition, not affected by chemicals. The plate is then supplied with foot and terminal pieces very firmly connected and also of special composition. The plate is then ready for formation. It is submitted to a forming process, which oxidizes the surface of the lead strips by attacking but slightly the surrounding support. The oxide taking more space, expands, and the process is continued until the interstices between the strips are filled solidly, thus leaving the alternate strips of lead and active material horizontally across the plate and surrounded by a solid frame of metal. This makes the plate solid and strong. The lead strips being not over $\frac{1}{10}$ of an inch apart, allow a large as well as intimate connection with all parts of the active material, thus giving heavy charge and discharge rates. The indentations in the lead strips effectively key the active material in place, preventing its falling out, while the strips, being placed edgewise to the plate solidly filled with active material, give the plate rigidity sidewise nearly equal to a solid lead plate of like thickness, thereby preventing buckling or bending. After the plates are formed, they are assembled together, the number of plates being governed by the size of cell desired, commencing with the outside plate and separated by rubber insulators called combs. Every alternate one is negative, while those between form the positive, there being one more negative than positive plates. The plates are then firmly united by fusing a T-shaped piece of rolled lead, positive to positive and negative to negative plates, this piece having a projecting strip for conveniently connecting cells. The elements are placed in a containing jar of glass or hard rubber which is filled with a solution of one part sulphuric acid to five parts water. The cell when charged is ready for work. The exhibit showed the battery, for lighting and for power. A plant of 52 150-ampere-hour cells connected in two series of twenty-six each was shown, furnishing a current of 30-50 16-c.p. lamps. The use of this battery for power was shown in the electric carriage. This carriage had a seating capacity for six people and was equipped with 24 150-ampere-hour cells placed under the seats. A 2½-h.p. motor wound for 50 volts was geared to the back axle of the carriage. The controlling apparatus consisted of a contact board and a reversing switch, both fastened to the underneath side of the carriage floor. The speed lever throws the cells into the following combinations and gives three speeds: Eight (8) in series, three (3) in parallel, first speed; twelve (12) in series, two (2) in parallel, second speed; twenty-four (24) in series, third speed.

A spring plug and foot switch was placed in the floor of the carriage directly in front of the driver, who had only to remove his foot from this spring switch and allow the small contact plug to rise from the cup

Peristyle by Night.—Search Light on the Quadriga.

of mercury. This action broke the circuit between the battery and motor. One of the most important features of the carriage is the arrangement of the front wheels to facilitate steering. Each of the two front wheels turns upon a bearing that is swivelled to swing horizontally and independently of the axle. The front axle is rigidly fixed in a position parallel to the back axle. A yoke bar carrying a rack couples the two front wheels, and the gears, working in this rack and operating through the steering wheel shaft, permits of the two wheels being swung to the right or left so as to guide the carriage. The ease and facility with which this carriage is operated is evidence that electric carriages will be in general use in the near future. The large current used from the batteries at times, without injury to them, fully demonstrated the great durability of the American battery as a source of power.

PUMPELLY-SORLEY STORAGE BATTERY COMPANY.

This battery is made on the Plante system, by electro-chemical process. The lead plates or electrodes differ from those of other batteries made by the Plante process. They are made from solid rolled lead, three-eighths of an inch thick, and the interstices in the plates are cut out by a gang of circular saws, leaving the surrounding rim solid lead, hence no joints, and no burning or soldering are necessary, thus giving great stability and durability to the plate. The plates are six by seven inches, three-eighths of an inch thick, and nine of them give the 150 ampere hour capacity of the battery exhibited. Storage batteries to-day are so well known to the public, and their uses and marked convenience for many purposes when stored energy is desired, ready to be drawn upon at any time, are so well appreciated, that it seems unnecessary to speak at length of their many applications. Among the peculiar features possessed by these cells are: 1st. The mechanical construction of the lead plate of which the electrodes are formed, already spoken of; 2d. The electro-chemical formation, obviating all of the many and serious disadvantages of pasted cells. The active material is solidly imbedded in the interstices of the plate, and neither by transportation, handling or use can it become detached from the plate. These cells can be discharged at any desired rate without injury or loss of storage capacity, hence it is believed that their strong point is found where the weakest point of all storage batteries has heretofore been developed, namely, in traction work, where the output required is irregular, and at times much above the normal rate of discharge of the cell. 3d. The electro-motive force is sustained under any rate of discharge, until the capacity of the cell is practically exhausted. 4th. The use of an absorbent fibre (patented), between the electrodes, absolutely prevents short-circuiting and renders it possible to arrange the plates compactly, preventing disintegration and loss of energy. 5th. The cell is of such comparatively recent invention that the test of time cannot yet be claimed for it, but its construction is such that theoretically it would seem indestructible, and the inventors and the company have the utmost confidence in its durability.

UNION ELECTRIC COMPANY.

This firm exhibited what was styled the Main high voltage cell. It possesses the advantage of high E. M. F. (2.6 to 2.4 volts), extremely light weight (9.25 watt hours per pound of complete cell), entire absence of sulphating and no buckling.

The positive plate of this cell is made of a number of thin sheets of lead enclosed between two stout outer sheets; these are all perforated simultaneously in a press and secured together by numerous leaden rivets. The plate is formed by the prolonged action of a charging current which converts the thinner inside sheets completely into peroxide of lead, leaving the thick ones, which are but superficially affected, as permanent retaining walls and supports. The transverse holes which remain open give free access of current and fluid to the interior of the plate, thus giving large active surface; no peroxide can deposit, and the plate is discharged uniformly throughout to maintain the distance between the positive and negative plates. The construction of the negative plate differs even more widely from the usual methods than that of the positive plate. It consists of an amalgam of zinc and mercury deposited electrolytically upon a number of perforations through it, giving a good hold for the zinc amalgam. Zinc gives a higher E. M. F. than the spongy lead usually employed, and but little is required. The copper support gives permanence and high conductivity to what would otherwise be a perishable plate, while a small amount of mercury suffices to prevent local action, as it spreads by capillary action over the whole surface. The method of connecting the cells is extremely simple and effective. The connecting wires are of copper, solidly coated with lead. The clamp is an oval collar of hard lead alloy, slightly tapered in one direction to correspond with a key wedge of the same material which is lightly rapped into place, or pressed in with pliers. The inner face of the wedge is hollowed to fit the wire, so that a considerable surface is driven firmly into contact. Rapping the collar downward, or the application of pliers to the small end of the key and the upper face of the collar, will loosen it in a moment and thus no brass or copper is exposed to the action of the acid. The plates are not liable to disintegrate by rapid discharge or from standing idle. They are not injured by over-charge and absorb current rapidly. In these plates the active material, instead of filling separate cavities opening outward, is solidly enclosed between continuous sheets so that it cannot and does not work out. The plates are very stout and strong mechanically, and for this reason, and from the fact that they discharge uniformly throughout the interior, and not from the surface, they do not warp. Their weight in proportion to their capacity is extremely light.

TYPE 2.—In this type of cell both electrodes are made of lead and in a manner similar to the positive plate in Type 1. From this cell an E. M. F. of two volts is obtained. The methods of connecting, etc., are identical with those described in Type 1. The batteries exhibited ranged from 15 to 1,000 ampere hours.

C. POLLAK & COMPANY, FRANKFURT A/M, GERMANY.

In the storage batteries made by this firm the plates are not cast, but laminated between special rollers, giving them a large surface. The plates are not coated but are electrolytically filled with pure metallic lead of great porosity. There is no dropping of the active material, but the porous lead is so strongly coherent with the core that plates may be bent or even hammered without bursting. There is no breaking of active material by the extension of the positive plates but these plates are allowed to expand freely in either direction by a special method of suspension on glass tubes. The Pollak accumulators have an electro-motive force somewhat higher than those of other systems, staying during a good time near 2 volts and giving an average of 1.93 volts. They may be occasionally discharged with very strong currents, and even short-circuited without any harm and yielding nearly the whole capacity. They are light and want but little space in stationary installations. They are extremely light in the special forms built for transportation purposes, as tramway locomotion, car lighting, etc. They are durable to a very high degree, because the plates consist throughout of metallic lead and therefore undergo no local actions. The Pollak system is especially adapted for manufacturing on a large scale, because all the operations—laminating, filling, electrolysis—are performed by machines or by the current, and therefore yields a good financial result.

PLANTE'S STORAGE BATTERY.

This exhibit was one separate cell, spiral type (1860 pattern) and a battery made of twenty cells or accumulators. The charge is accomplished when the cells are in parallel, and the discharge when they are in series; this is done by a special switching device which has been adopted in accumulator plants.

M. PEYRUSSON.

STORAGE BATTERY.—The object of M. Peyrusson was to obtain a greater amount of active lead in the cell for the same total weight. The lead is made in plates one-half millimeter thick. The positive electrodes are made of thin plates welded to a lead axle and firmly held between two horizontal lead plates. In that way the wearing of the plates is from the circumference to the center, which is the conductor part of the system, and the electrode can be used to its complete wearing. In other cells the conductive part of the electrode is, on the contrary, the first to be worn. The negative electrode is a hollow cylinder made of thin plates and held by two lead rings. The capacity is 40 ampere hours per square meter of surface, much more than for any other cell. The surface is 10 square meters for a cell of 33 centimeters high and 25 centimeters in diameter, giving thus a very small resistance.

CHAPTER XVII.

ELECTRIC HEATING, WELDING AND FORGING.

ELECTRIC HEATING.—The latest application of electricity is in the generation and utilization of heat. The discovered trinity in this modern force is light, power and heat. Even more extended and wonderful than the applications of light and power is the application of electricity in the form of heat. Light is required only by night, power is required chiefly by day, while heat is needed throughout the twenty-four hours. So great have been the successes of electricity in its former fields that there is no longer surprise, suspicion, doubt, or criticism offered regarding electricity applied as heat. Because of irrefutable demonstration in these former fields, the popular mind does not repeat its old objection that electric heating and cooking can never be anything more than a luxury. So complete has been the success of electricity in the domain of light and power that the people are ready to accept electric heating without a question. Wherever the convenience, efficiency, and economy of electric heating and cooking are known it is preferred and used. Wood and coal stoves, petroleum and gas-heating mechanisms, are giving way to electric-heating appliances, just as the older fashioned fire-places and brick ovens gave way to their predecessors. The practicability of electric heating has been conclusively demonstrated. The mechanical and domestic arts are opening their doors to receive and prove the advantages of this most recent application of electricity. Nearly all the systems of heating by electricity, especially for domestic purposes, are made to work off the ordinary incandescent circuit, since this form of the current is the commonest where this class of heating is desired. The heat generated by these systems is applied in a variety of ways; cooking, house heating, carriage and car heating, chicken hatching and the like, are a few uses to which it is put. All of these forms were shown in operation at the Exposition.

View Looking South From Electicity Building.

WELDING AND FORGING.—Johnson, in 1865, obtained a patent for heating and annealing wire by a current of electricity. Williams, in England, in 1881, obtained a patent for working metals by electricity, using a system of accumulators. Other inventors procured patents about 1870 for the process of working metals by a current of electricity. That the methods then employed for producing the current were too expensive was no doubt the reason why they were not successful. Since that time some of the ablest electricians in this country and Europe have employed their talents and time perfecting commercially this branch of electricity, until now the shaping of metal, welding and forging, are commercial and are in practice in many factories, car shops and special machine shops all over the land. At the Exposition, one of the most interesting operative exhibits was a plant of welding and forging machinery—that of the Electrical Forging Company, of Boston. Other companies showed metal worked in various shapes, also machinery for doing the work.

THE ELECTRICAL FORGING COMPANY.

This company maintained in operation, from the opening to the closing of the Fair, a plant requiring constantly about thirty horse power of current at 210 volts, and periodically about 100 horse power at 500 volts. The operation consisted in melting and shaping iron, steel and copper bars in the presence of visitors, forging and shaping miniature horseshoes, and in melting, for welding and shaping purposes, bars of iron, steel and copper in a water bath containing sal-ammoniac. This bath process attracted much attention on account of its novelty and apparent mystery. One line of the current was connected to a plate in the bottom of the bucket of water, the other being attached to the end of the tongs used for holding the article desired to be heated. It looked very strange to visitors to see all attendants hold a bar of copper by a pair of tongs in a pail of cold water for a moment until it fused ready for welding, then see the attendant take the end of the bar in his hand and dip it into the same pail of water to cool it off. The miniature horseshoes made were sold for ten cents each, the Exposition company taking the profit in return for the current furnished for the operation of the machinery. The plant used by the company consisted of two alternating generators, each of one hundred horse power capacity; two one-hundred horse power moters for driving the generators ; one forty horse power motor for driving metal working machinery, two exciters for exciting the fields of the alternating generators, one metal rolling mill, two special rolling machines, one punching press, one hot punching press, one six hundred pound drop press, one eight forge heating apparatus, capable of heating, or adapting to heat at different or the same time, metal rods one-quarter inch square by three feet long; two bars three inches square and three feet long; one apparatus for working metal by the electrical arc; one complete apparatus for working metal in a liquid bath; samples of forging, welding and brazing by electric heat; samples of brass, copper and steel, worked by electricity; samples of ornamental metal work, built by electricity; machines for forging by the rolling process and by electric heat, articles of regular and irregular shapes. In this connection, some figures showing the relative cost of operating an electric plant for heating, welding and forging, as against the cost of a similar plant operated by coal direct, will be interesting.

ELECTRIC PLANT.

Cost of plant,	$22.500.	
Interest, 6 per cent,		$1,350.00
Repairs, 4 per cent,		900.00
Coal at 2 lbs. per h. p. hour at $1.25 per ton,		2,250.00
Man to run generator,		900.00
Helper for heating on transformer at $1.50 per day,		450.00
Total yearly cost,		$5,850.00

To this must be added the power plant for running the main generators.

Engines, 600 h. p. boilers, pumps, heaters, piping stack, shafting, power house,	$39,000.00

YEARLY COST.

Interest on $39,000,	$2,340.00	
Engineer,	900.00	
Fireman and helper,	1,800.00	
Assistant engineer,	700.00	
Insurance,	400.00	
Ground lease,	(?)	
Repairs,	2,000.00	
	$8,140.00	
Power plant,	5,850.00	
Total yearly cost,		$13,990.00
Total daily cost,	46.66	
Or for one billet at one billet per minute for 600 minutes, viz., a day,	.0778	

PLANT FOR COAL.

The cost of heating a billet by coal in a shop would be as follows:

Billet 1x8x36.288 cubic in.; weight 80 lbs. Coal required, 36 lbs. at $1.65 per ton,	$ .0252

The time required for heating this bar will be about ten minutes longer than the electric process, which will require, as claimed, one minute. Then must be added to cost of heating by coal, ten minutes.

One smith at 30 cents,	$ .05
One helper at 15 cents,	.025
Total cost for heating by coal one 80 lb. billet,	.10
Total cost by electricity,	.0778
	$ .0222

If two helpers were required, the cost of heating billets would be:

By coal, at $1.65,	$ .125
By electricity,	.078
Saving,	$ .047

THE THOMSON ELECTRIC WELDING COMPANY.

This firm made a large and comprehensive exhibit in the Electricity building. The smaller types of machine were the only ones shown in actual operation, but it was the aim of the company to show as great a variety as possible of special apparatus and a large number of specimens of work done commercially by those using the machines throughout the United States. Twelve machines of various forms were exhibited, arranged in such a way that three different kinds of plants were so shown that the method of operation could be readily traced and explained. The largest dynamo was of 60 kilowatt capacity, generating an alternating current of 300 volts. This was a constant potential, separately excited machine. Another dynamo was of 15 kilowatt capacity, also generating an alternating current of 300 volts. The larger of these two dynamos was wired to a welder of the Universal pattern, which will weld steel or iron having a section of three square inches to be welded, also smaller sizes. The dynamo was connected to a switchboard as used in ordinary practice. The smaller dynamo was connected to a Universal welder built to weld steel or iron having a section to be welded of 1.2 square inches, and smaller sizes. This dynamo was also connected to the regular switchboard. Both of these welders were intended to illustrate the form of machine used where any company may have a variety of work to be done. The machines are so designed that by a simple changing of the clamping device, which can be very quickly accomplished, an infinite variety of welding can be done upon them.

In contrast to these two Universal machines were shown a large pipe welder designed to weld extra heavy pipe, 6 inches in diameter, and other smaller sizes. In this machine was shown the special arrangements used by the Thomson Electric Welding Company for procuring a perfectly equal distribution of the electric current throughout the piece to be welded. This pipe welder is operated almost entirely by hydraulic devices, and by repeated tests has shown unusually high efficiency. The second special welder was one designed for welding tires of a section 3½ x¾ inches, and smaller sizes. This was a machine similar to those used by many of the large wagon and carriage manufacturers throughout this country. A plant for small welding was shown in operation. This consisted of a 1000 Watt dynamo, self-excited, generating an alternating current of 300 volts. It was connected directly to a welder intended to weld ¼ inch rod, and smaller sized to No. 16 wire, and to a second welder for welding copper wire from No. 6 to No. 16 gauge. The dynamo was belted directly to a Thomson-Houston motor operated from the 220 volt circuit. This latter apparatus was operated regularly at the Fair and proved of great interest to those connected with all forms of metal working, as well as the public generally.

Electrical forging is a branch of work to which a great deal of attention has been paid. To illustrate this class of machines an electric upsetting machine was shown, used at the works of Niles & Scott, of La Porte, Indiana, where it has been in successful operation, upsetting the spokes into the rims of their agricultural wheels. The whole machine is automatic throughout, the spokes being clamped in position, the current turned on, the necessary upset given, and the current thrown off by the operation of the apparatus. Another device illustrating electric forging consisted of a small heater and a mechanical hammer operated by a flexible shaft. This is especially designed for reducing the burr formed by electric welding on rods and wires, but could of course be used for regular welding purposes. While the Thomson Electric Welding Company have devoted a large part of their space to the exhibits of their various forms of machines, it was their intention to show in a thorough manner the various forms of work actually accomplished by their machines as performed in many industries throughout the country, and to enable them to do this many of the companies using these electric welders furnished specimens of their work, which were shown to advantage on specimen boards in the exhibit. Some of these companies sent specimens of large axles and tires, as well as all the sizes of hub bands used by them. Two axles were shown of each size used; the first left exactly as it comes from the welder; the second finished ready for use, the burr formed by welding having been removed by hammering at the original heat. Tires were shown in a similar way, two of each size—one with the burr on, the second with the burr off. Some of these specimens illustrated the welding of channel iron, which can be easily accomplished by the electric process, which it is exceedingly difficult to do in the forge. Specimens of pipe work as used for ice machine and refrigerating purposes were shown in which many electric welds were made. Return bands were shown welded to various forms of couplings and flanges; straight pieces of pipe also welded to flanges and couplings, and pieces of pipe welded directly to various forms of return bands made of cast steel, wrought steel and other forms impossible to weld by other means. Welding, as done in carriage hardware, was illustrated. Also specimens of steel and iron wire and rod welding were exhibited, the latter showing spring wire containing 60 to 80 per cent carbon, which can be readily welded by a special process recently developed by the company. Copper welding was illustrated by specimens in which were shown copper rods ⅜ inch in diameter, containing eight welds drawn down to No. 16 wire without any breaks occurring in the weld; also ¼ inch rod drawn down to about No. 30 wire without any welds being defective. Specimens of welds were shown in aluminum, lead, brass, copper and other metals impossible to weld by any other process. Besides these specimen boards described were shown many forms of large welds, the largest collection being of splice bars welded electrically, welded chairs, rails to which the chairs are welded to the rail proper, various forms of switches and crossings in which one rail is welded directly to another at any desired angle, and many other forms of work. Several of the specimens shown had a total cross-section welded of over 12 square inches, one piece having a welded area of 24 square inches.

THE AMERICAN ELECTRIC HEATING CORPORATION.

The American Electric Heating Corporation exhibited among other things the following systems of electric heater: The Dewey system of Syracuse, N. Y.; the Ahern system of Ottawa, Canada; the Butterfield and Mitchell system, and the Morford system of Chicago. The Dewey system as shown consists of a wire resistance wound continuously around an iron frame from end to end. Various metals are used for the resistance wire, according to the work to be done, and according to the size of heater and voltage of current on which it is to be used. The wires are separated and insulated from the iron frame by means of porcelain insulators. This heater is very largely used at present in electric street car heating. The heater is placed in a cast iron box with openings along the upper and lower edges to allow a constant current of air to pass through for heating. The air enters at the bottom, comes in contact with the heated wires, and passes out as hot air at the upper openings. The Ahern system consists of resistance wires held clear of the outer surface of a tube by means of porcelain or other insulators. Outside of this tube is placed another tube or outer shell of the heater, and the intervening space is filled with a loose composition of whiting, fire clay and sand. The heater is supported on

ADMINISTRATION BUILDING.—LOOKING FROM ELECTRICITY BUILDING.

legs, allowing a constant current of air to pass in at the bottom and pass up through the inner cylinder and out at the top, the air coming in contact with the heated tube passing out at the top for service. This system is also applied very successfully to hot water heaters and various forms of cooking utensils. The packing of whiting, fire clay and sand serves to retain the heat for a long time and prevents cold air from causing a rapid decrease in the temperature of the heater. The Butterfield & Mitchell system consists of numerous coils of wire or thin, narrow bands of resistance metal wound continuously around insulating stems. The insulated rods are fastened at the ends into any convenient bracket. The whole is placed within a shell for atmospheric heating and the current of air passes in under and up through the cylinder and comes in contact with the hot wire or band resistance. In cooking utensils and irons constructed by this method the wires are wound around cast lugs which are made in one piece, without the heater plate. Asbestos is used as insulation in this construction, the lugs themselves being covered with it and the layers of wire being separated by pieces of asbestos. The Morford heaters are made either in tubular form or on flat plates. Enamel is used as the insulating medium and hermetically sealed in the insulation. This form of heater is very desirable for cooking utensils, flat-irons, etc., on account of the compact manner in which it can be constructed, and its instantaneous action. The film of insulation between the metal body and resistance wire being only about a sixty-fourth of an inch thick and having a resistance of over 10 megohms. The insulation is burned on at a bright red heat and adheres firmly to the metallic. body, at the same time holding the resistance close to the body to be heated.

Following is a test made at the Algonquin Club at Boston to determine the expense of cooking a meal by electrical heat, a dinner for twenty persons, consisting of 12 lbs. beef (sirloin roast), 1 large loaf of bread, 1 doz. rolls, 3 doz. potatoes, 4 lbs. of steak, 1 doz. mutton chops, peas for 20 persons, fried fish for 20 persons, coffee for 20 persons; current for doing this was supplied for five cents per h.p. hour and the amount used was 1968.5 ampere minutes at a voltage of 108, giving 3543.3 watt hours, or 4.75 h.p. hours, making the total cost for fuel for cooking this dinner $0.2375, or a little over one cent for each person.

CARPENTER ENAMEL RHEOSTAT COMPANY.

PRINCIPLE OF THE RHEOSTAT. — The principle of this rheostat depends upon the conduction and dissipation of heat by holding a resistance wire in intimate mechanical contact with a supporting plate, so that the heat generated in the resistance wire, by the passage of the current, is rapidly conducted to the plate, which becomes the radiating surface.

This is accomplished by the use of enamel, which attaches the wire to, but insulates it from, the radiating surface plate, and which also completely surrounds the wire and protects it from chemical action. In this way the capacity of a resistance wire for current is greatly increased. In practice it is found that a wire that will carry a certain current when exposed to

A MODEL ELECTRIC KITCHEN.

the air will carry ten times that amount of current safely in this rheostat. On this principle, since the cross-section of the conductor can be greatly reduced, it follows that its length can also be proportionately shortened for any required resistance, and no consideration of mechanical strength of the wire enters into this construction, since it is so perfectly protected and supported on all sides. For example, a No. 20 B. & S. German silver wire will, in this rheostat, carry 25 amperes, and thirty feet of the wire, coiled upon the surface of a plate 10x10 inches, will absorb and dissipate five kilowatts of energy in starting a five kilowatt motor of 220 volts under full load. To increase the radiating surface the back of the plate is provided with ribs for extending its radiating surface. The plate

ELECTRIC SAUCE PAN.

will dissipate continuously eight watts per square inch of one surface, i.e., 10x10 inch plate, or 100 square inches, will dissipate 800 watts. The temperature of the plate dissipating energy at this rate will be entirely safe, as it will not char wood or similar materials when operating under this condition. In order to prevent the expansion of the wire, due to

ELECTRIC CURLING IRONS.

the heat generated in it by the passage of current, from breaking the enamel, the wire is crimped, or bent in a zigzag manner, so that no material length of the same runs in one direction. This so distributes the strain that the wire may expand two or three times as much without breaking the enamel as would be the case were the wire to run in one direction. The four dynamo field rheostats used on the switchboard of the

Weston Electrical Instrument Company in their exhibit were 15 inches in diameter, each intended for 60 kw. generators. They have a total resistance of 20 ohms, divided into 60 graduated or tapered steps, and have an initial and final current capacity of 8 amperes and 3.5 amperes respectively. The body of the rheostat is cast iron and provided with circular ribs, on the side opposite the enamel, for extending the radiating surface. As before stated, the resistance wire is mechanically attached to, but insulated from, the iron plate, which enamel not only serves to secure the wires, but also protects them from oxidation and deterioration. The terminal points of the resistance are copper pieces projecting from the enamel and are connected by means of screws of good conducting metal to the switch piece or buttons on the opposite surface of the plate, which buttons and screws are insulated from the plate by fire-proof insulating bushings. These field rheostats are so designed that they do not exceed 200 degrees F. maximum capacity when operating under full load.

CONSOLIDATED CAR HEATING COMPANY.

In the construction of the Consolidated Car Heating Company electric heater, the resisting conductor is preferably divided into a number of parts, so that these parts may be assembled in different combinations, offering different resistances to the passage of the current, so as to obtain different intensities of heat. For convenience, the resisting conductor on each car may be divided into twelve equal parts, two of these parts being combined into one heater, so that six heaters are placed in each car, three being on each side of the car. The resisting conductor wire is coiled in the form of a close and continuous spiral. A cylindrical glazed porcelain insulator is provided, mounted upon a five-eighths inch square iron rod. A spiral groove is formed upon the surface of this cylindrical porcelain insulator and extends continuously from one end of the heater to the other. The resisting conductor above described is placed in the spiral grooves after being drawn out sufficiently to prevent neighboring coils of the spiral from coming in contact with each other. At the end of each division of the heater a glazed porcelain plate is provided, in which binding screws are placed, to which the ends of the spiral are connected. The two parts of the resisting conductor placed in each double heater are separated by a glazed porcelain plate, so that the parts may be connected independently where the plan of wiring so requires. The spiral resisting conductors are placed on the porcelain insulator, which is supported by the iron rod above-mentioned and carried at the ends of the heater upon two cast iron supports. By these supports the heater is fastened to the floor of the car, and these supports also furnish the means of securely attaching the covering box, of the heater itself. All parts of the apparatus are thoroughly insulated and are so arranged that parts near each other can have but a slight difference in potential. Dangers from short-circuiting are, by the construction of the heater, easily removed. The heater is covered with a wooden box which is lined throughout with asbestos mill board, in which is made a screened opening through which the circulation of air from the heater is maintained. The covering box may be made in wood of any desired kind and finish. To prevent the oxidation of the resisting conductor a galvanized wire is used which has the advantage of not oxidizing readily, even when a high temperature is maintained. An absolute assurance against any dangers from the oxidation of the conductor is secured by carrying in the resisting conductor a comparatively low temperature. To accomplish this, it is necessary to use a wire of considerable length, so that this wire may present a large heating surface and give off the desired amount of heat without the necessity for a high temperature.

H. W. AXFORD.

Electric Incubator.—This exhibit was an ordinary incubator, the eggs made to lay in warm, damp sand, the heat being furnished by an electric current, controlled and kept at a proper temperature by a specially devised thermostat—the principal device in the system. This thermostat is made sensitive to one degree, and is set for 102 degrees, the temperature required to hatch eggs. An automatic switch is the controlling device, which, when the mercury rises above the normal point, closes off the current and again throws it on when the mercury falls below 102 degrees. The exhibit proved a very popular one among the masses at the Exposition, and was continuously surrounded by great crowds. One of the peculiarities about the incubator was that chickens were hatched in nineteen instead of twenty-one days, probably on account of the temperature, 102 degrees being a degree or more above the temperature of the hen. It is a question whether this forcing process is not at the expense of the chicken's development and strength.

GENERAL ELECTRIC COMPANY (LIMITED), LONDON.

The apparatus for heating exhibited by this firm does not employ the principle of enamel imbedded conductors favored in this country. The heating resistance, which can readily be attached to any cooking utensil, is imbedded in a special patented composition which will stand the heat, as experiments have demonstrated, without cracking and dropping off. On exhibition were shown a variety of utensils, such as kettles, frying pans, fish kettles, coffee pots, flat irons, curling irons, etc., etc.

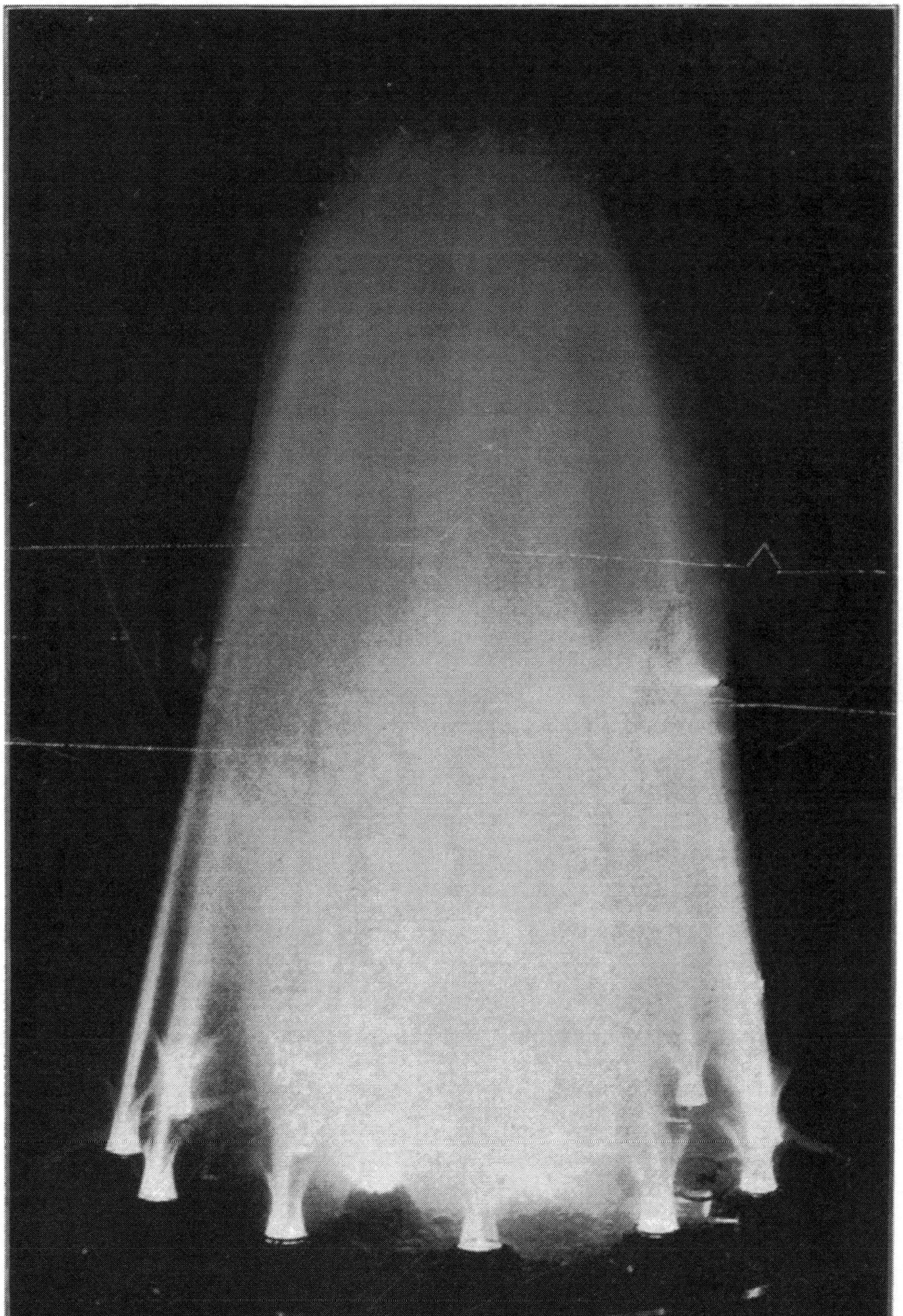

View of Electric Fountain.—Machinery Hall in Distance.

ALLGEMEINE ELECTRICITATS GESELLSCHAFT.

Electric Heating Apparatus.—Heaters for hair dressers' tongs consist of a metal box, in which the tongs to be warmed must be placed. This box, being open on one side, is surrounded by the heating coil of nickel wire covered with a fire-proof asbestos insulation. The whole being covered by a protecting case, the small oven has the appearance of a box having on one side an opening for the tongs. On the same system are the heaters for cosmetics, irons, water-heaters, tea-kettles, glue-boilers and electric cigar lighters, consisting of a platinum wire passing several times through a sheet of asbestos. The current passing along this wire produces a small, glowing surface of the size of the end of the cigar to be lighted. The apparatus is thrown in or out of action merely by taking it off or hanging it on a hook. There were also electrical pipe-lighters.

CHAPTER XVIII.

ELECTRICITY APPLIED TO MINING.

AS applied to mining, electricity is comparatively new, and its advent is in consonance with the commercial success of the alternating current for power and light work. Heretofore the difficulties have related to cost alone.

In many of the western mining districts, there is no coal and very little wood for fuel, hence the objection to steam apparatus. The direct current requires so much copper that its employment over great distances is practically out of the question; the water courses in mountainous countries are usually only capable of utilization in the valleys, consequently remote from the mines.

The use of the alternating current with water wheel solves the difficulty and permits the transmission of the current at high voltage over long distances, and by use of transformers the current may be reduced to suit the various purposes. Under present applied methods power to operate pumps, drills, tramways, elevators and hoists, as well as current for lighting the mine, may be taken from the same high voltage line, transformers being used to control the current and adjust it to meet any required voltage. Auxiliary motors and generators are also employed to transpose an alternating to a direct current if deemed desirable. At the Exposition, there was a complete demonstration of the possibilities of electricity for mining work.

THE GENERAL ELECTRIC COMPANY.

The mining and power transmission display showed all apparatus for use in mining and kindred work and for long distance transmission where the usual system of continuous current and comparatively low voltages cannot be made commercially successful on account of the great first cost. The exhibit demonstrated how rapidly electricity is gaining a foothold in mining work, and showed how easily the energy of a water power located within, say 25 miles, of some commercial or mining center may be made available there in its most flexible form. For mine uses there were shown many styles and capacities of electric pumps doing actual duty. A centrifugal pump throwing 1250 gallons per minute was mounted with its motor on a common bed plate, the pump disc and armature shafts being direct connected by an ordinary coupling, while another pump of similar style was belt-driven. For deep mine work there were several styles of three-cylinder pumps, designed for duty under heads of 200 feet to 800 feet, the motors in all these cases being on the same bed plates as their respective pumps. The motors themselves varied in design from the high speed bi-polar motor with double reduction gear to the very slow speed multipolar motor direct connected. One ot these pumps was mounted on a truck for convenience in shifting to various parts of the mine. A sinking pump was shown which is an ideal one for mine work, being at once compact, light, efficient and proof against damage from water. The motor and gearing are enclosed in water-tight steel cases, so that the pump may be operated, entirely submerged in water, without damage. This pump has a capacity of 125 gallons per minute under a head of 400 feet. A most interesting part of the exhibit was the long distance power transmission system, in connection with which a number of interesting transformations of energy were shown; a very slow speed motor was connected to the crank shaft of a triplex pump by a single reduction gear. The pump, which is double acting and has a capacity of 500 gallons per minute under 650 feet lift, drove a Pelton water wheel, the shaft of which was directly connected by the usual coupling to the armature shaft of a tri-phase alternator of 35 kw. capacity, running at 750 revolutions per minute. This generator delivered current at 500 volts, which in turn was increased in the step-up transformers to 10,000 volts, transmitted across the space on three No. 18 B. & S. gauge bare copper wires supported on oil insulators, reduced in the step-down transformers to 110 volts, when it was delivered to the motors, incandescent lamps, and arc lamps run two in series. There were operated from the secondaries of these step-down transformers 270 16-c.p. incandescents, 6 2000-c.p. arcs and a number of motors. The three-phase motors exhibited were all of the induction type, without brushes, collector rings or external resistance, and ranged in size from the small fan motor to one of 15 kw. capacity. They run at constant speed, starting under load, are highly efficient, and may be reversed even when running at speed by throwing a double-throw switch, reversing the current instantly, There was also shown a machine which, with but a single winding on its armature, may be used in any of the following capacities: as a direct current motor; as a three-phase alternating motor; as a transformer receiving direct and delivering tri-phase current; and as a transformer receiving tri-phase current and delivering direct, the latter being either constant potential or over-compounded 5 per cent. This machine demonstrates the extreme flexibility of applied electricity, and shows how the alternating current may be used for transmitting the energy of distant water powers to desirable markets and there transforming it for use on continuous current systems already installed. A single-phase synchronous power transmission was exhibited, a 60 kw. generator, composite wound for 10 per cent. line loss, driving a motor of similar capacity at 1000 volts, generator and motor being on opposite sides of the space. The armatures of both are ironclad, with no binding wires. The motor is separately excited, the exciter being used as a motor to bring it into synchronism, it being a special form of resonator which indicates in an unmistakable manner when to close the switch throwing the motor on the generator circuit. In the line of electrical drills there were two classes shown, one for percussion rock drilling, and the other for diamond core drilling. The percussion drills are of the same general appearance as steam or air drills for similar work, but they strike a much harder blow, drill faster and consequently are made slightly heavier. The drill consists of two coils of copper wire enclosed in an iron tube, and a wrought iron core moving within them. To one end of the core is fastened a rifle bar rotating the drill, to the other end a rod carrying the drill chuck. Current is brought to the drill from the special generator with rotating brushes, through a flexible cable with three insulated wires. These percussion drills were shown at work drilling 2½ inch holes through limestone rock at the rate of 1 foot per minute, there being two forms of the drill exhibited. At one time a hole 3 inches in diameter and 1 foot deep was drilled for a visitor, who took the time on his stop watch, the hole being completed in 38⅛ seconds after the drill was started. Of diamond drills there were three different types at work, and cores were taken out of limestone rock for visitors. Two drills were for under-cutting, and the third for prospecting. With the latter a core can be taken from a depth of 600 feet, a drum hoist being provided for raising the cores and drills. The prospecting drill is provided with a hydraulic feed, while the under-cutting drills have the usual screw feed. Each of the drills has a triplex pump run by a separate motor, all mounted on the same truck, for forcing water to the cutting points. The mine hoists were illustrated by two examples. The smaller one was a single drum hoist with a 40-h.p. motor of the waterproof railway type, series wound, the average weight hoisted being 3000 pounds, at a speed of 400 feet per minute. The other hoist was also a single drum hoist, but connected with a multipolar motor of 110 h.p., giving a hoisting speed of 550 feet per minute with a load of 6500 pounds. The motors and drums are reversible, and may be separated by a clutch when desired, so that the motor need not be run when lowering unless desired. The drum is provided with a powerful hand-brake, and the brake and other controlling levers are conveniently located in a manner to be most easily handled by the operator. These machines are representative of the hoists built by the General Electric Company, in sizes varying from a capacity of 600 pounds lift under a speed of 200 feet per minute, with double reduction gears, to the hoist with a capacity of 32,000 pounds under a speed of 3000 feet per minute, with direct connected motors. For mine haulage there were three types of locomotives exhibited, designed for very low and narrow entries, and to withstand the rough usage of mine work. One of these was a 25-h.p. locomotive for a 24-inch gauge track, with a motor on each axle connected by a single reduction gear. The extreme width and height of this locomotive are 40 inches and 24 inches respectively; it has a speed of 10 miles per hour and exerts a draw bar pull of 800 pounds. The motors are of the water-proof type, equipped with a controlling switch designed especially for mine work. Another of these was a 30-h.p. locomotive for a 36-inch gauge, with 2 15-h.p. water-proof motors connected to each axle by single reduction gear. The width and height of this motor are 57 inches and 34 inches respectively; it has a speed of 6 miles per hour and a draw bar pull of 1500 pounds. The third was a 60.h.p. locomotive for a 30-inch gauge. It was equipped with one motor, its armature being guarded through two intermediate shafts to the two axles, doing away with connecting rods. The motor is of novel construction, its armature, field coils and all parts being entirely protected from outside injury. The armature is iron clad and of the ring type. The axle gears are set in fixed bearings, and fastened to the axles by flexible couplings, allowing of a spring suspension for the entire machine, motor and all, the advantages of which are obvious. These locomotives are all provided with head lights, sand boxes and powerful brakes, have all their working parts thoroughly protected, and are proving themselves in actual service equal to the most severe demands made upon them. For ventilating purposes a No. 5 Baker rotary blower was shown in operation, driven by a 20-h.p. Edison type motor through a double reduction gear. An air compressor of the Ingersoll-Sargent make, driven by a 25-h.p water-proof motor with double reduction gear, was shown in operation, together with compression tank. For coal mining there was exhibited a pick machine for under-cutting, which in operation strikes 150 blows per minute. A motor of the water-proof type draws a projectile holding the pick against a powerful compression spring and then releases it, the direct blow being struck by the spring and the motor relieved of the force of the dead blow by a

Swiss Alpine Scene.—Western Electric Theater.—Electricity Building.

rubber cushion. In connection with the exhibit were shown a horizontal triplex mine pump of the Gould Company's manufacture, and a Knowles triple, double-acting outside packed plunger pump coupled to a T.-H. motor.

The Gould pump is built with bronze plungers, bushed cylinders and glands, mounted upon iron frame, forming side bars or frame of four wheel truck. This frame is strongly tied together and may be furnished without wheels, adapted for any mounting. It is especially designed for operation in horizontal drifts of mines for pumping water back to base of mine shaft. The Knowles pump is built for a T-H. multipolar motor, 75-h.p., making 275 revolutions per minute. Space occupied, including motor: Length 17 feet; width 10 feet 8¼ inches; height 6 feet. There are three pump cylinders, each having two single acting outside packed plungers, which plungers are connected together by means of rods on the outside of the cylinders. Each end of each cylinder has a number of small valves having an aggregate effective area of ninety per cent of the cross-section of pump plunger. These valves have heavy leather faces. There are six bearings for the main shaft, each 6½ inches diameter and 10 inches long. The pinion shaft has three bearings each 6 inches diameter and 10 inches long. The crank pins are 6¼ inches diameter by 4 inches long; cross-head pins 4½ inches diameter by 4 inches long. The main gear has a pitch diameter of 76 inches and the pinion 114 inches; both have a face of 7½ inches. The main gear cuts with involute teeth and works in a pinion of raw hide. The outside plungers are supported by a slipper working on a slide bolted to the outer end of the bed plate. All irregularities of pressures in the discharge pipe are absorbed by means of three spring accumulators, which act on plungers moving through boxes on the discharge pipe. All the irregularities on the suction pipe are absorbed by the air in two chambers at the end of the pipe. The whole machine is firmly secured to a heavy bed, making it a self-contained machine. Ample means are provided for charging and draining all the cylinders by means of very simple pipe connections. There was also in the exhibit a stone channeling machine made by the Steam Stone Cutter Company. The exhibit purposed to show the application of an electric motor in place of engine and boiler, and to demonstrate the practicability of the application of electricity in place of steam whenever it is desired to do so. And it is fully proven that it can be worked by either steam or electricity without material change, except the substitution of the electric motor in place of engine and boiler. There was also exhibited an air compressor made by the Ingersoll-Sargent Drill Company. The principal features of the air compressor consist in the manner of receiving the free air into the cylinder through a tube (which acts also as a piston rod), creating a uniform draft of air in one direction only, this uniform movement giving a certain momentum to the air, and causing it to fill the cylinder to its fullest extent at each end of the stroke. The air inlet valves are large metallic rings which open and close by the natural momentum caused by the movement of the piston, admitting of a large area of inlet, with but a small throw of the valve. The cooling is effected by means of a new form of water jacket, the construction of the cylinder admitting of a larger cooling surface being presented to the compressed air near the end of the stroke, where the air is hottest, than any other form of air cylinder. The swivel cross-head on this compressor is the most perfect device as yet applied for equalizing unequal wear, and any irregularity in setting up the connecting rod braces.

THE JEFFREY MANUFACTURING COMPANY.

ELECTRIC COAL MINING MACHINE.—The machine consists of a bed frame occupying a space two feet wide by eight feet six inches long, composed of two steel channel bars firmly braced, the top plates on each forming racks with their teeth downward, in which the feed wheels of the sliding frame engage. Mounting upon and engaging with this bed frame is the sliding frame, similarly braced, consisting mainly of two steel bars, upon which are mounted at the rear ends one electric motor, from which power is transmitted through straight gear and worm wheel to the rack, by means of which the sliding frame is fed forward. Upon the front end of this sliding frame is mounted the cutter-bar, held firmly by two solid steel shoes, with suitable brass boxes. The cutter-bar contains bits, made of tool steel, held in place by set-screws. When the cutter-bar is revolved, these cutters, or bits, cover its entire face. The cutter-bar is revolved by an endless curved link steel chain from the driving shaft, and, as it is revolved, is advanced by the above mechanism into the coal or other

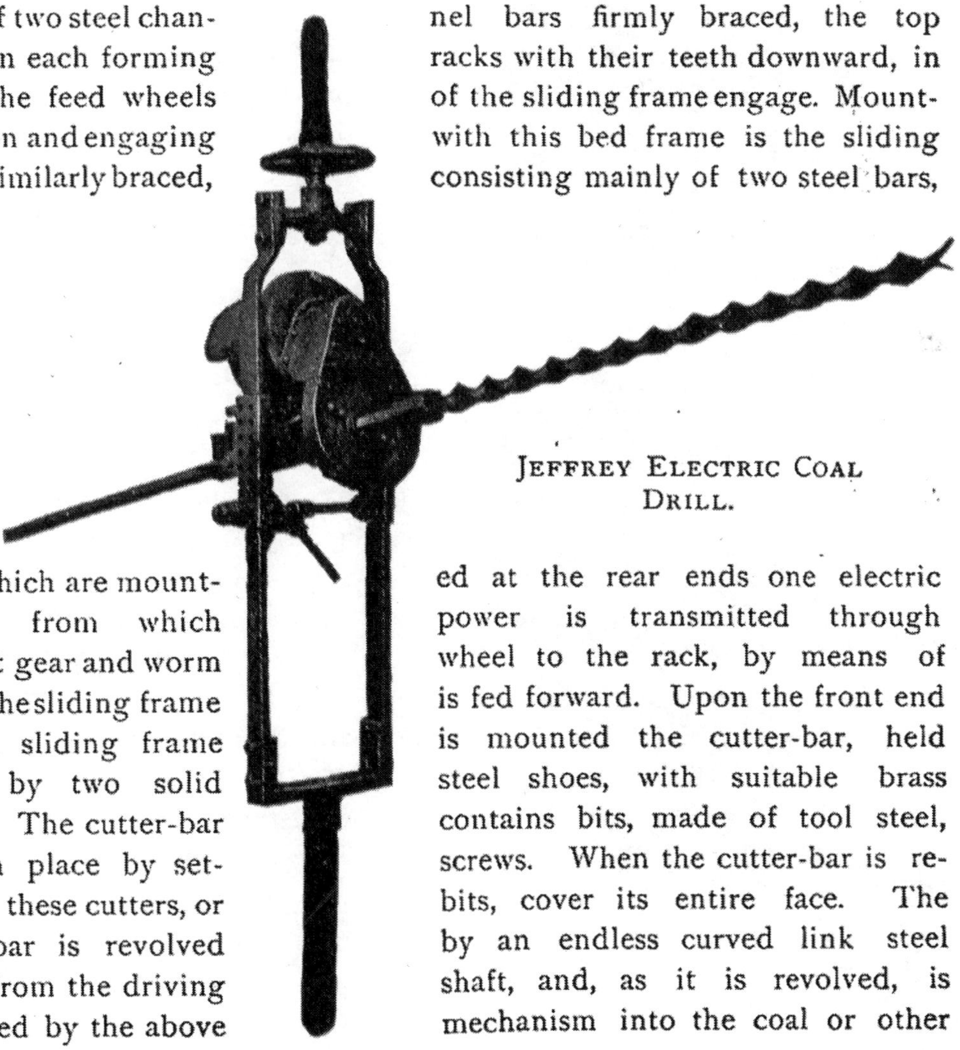

JEFFREY ELECTRIC COAL DRILL.

material, to be under-cut to the desired depth. The electric motor occupies a space about 20 inches square, and is built in the most workmanlike manner, both mechanically and electrically. The current required is from 30 to 50 amperes at a pressure of 220 volts; each motor is wound to develop fully 15 h.p., though frequently, in some veins of coal, the machine only uses 30 amperes or 7½ h.p. in making cuts.

JEFFREY TRUCK FOR HANDLING THE JEFFREY ELECTRIC COAL MINING MACHINE.

The machine is started by means of a switch, located on a suitable resistance box, on the rear end of the motor, the same being arranged with buttons. The current is gradually turned on by simply passing the lever over these buttons. The armature of the motor is calculated to run at a speed of 1,000 revolutions per minute, from which the speed is reduced so as to run the cutter-bar 200 revolutions per minute. The momentum of the armature is such that ordinary obstructions met by the cutter-bar in the coal are not perceptible, causing the machine to run steadily and comparatively quiet. The machines are operated by two men, one man in charge and the other as helper. The machine is taken into the mine upon a truck and run into the room to be under-cut; it is then placed on two boards in front of the coal at one side of the room, and fastened firmly by means of the front and rear jacks, which are braced against the face and roof of the coal—this prevents the machine from moving while in operation; the power is then turned on by the machine runner, and the machine proceeds to its work. The cutter, which is revolved by an endless chain, is fed forward by means of rack and pinion wheels to a depth of five or six feet, according to the size of the machine (the usuai length of the cutter-bar is 39 and 42 inches), and when the full depth has been reached, the feed is thrown off, and, by means of a reverse lever, the cutter-bar is withdrawn to its starting place. This completes the cut, and the machine is moved over the length of the cutter-bar used, and another cut is made in the same manner. Thus the work is continued until the entire width of the room has been under-cut, after which the machine is again loaded on the truck and taken into another room. These cuts are made in an average of from four to six minutes each. The amount of coal under-cut, or the lineal feet face for each machine, depends upon the quality of the coal, and the skill of the men handling the machines. In some coal veins, the machines have cut at the rate of 130 to 150 feet face in ten hours, to a depth of six feet. Twice this amount can be cut if the machines are run on double shift. The construction is very simple.

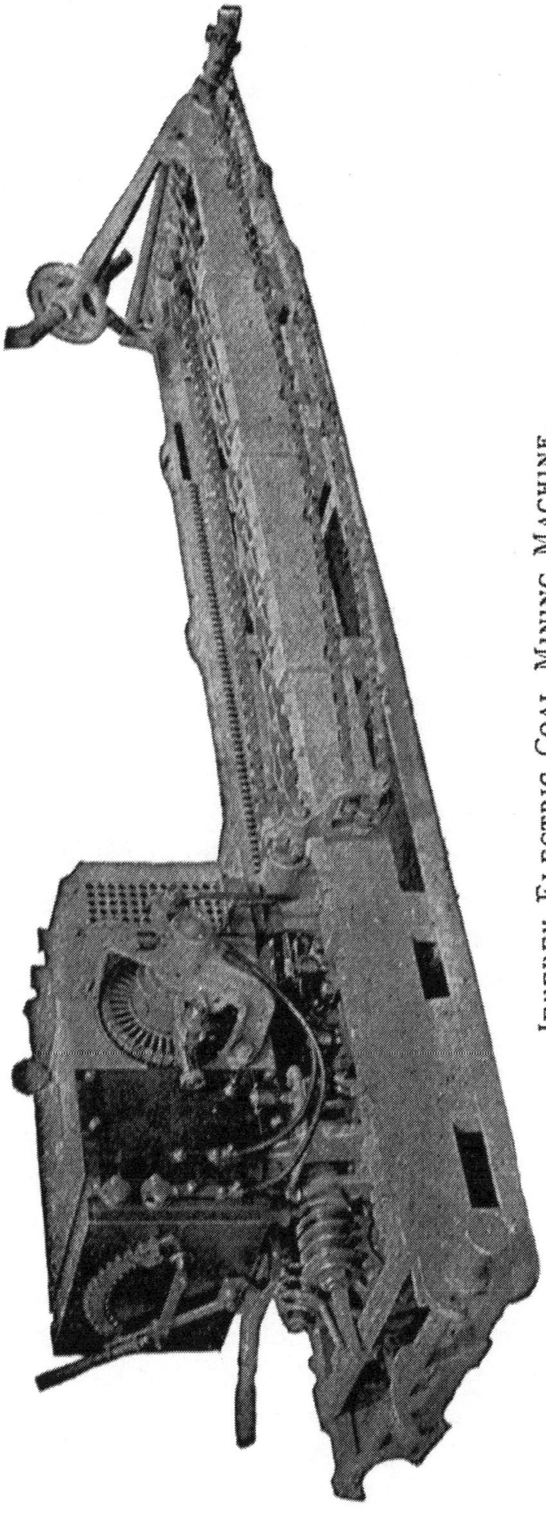

JEFFREY ELECTRIC COAL MINING MACHINE.

JEFFREY ELECTRIC DRILL.—This drill has a small electric motor hung in an upright frame, having projections at top and bottom, arranged with adjustable screws, by means of which it is fastened to the roof and floor of the mine. This is supported by a brace to stiffen and hold the frame rigid while the drill is in operation, although in low veins, with short posts, this brace would not be needed. The motor is of the iron clad type, and will develop 1½ h.p. at a pressure of 220 volts, but can be wound for a lower or higher voltage to suit plants already installed. The working parts, including the field coils and armature, are inclosed, with exception of the commutator, brush holders and terminal connectors, the latter being covered by a sheet-iron hood, thus preventing dirt and pieces of coal from damaging the armature and field coils. There are but few parts to the motor, and these are easily replaced. The armature and field coils are insulated in the best manner, and all danger of a burned or grounded coil is obviated. The motor is not encumbered with a starting resistance or rheostat, as one is not required. The auger runs at about 300 revolutions per minute, and bores a two-inch hole six feet deep in from two to four minutes.

Jeffrey Rotary Electric Pump.—The electric rotary pump is a light, portable combination of rotary pump and electric motor, for use in mines where there is no need for a large or powerful pump. The combination consists of a double rotary force pump, all parts exposed to mine water being made of bronze. The pump has a capacity of from fifty to seventy-five gallons per minute. The motor is standard type, wound for 220 volts, but could be wound for higher or lower voltage; is covered so as to be protected from water. The whole arrangement makes a very efficient and light combination and one that can be easily moved from place to place in the mines, by one man. Wheels and axles can be fitted to suit any gauge of mine track.

Jeffrey Electric Mining Machine Power Truck.—The power truck consists mainly of a well-built frame mounted upon axles, which are fitted with wheels 14 inches in diameter. This forms a carriage for transporting the machine about the mine. A special spool with ratchet wheel, pawl, lever and chain is mounted at one end, supported on a shaft which is mounted on suitable bearings.

The spool, operated by means of chain and sprocket wheels, engages with the machine after it is upon the truck. This drive is quickly adjusted and transmits power to the spool shaft, which in turn transmits the power to the track shaft by another drive chain. The mining machine is arranged with sprocket wheel and clutch, so that it can be thrown in and out of gear. In operating, the machine is drawn upon the truck by means of the spool and is guided into position by means of angle guides which always bring it into a proper position. The chain drive is then adjusted and, by a forward movement of the inside frame, is lightened sufficiently to operate the truck. The machine is thrown out of gear when motor is used to propel truck, and, by means of a reverse switch

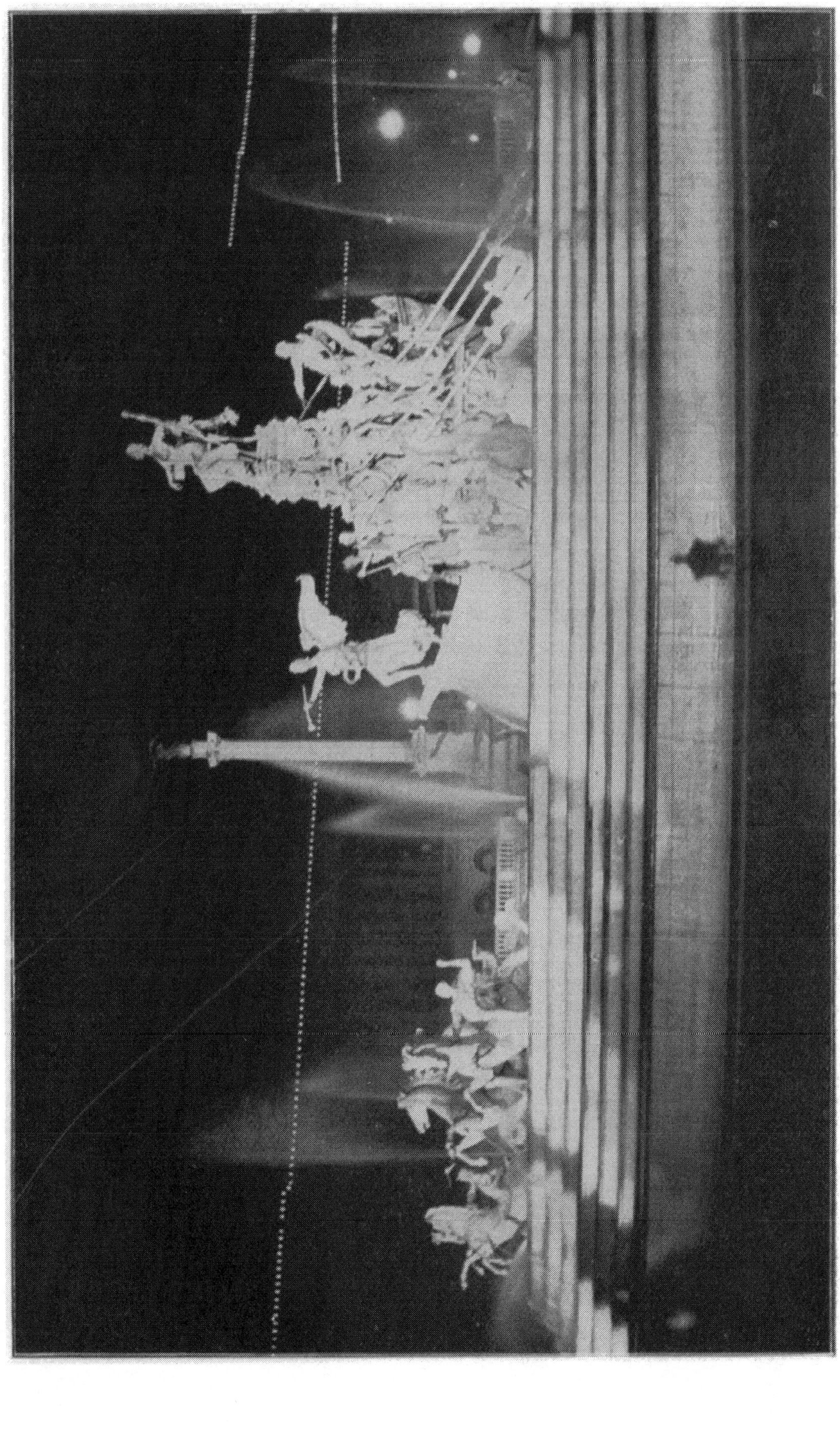

Columbian Fountain.—Under the Great Search Light.

on the motor, it can be run in either direction. The electric current is transmitted to this truck from the wire in rooms or along the entries by a trolley, brush or other device. The regular mining machine cable can also be used. The truck is arranged to travel from 200 to 300 feet per minute.

CHAPTER XIX.

ELECTRO-CHEMISTRY.

ELECTRO-CHEMISTRY is so closely allied to all other branches of electricity that it has been difficult to draw a line of separation between them. There have been left in this class only the electro-deposition of metals, electrolysis, and a few special cases like the manufacture of illuminating gas. Electro-plating, brazing, etc., also come under this head, also the special machinery for performing this work. It might have been proper to treat primary and storage batteries under this heading, but these seem to be so well suited to special classification that it was deemed best to so treat them.

The old Greek philosopher Thales seems to have been the earliest experimenter in this department of science, though his work served merely as a stimulus for others who came after him. It has been said that the scientific electrician must of necessity be a good chemist. This is perhaps true. At least Edison relied upon chemistry almost entirely in his early work upon the incandescent lamp, and other of the most important acquisitions to the science have been along the lines of chemistry. Volta, Ampere, Faraday, and many others who have contributed to electrical development and progress were masters of the art of chemistry, and if the entity called electricity is ever to be defined and analyzed, and given a specific place among the primary elements, or proven to be a combination of some of those already segregated, the result will depend upon chemistry and its laws. At the Exposition, not a large but, a complete, exhibit in this class was made—not so nearly complete in a commercial aspect as some of the other classes, yet, from a scientific standpoint, thorough and exhaustive.

ZUCKER & LEVETT CHEMICAL COMPANY.

Prominently located in the Electricity building, on the main floor in the northeast hemicycle, was the exhibit of the Zucker & Levett Chemical Company, importers and manufacturers of nickel and electroplaters' supplies and polishing materials, also manufacturers of the American Giant dynamos for electroplating and electrotyping, and the Hercules electric motors and dynamos for light and power purposes. The exhibit was not a mere model, but consisted of a complete polishing and plating plant in actual operation. The entire plant was operated by electricity, the current being supplied by the Exposition Company from the power plant in Machinery Hall and transmitted through the subway by one of the contractors in Electricity building to the several Hercules motors in the company's space; these motors in their turn driving plating dynamos and secondary motors for lighting and power purposes. The detail of the exhibit of the Zucker & Levett Company consisted of five American Giant dynamos connecting and charging tanks, for plating purposes, containing copper, nickel, brass, bronze and silver solutions, with a total capacity of five hundred gallons. A two hundred light incandescent dynamo and a twenty-five horse power Hercules motor were shown without load, while several small motors were in operation illustrating some of the numerous applications to which motors of this pattern are peculiarly adapted. A three h.p. motor drove a plating dynamo and at the same time was the motive force to a dynamo of thirty-two 16-c-p. incandescent lamp capacity. A two horse power motor drove a blower, a one horse power motor a buffing lathe and a small one-half horse power machine operated a duplex lamp. The machines are of the inverted horse-shoe type. They have laminated fields in the smaller, and a special cast iron equal to wrought iron, in the larger sizes. The field coils are wound on brass spools which are slipped over the cores. The armatures are of the modified Siemens type, thoroughly laminated to prevent all possible heating due to Foucault or Eddy currents. The armature heads are covered to prevent any dirt or oil from entering into same, also to prevent the wire from mechanical injury. The proportions of the machine are such that the resistance of the field coils is very high, consequently the current is very small, which, together with a strong field, a weak armature—that is, one having few turns and low resistance, with the absence of Eddy currents, give the highest possible commercial efficiency. To make still more attractive and practical to the electroplater, in this exhibit was handsomely arranged, in pyramidal forms and elegant show case, a complete outfit of polishing materials and buffing wheel, scratch and bristle brushes, imported walrus and bull-neck hides, felt wheel, plating batteries, nickel anodes, rouge and chemicals.

THE ROESSLER-EDELMANN PROCESS OF DESILVERING LEAD BY MEANS OF AN ALLOY OF ALUMINUM AND ZINC.

This exhibit, which was in the Mining building, was also of far-reaching interest to the electrician on account of the application which electricity is destined to find in it.

The following is a synopsis of the process, which has now been in operation for two years at Hoboken, near Antwerp, Belgium. It works very satisfactorily because it is simple in practice, and compared with the old methods it saves time, labor and material. It further reduces the losses in lead, silver and zinc, and increases the capacity of the plant. Besides, the zinc produced is of a highly refined quality, approaching chemical purity. The process comprises two parts: (1) desilvering the lead by means of an alloy of zinc and aluminum, whereby refined lead on the one hand, and a homogeneous alloy, consisting chiefly of zinc and silver, on the other, results; (2) working up the zinc silver alloy to refined silver and refined zinc. The new desilverization of lead, which does not call for any modification of the usual plant, differs from the process hitherto applied by the employment of an alloy of zinc with about 0.5 per cent of aluminum, instead of using pure spelter. The advantages of employing this alloy are explained by the following facts: (1) The quantity of zinc required for complete desilverization depends on the quantity of silver present in the lead, on the understanding that comparatively pure (softened) silver lead is to be treated. (2) The quantity of spelter which the molten lead is capable of absorbing depends on the temperature of the latter—the higher the temperature the greater its capacity for taking up spelter. (3) In order to convey to the silver lead the quantity of spelter required for complete desilverization at one operation, without much stirring, the temperature of the lead ought to be a sufficiently high one. (4) At such a temperature the ordinary spelter unavoidably oxidizes; the oxide of zinc covering the spelter prevents the complete solution of the latter in the lead bath, the result being the production of a dirty conglomerate of zinc oxide, zinc, lead and silver (known under the name of zinc scum, being comparatively poor in silver and rich in lead, and offering many difficulties to the subsequent operations for extracting the silver. (5) An alloy of zinc with 0.5 per cent of aluminum does not oxidize at the temperature required for the desilverizing process. A complete desilverization of the silver lead is therefore possible at one operation without much stirring. An excess of zinc is not required. The resulting alloy is perfectly free of oxide of zinc—a liquidation of the excess of lead does not offer the slightest difficulty; an alloy rich in silver is produced, *i. e.*, the whole of the silver is brought to a concentrated state, therefore the bulk of the stuff to be treated for getting out the silver is essentially diminished, and, besides, its treatment is much simpler. Concerning the modus operandi, the zinc aluminum alloy, which has to be prepared previously, is thrown upon the lead bath when the latter has acquired the necessary temperature. The temperature varies somewhat according to the contents in silver, but is nearly 400-500 deg. C. Then the whole is stirred, and allowed to cool, whereupon the molten lead, which at the low temperature is no more capable of keeping the zinc absorbed, releases the latter, which meanwhile has taken up the silver and rises to the surface, whence it is, together with some lead, ladled off. In order to get rid of the excess of lead, the alloy is changed into a cast iron pot with an outlet at the bottom, and slowly heated for the purpose of draining the greater part of the lead. Subsequently the temperature is raised to red-heat for smelting the zinc silver alloy as well as for separating it from the remainder of lead still present. This alloy will not be dissolved in the surplus of lead, but floats on top of the latter as a liquid and homogeneous bath. This alloy is now ladled out, care being taken to leave the lead underneath, untouched. The latter is drawn off subsequently, This alloy consists of 20-40 per cent silver, according to the richness of the silver lead treated, of 5 per cent lead, 2-4 per cent copper and 70-60 per cent zinc. The quantity of this alloy amounts only to about 2 per cent of the silver lead treated, while by the old process about 15 per cent zinc scum, consisting of 4-6 per cent silver, 70-80 per cent lead, 6.5 per cent copper and 10 per cent zinc, was produced. It will be seen that there is a great advantage in having to deal with such a small quantity of homogeneous alloy, in which nearly the total contents of silver from the silver lead is concentrated, and, on the other hand, the direct production of refined lead is increased to 95 per cent. For working up the zinc silver alloy there are two ways: The first way is to treat the granulated alloy by hydrochloric in diluted sulphuric acid, whereby the zinc is secured as a salt, and the silver in the shape of slime. The second way is by electrolysis. Thereby the spelter is obtained as in metal of high purity. This pure metal, of course, commands a price much superior to that of ordinary spelter. The cost of electrolysis is nearly covered by the high value of this electrolytic zinc. The electrolyte consists of a solution of chloride of zinc in chloride of magnesium. The cathodes consist in vertical circular sheets of metallic zinc fixed upon a horizontal spindle, the latter revolving just above the surface of the electrolytic bath in such a way that the rotating dies immerse nearly half way into the electrolyte, while the other half is outside. The spelter is thereby obtained in sheets. The residue of the

The Quadriga on the Peristyle.—Search Light Effect.

anodes left after the electrolytic extraction of the zinc in the shape of slime, consists of 75 per cent silver, 12 per cent lead, 5 per cent copper, 6.5 per cent zinc, with some impurities of arsenic, antimony, iron and chlorides. It weathers on the air very rapidly, forming oxides of copper, zinc, iron and lead. A small quantity of chloride of silver is also formed. A solution of highly diluted sulphuric acid dissolves the oxides of copper, zinc and iron, while shavings of metallic iron will reduce, at the same time, the chlorides of silver to a metallic state. The silver slime now contains nearly 15 per cent of lead, with some copper, and 80-85 per cent of silver. It is then smelted upon a test, whereby the remainder of the lead results as litharge or is absorbed by the furnace. 200 kilos of the silver slime are refined in 8 hours, and it is possible to refine three charges in 24 hours. The cupellation, therefore, is done away with by the new process, and with it the reviving of litharge and other products of cupellation. In lieu of it the short refining process on the test is substituted. As there is only a very small quantity of litharge produced, practically the whole of the silver lead is worked at once into refined lead, so that no subsequent desilverization of the revived bullion as hitherto is required.

THE PITTSBURG REDUCTION COMPANY.

Aluminum and its Reduction.—This company had an exhibit of aluminum, showing: (1) The ores of the metal, the minerals bauxite, cryolite, corundum, Gibbsite, clay, etc.; (2) exhibits of the processes used in the past with metallic sodium as the reducing agent, and description of the older method of manufacture of aluminum by the sodium process; (3) a model showing the use of the Hall process for the reduction of aluminum by the electrolysis of the oxide of aluminum dissolved in a bath of molten fluoride containing the fluoride of aluminum and some metals more electro-positive than aluminum. The pot had the following date, describing the process with it: "Model of the pot in which the Hall electrolytic operation is conducted in the manufacture of aluminum by the Pittsburg Reduction Company in accordance with United States patent No. 400,766." The pot consists of an iron box with a lining of carbon and contains the materials from which metallic aluminum is produced. A glass plate is put in to show a section of these materials and reduced aluminum, in the way in which they are arranged when the operation of producing aluminum is carried on. In the bottom is a layer of accumulated aluminum, which, in the actual operation, is in a molten condition. Floating above this aluminum is the molten strata in which the reduction of aluminum takes place. This consists of a bath of the molten fluoride of aluminum with the fluoride of sodium or other electro-positive metal, represented by the white powdered cryolite. This layer also contains dissolved alumina. On top of this molten bath is a layer of carbon dust, which is employed to retain the heat and prevent volatilization, and a layer of alumina. The latter—alumina—is stirred down into the bath beneath, from time to time, as required, and being dissolved, is the material used up in the process from which the pure aluminum is produced. To the iron pot are attached copper bars which serve as one of the connections for the electric current, the pot itself, with the carbon lining, acting as the negative electrode in the electrolytic operation as ordinarily carried out. The carbon cylinders, immersed in the molten bath and connected by copper rods to the copper bar above, form the positive electrode. The electric current passes along the copper bar down the copper rods to the carbon cylinders, from the carbon cylinders through the bath, when it performs the electrolytic work and reduces metallic aluminum to the carbon lining and iron box and back by the copper bars. In practice the copper bars are connected respectively to the positive and negative poles of a dynamo. The process of the Hall patent, is carried out in substantially the way shown in this model of the only process by which aluminum has been made in any commercial quantities, in the United States. The exhibit showed some of the first aluminum made by the Hall process by the inventor, Mr. Hall, himself; also the first metal made by the Pittsburg Reduction Co. In contra-distinction to this, some of the best metal ever made was shown, running 99.93 per cent. pure aluminum. There were also ingots of the various shapes and dimensions as sold by the Pittsburg Reduction Co.; also ingots of aluminum alloyed with various metals, the whole range being shown of alloys of aluminum with almost all the known metals. Sheet, plates, bars, wire and castings of aluminum were also shown. Besides these, was shown quite a large collection of articles fabricated by different parties, the aluminum in all of which was made by the Pittsburg Reduction Co. This company was the only commercial manufacturer of aluminum that had an exhibit at the Exposition.

F. A. RINGLER COMPANY.

PROCESS OF ELECTROTYPING FOR PRINTING AND DECORATIVE ART. —This is the art of taking one metal and, after placing it in a prepared solution, causing it, by electric or galvanic action, to deposit itself over the surface of a mold or matrix of whatever design, and there being deposited in a film or sheet of any desired thickness. The process is as follows: After the type-pages, wood-cuts or photo-engravings which are to be electrotyped are locked up in a form, they are perfectly cleaned, then pressed by means of an hydraulic machine into a sheet of beeswax. After the impression is made in the beeswax, the surface is thinly covered with powdered plumbago and suspended in a galvanic bath and, by the action of chemicals, electrical or galvanic action is excited, and copper which has previously been dissolved is taken up and spread in a thin and even layer over all the holes and interstices of the wax mold, becoming thicker the longer it is kept in the battery. When deposited to the required thickness it is removed from the wax mold; the back is then covered with a thin coating of solder and backed up with electrotype metal to the required thickness.

ELECTROTYPING FOR INTAGLIO PRINTING.—Under this heading duplicates from fine steel and copper engravings, etchings and photogravure plates are manufactured. This process is as follows: A mold is taken from the original plate in a composition of gutta percha on which the face is metallized, then placed in a copper solution where the copper is deposited to the desired thickness. The face of the copper deposit is then steel-faced in order to harden the printing surface of the plate. By this method a costly engraving can be saved, as the prints from the duplicate are the same as if taken from the original.

ELECTROTYPING FOR THE ART OF RELIEF DECORATING.—This process has given a great extension to the employment of galvanic bronzes for the art of decorating—it is the most effectual and perfect process for popularizing ancient and modern types which may serve to guide the artist and form the public taste.

ELECTROTYPES FROM PHOTOGRAPHS BY THE HALF-TONE PROCESS.—This heading relates to half-tone photo-engravings made direct from nature or photographs, and are used for relief and intaglio printing. The chief points of this process are as follows: A negative of the painting, engraving, photograph or work to be produced is first obtained. As in any branch of photographing, a picture in relief is then made on copper. The copper plate, being finished, is then covered with an infinitesimal coating of steel by means of an electric current, so as to enable it to withstand the wear of printing and keep up the sharpness of the printing surface to an unlimited extent.

TITUS ULKE.

The exhibit consisted of samples and diagrams illustrating improved practice in the electrolytic refining of copper; blister copper, containing 98 per cent copper and 80 oz. silver per ton; electrolytically refined copper, cathode deposit, containing 99.98 per cent copper and 0.3 oz. silver per ton. Blister copper, resulting from the "Bessemerizing" process, is cast into anode plates and hung in vats constructed of well-seasoned wood coated with insulating paint. The electrolyte consists of an acidified aqueous solution of copper vitriol containing 16 per cent bluestone and 5 per cent free sulphuric acid. One side of the anode is coated with molten paraffin and dusted with graphite, to receive the deposit of pure copper. The plates are connected in series. Upon closing the circuit, the anode copper dissolves and pure copper is deposited, while the silver, gold and other impurities in the anode are not dissolved and settle in the bottom of the vat. Samples illustrating an improved process for refining base slimes were shown: base slimes (pulp settling in the electrolytic vats) containing 35 per cent copper, 45 per cent silver and small quantities of antimony, lead, arsenic, sulphur and other elements; dore silver, obtained by refining the slimes (silver mud); gold (99.9) fine, obtained by parting the bullion with sulphuric acid. Base slimes from the electrolytic vats, containing much metallic copper, are refined very cheaply by subjecting them in a vat to the action of air and a small amount of dilute sulphuric acid, whereby the copper is completely dissolved, leaving the gold and silver thus freed from copper in a residue. As dilute acid is used instead of strong, obnoxious fumes and losses of acid are obviated and the process is made very much cheaper. The acid is heated by steam supplied by means of a Korting injector, which also furnishes the air or oxygen necessary to the process. The resulting copper sulphate is either added to the electrolyte or made into crystals of blue vitriol. The gold and silver residues by thorough washing are freed from the little remaining copper. They are then cast into bullion plates or bars and may be parted electrolytically or by the ordinary methods.

THE HANSON–VAN WINKLE COMPANY.

In addition to the exhibit of dynamo-electric machines, the exhibit consisted of a full line of electro-chemicals, comprising the various salts and anodes of nickel, copper, zinc, and their alloys, and the various grades of cyanide of potash used so largely in electro-plating. This company is a manufacturer of these chemicals, anodes, etc., consisting of sulphate of nickel and ammonia, sulphate of nickel, chloride of nickel, carbonate of copper, carbonate of zinc, cyanide of potash, anodes of nickel, copper, brass, zinc, bronze, etc. Processes were shown of the applications of electricity to the refining of metals—bleaching, disinfecting, storage batteries—dependent on the decomposition of the various metallic and other salts, by means of the electric current of the dynamo. In this exhibit, also, was the original Woolwych electro-plating dynamo, the first of its kind, owned by the city of Birmingham, England, and loaned for exhibition as a part of the Hanson-Van Winkle exhibit.

THE ROESSLER & HASSLACHER CHEMICAL COMPANY.

The exhibit consisted of Isolatine, shown in three different forms, as the Isolatine insulating material, as Isolatine sticks and as Isolatine paint. Isolatine proper was shown in two big glass jars in its hardest and softest consistency. In this form it enters directly into the manufacture of insulation of every kind of cable and of most grades of insulated wires, of which samples and coils, insulated with the material, were also shown. It forms either a concentric coat around the metal, with suitable mechanical means for keeping the metal in position, or it is used for the impregnation of the braidings and wrappings of the wire. The Isolatine sticks were of the same material, put up in specially prepared wrappers. They offer a convenient form for applying the material to small objects, such as the covering of splices, etc., and for work in man-holes and on poles. Isolatine paint was exhibited in high glass cylinders and in various kinds of packages showing the style in which it is brought into the market. The paint is prepared from Isolatine, which, with its solvent, forms a perfect solution—not a mixture. As the solvent in drying out leaves no residuum, the dried coating of the paint partakes of all the qualities mentioned under Isolatine. Through its flexibility, permanency and insulation it furnishes one of the best electrical paints, and at the same time it is preferred for the painting of iron and wood, as the usual protective coating. It therefore enters into almost every branch of electrical construction work, from the dynamo to the pole. The exhibit also contained cyanide of potassium, containing 98–99 per cent. pure cyanide. The advantage in the use of the pure cyanide is that the baths are not loaded with carbonates and other ingredients, with which the fused cyanides are loaded, so that they give much better and more reliable results, and can be exhausted more completely. Besides cyanide of potash and Isolatine, of which articles they make a specialty, and on which they report separately, the company exhibited sal-ammonia granular, prime white, containing 99¾ per cent. pure muriate of ammonia, free from any metallic impurities; it evaporates from the platinum plate without showing any residue; chromic acid crude, oxide of manganese black granular, both being used in batteries.

THE ELECTRIC GAS COMPANY.

A peculiarly unique exhibit in the southeast gallery of the Electricity building, and one which excited much favorable comment, was the machine and appliances for the generation of illuminating gas by the chemical action of acid on zinc by methods similar to those employed in an ordinary primary battery. This machine—shown in part by a working

Northwest Side Manufactures Building.—Tiger Bridge.

model—was instant in its action, perfectly automatic and thoroughly safe. The construction showed the following essentials: A main copper tank for acidulated water; a perforated copper pan, containing zinc; a submerged pan of carburetting oil; bell, gas holder. When these are placed together the result is the immediate production of a fine gas light without the aid of heat or fire, the process being as follows: the action of acid on zinc produces a hydrogen gas which is passed, by its own force, through the carburetting oil, resulting in a superior commercial hydro-carbon light of intense power and brilliancy. For all ordinary house or car lighting, or for cooking, heating, etc. and more especially for isolated plants, this system is most applicable and exceedingly effective.

HENRI MOISSON.

ELECTRIC FURNACE.—This furnace is made of two bricks of quick-lime. The lower one has a longitudinal groove wherein to put the carbons and in its center a little hole in the form of a crucible wherein to put a carbon crucible. This crucible contains only a few centimeters of the substance to be reduced. The furnace contains, of course, one electric arc only, and the diameter of carbons is fixed in accordance with the amount of current used. Further, with an engine of 8-h.p. there were produced 100 amperes 45 volts and 2500° Cent. temperature. The first experiments were conducted with a small Edison dynamo and a gas engine of 4-h.p. The current was in general 30 amperes 55 volts and the temperature 2250° Centigrade. Then in the Conservatoire des Arts et Metiers, there were produced 450 amp. 70 volts and 3,000°. Great precaution must be taken in insulating the wires and dark glasses to protect the eyes of the operator must be used. Near 2500° lime, strontian and magnesia crystallize in a few minutes. At 3000° quick-lime is as fluent as water and is reduced to calcium which forms with the carbons a liquid composite. It is easy at this temperature to produce in ten minutes 120 grammes of metallic uranium from its oxide.

PREPARATION OF CARBON UNDER HIGH PRESSURE.—The pressure is realized by the diminution of volume produced when cast iron is solidified. M. Moisson takes pure carbon made with sugar, and puts it in a soft iron cylinder with a screw cork of same metal. In the electric furnace he melts 50 to 200 grammes of soft iron, and throws in it the cylinder. Then he takes the crucible out of the furnace and puts it in water until the surface becomes solid iron, takes it out of water and lets it naturally cool. Afterward the metallic particles are dissolved in hydrochloric acid and the carbon remains in crystals. Some are black, the others transparent. They are to be seen with a microscope. The transparent atoms, where burned in oxygen, give exactly the same result as impure diamonds, but they were too small to allow any weighing of the carbonic acid produced. In the same furnace boiling silver is able to dissolve carbons, and, after cooling, gives also black crystals of carbonate. M. Moisson had in that way a form of carbon crystals whose density runs from two to three and more, very similar to the black diamond.

Preparation of Metallic Uranium.—In the lime furnace some magnesia is put and upon it a carbon crucible containing the mixture of carbon and of oxide of uranium. The arc must be 60 volts and 450 amperes and in a few moments a metallic pig of uranium, exceedingly hard, is made. This mass is not absolutely pure and contains some carbon. The same preparation obtains for manganese and chrome.

HENRI BERTRAND, PARIS.

Process for Rust Proofing, Tinning and Enameling Iron.—Among the various means in use to protect iron from oxidation the Bower-Barff process, or some modification of it, has proved the most successful. It is well known that by this process a coat of permanent magnetic oxide, if deposited on wrought and cast iron, if proper care has been taken, will protect the coated surfaces for all time. This method has been in use in this country for many years and has given very satisfactory results, but, owing to the necessity of a rather costly special form of furnace and the limited output of such furnace, its use has been restricted and has never reached the development that a cheap and rapid method of producing a coating of permanent oxide upon iron articles of any size or weight should have reached. Researches in this direction have been going on for some time, and lately Pierre Henri Bertrand, of Paris, has brought out a new process by means of which iron can be coated with magnetic oxide rapidly, cheaply and effectively, in any furnace of ordinary construction. M. Bertrand discovered the fact that, when an iron article previously coated with a thin layer of a volatilizable metal or alloy is heated to a temperature at which the metal or alloy is volatilized, the protection afforded by this layer prevents the free access of oxygen to the iron surfaces, and the resulting oxide, instead of being a sesquioxide, is a magnetic oxide, of a fine blue slate color. M. Bertrand, perceiving that the ultimate success of his process depended upon a cheap and rapid way of placing upon the surface of the articles to be blued, the first thin layer of alloy, developed a new method by which this can be done in a few seconds. It consists in the preparation of a bath or mixture in which the salts of the metals to be deposited are dissolved. Upon simply dipping the articles in this bath, without the aid of electricity, the metals of the salts are quickly precipitated upon the surfaces, so that the whole operation of alloy, coating and oxidizing is completed in thirty to forty minutes. No special furnace is needed, and all there is wanted to form the coat of oxide is a stay of about ten minutes in an atmosphere of about 1000° C. With proper charging arrangements the furnace can run continuously at the rate of two charges an hour. The operation is very simple, and requires no specially trained men, so that the process can be installed in any foundry at a trifling expense. Pursuing the problem of coating articles with a thin layer of volatilizable metal or alloy a little further, M. Bertrand has succeeded in causing a durable coat of tin to be deposited on cast iron. The method is simple and inexpensive and is used extensively now in coating the inside of cast-iron cooking utensils, which are thus kept from oxidation and preserved from the usual inherent iron taste. Further experimenting with articles coated with magnetic oxide by his process led M. Bertrand to discover a very cheap method of covering such articles with a layer of enamel, which may be varied in design and color, as enamels usually are. The interesting part of this method is that any article of cast or wrought iron or steel may be enameled in any color at one single firing. Articles thus doubly coated with the magnetic oxide and enamel are non-conductors of electricity, and this property has led to their use in various electrical devices, such as parts of dynamos, isolating plates, conduits, etc. The oxide coating and enameling occupy less than an hour. These processes have been in practical use at M. Bertrand's works in Paris for nearly two years, and the results have proved highly satisfactory. The usual products of the works are blued and enameled soil pipes, enameled parts of electrical devices, blued railings and stable fittings, enameled mantel-pieces to imitate various marbles, blued and enameled drinking fountains, etc. The pipe foundry at Pont-a-Mousson, the largest of the kind in France, has acquired the right for bluing and enameling gas and water pipes, and the cheapness with which these operations can be carried on bids fair to develop a large trade, not only in that line, but in coating of all iron exposed to the weather. The exhibit at the Columbian Exposition was an operative one, and samples of work were shown in great variety.

CHAPTER XX.

ELECTRO-THERAPEUTICS AND ELECTRO-SURGERY.

THE electrical current was generated by means of friction, in 1740, and, beginning with that date, static electricity was employed with more or less success in the curing of disease. That there was not pronounced success in its application at that early date was due probably to the fact that very little was known of the science, and, when electricity was used in medicine, it was without special reference to the class of disorder to be treated. It was as though a doctor's only medicine was calomel, and he gave it on all occasions and for all ailments. Under those circumstances it may be readily supposed that electricity did as much, or more, harm than good.

Within the past few years, however, since the subject of electricity has been so thoroughly studied and since our knowledge of the nature of electricity—or, rather, its peculiar properties—has so materially advanced, electro-therapeutics has taken on new zeal, until now nearly every well informed man appreciates the probability of almost miraculous progress in the immediate future.

We are not only bringing intimate technical electrical knowledge to bear on the subject, and are thereby advancing our ability to differentiate between the various forms of current to be employed, but are also making very great strides in another direction—namely, in the fields of biology, botany and physiology. Students of these special sciences are arriving at definite results in the study of the nature of the animal and vegetable nervous system, and have, I believe, already established an intimate relationship between the vital fluids and what we call electricity. The next step forward will be to establish definite currents or waves to or from the nervous centers of plants and animals. When this is done it will be a comparatively easy task to select normal from abnormal conditions, and, when the latter exist, to supply deficiencies, detonize superfluities, or to reverse conditions antagonistic to normality. Doctors agree that specific diagnosis is the essential in the cure of disease.

At the Exposition, no great or new principles in electro-therapeutics were demonstrated. The improvements lay in the direction of mechanics. There were only improved forms, and better construction of apparatus, most of it older than a generation.

There were some quite unique and valuable devices shown. Among these were several varieties of dental motors, mouth lamps, mallets and perforating needles; cautery apparatus for surgical purposes, and, last but by no means least, several devices demonstrating the use of the incandescent lamp in the exploration of cavities and the viscera. One of these applications was a tiny lamp at the end of a catheter, for exploring the bladder in order to detect stone or ulceration. A number of properly focussed mirrors placed along the course of the catheter made it possible to show to the eye the walls of the bladder.

Another instrument, similar in construction, was for examination of the stomach walls, another for the rectum, etc.

Static and Faradic machines were shown with many improvements, also the machines that are used to replace the old frictional devices, notably the Holtz induction and the Wimhurst influence machines.

WAITE & BARTLETT MANUFACTURING COMPANY.

The Holtz Induction Machine.—Static electricity, after having fallen into disuse, was revived in 1856 by the introduction of the Holtz machine. This machine is the most powerful of any ever presented. By those who have had experience with static machines, it is generally conceded that they are frequently charged with difficulty, and give but a feeble spark during damp weather, and that they are particularly unsatisfactory and of little service during the summer months, when such conditions are liable to prevail. It is also conceded that some form of motor is generally required to run a machine of large size, because the hand is soon fatigued in overcoming the friction of the plates upon the rubbers, in addition to that produced by the bearings of the axle and the belt which connects the driving-wheel with the axle. Furthermore, it is now well recognized that plates of large size (26 inches in diameter), and several of them, are absolutely essential to a machine which is intended for medical use. Small single-plate machines do not give sufficient quantity or length of spark to be of any practical benefit as a curative agent. Again, it has been found by experience that building a glass case over a static machine does not thoroughly protect the plates of the instrument from dampness. No cabinet maker can make joints of wood which will

Hemicycle Entrance of Electricity Building.—Morning Scene.

not admit of much moisture when the outside air is impregnated with it; and chloride of calcium, if placed within the case, will not absorb all the dampness that enters and collects upon the plates and metal of the machine. It was found by experience that all the electrodes of the static machine (being insulated by glass) were liable to be easily broken, and that the poles of the instrument, when by accident exposed to outside violence or a blow, were also liable to cause a breakage of the glass windows in the case of the machine (which they perforate in most of the later models). In some machines the case is, moreover, too small for the plates, and allows an escape of more or less of the electricity generated. The first machine made upon the present plan had this objection. It would give the operator an occasional shock in consequence of this defect when the hand was used as the motor, and it lost a large percentage of the volume of electricity generated by "grounding" that portion which jumped to the metal parts of the case. A summary of the new machine made to obviate these difficulties, is as follows: The case is so constructed that all its joints are packed with soft rubber before its screws are tightened. This prevents the entrance of moisture from without, and makes the machine a useful one at all seasons of the year. All imperfections of the joints of the case are perfectly sealed in this instrument. The doors of the case are so arranged as to be drawn tightly in contact with a frame covered with soft rubber by means of milled screws. The glass in the case is put into frames with putty, making it impervious to air. The poles of the machine perforate the wooden portion of the case, instead of the glass windows. They are insulated with hard rubber, and the apertures are protected by soft rubber packing placed between hard rubber buttons and the wood. Thus the danger of breakage of the case by accident is decreased without impairing its impermeability to dampness or allowing of the leakage of the electricity generated when the machine is in use, while the strength of the case is materially increased by this modification. The axle is so built as to reduce the friction to a minimum and to allow of its being oiled without opening the case. The driving wheel is very large, and rests upon a cast iron support. This insures both ease of motion and durability. The glass plates of the machine are nine in number. Six of these revolve and three are stationary. The stationary plates are of peculiar shape. The revolving wheels are made of carefully selected glass, so as to be as true as possible. The machine is charged with a small Wimhurst machine. By so doing it obviates the necessity of using cat-skin and continuously rubbing against the revolving plates. This is a great advance, as it prevents the plates from becoming scratched or marred, and the machine will last indefinitely. The excitants consist of cat-skin rubbers so arranged as to touch the outer plates when the machine loses its charge. This charger is a great improvement over all others previously employed. Metal buttons are also placed upon the outer plates, which as materially assist in charging. The collectors have tinsel attachments which aid in gathering the electricity generated; the electrodes are made with handles composed of hard rubber instead of glass. They are, therefore, less liable to be broken, and are perfectly insulated. Each machine is provided with three pairs of Leyden jars of different sizes. It is arranged, also, with hooks upon which the electrodes may be hung when not in use. The exhibitor claims to have produced with the machine described (26 inch plate) a spark of eleven inches in length during a muggy day in August when most static machines would fail to charge. It runs, after a few turns by the hand, for nearly a minute without any power, and generates without interruption.

The Ranney Cabinet.—For use of the constant current there are 40 cells of Axo Leclanche battery. The current selecter is universal, and any cell or cells from the entire series may be used, and thus a great saving of the battery, and uniform wear of same, are obtained. It contains also an automatic rheotome for interrupted galvanic current. The polarity changer or commutator is of substantial make, has rubbing contacts and keeps in order. The German-silver wire rheostat has coils measuring from 5 ohms to 5000 ohms resistance each, the entire resistance of all the coils being 16125 ohms. By means of the rheostat and the milli-amperemeter, the resistance of the patient's body may be measured or the condition of cells may be tested. The graphite current controller is used to increase or decrease the current gradually. This cabinet includes a set of Engelmann coils, 3 in number, with fine, coarse and medium wire; the coarse wire coil is 66 meters long, the medium coil 200 meters long and the fine coil 600 meters long—the coarse, for muscular contraction, as for contracting subinvoluted uterus; fine wire coil to relieve pain, as in chronic pelvic cellulitis, and medium wire coil for a higher electro-motive force to overcome resistance for muscular contraction. It is provided with a slow and rapid rheotome, also with a contract key, to be operated by the finger. This is used for single impulses, to contract single muscles or groups of muscles, a great help in diagnosis. This instrument is also furnished with milli-amperemeter for measuring the current strength. The case contains two small closets with bevel glass doors, two large drawers and a sliding shelf available as a desk. The castors are of large size, making it roll easily. The lower part has doors at the back as well as the front, making the cells accessible, and of easy inspection.

Improved Faradic Apparatus.—The improvements of this apparatus consist, first, of several coils wound with different thicknesses of wire, and different lengths—the fine wire for relief of pain; the coarse wire for causing muscular contractions and stimulating circulation. This apparatus is of the Du Bois Raymond type, and is arranged so that it can be adjusted by a screw, which increases the current almost imperceptibly. There is also an interrupter, which is caused to rotate by a motor run by an independent circuit. The number of interruptions can be increased from one up to one hundred thousand per minute, thus giving a wide range of electrical impulses. There is also a rheostat in the primary circuit, by which the amount of current flowing through the primary coil can be controlled. The intensity of the current can be increased or diminished by this means. By the above arrangement the physician has at his hands the means of controlling the Faradic current in all of its variations. This is acknowledged by leading physicians to be the only advance that has been made in the Faradic apparatus for therapeutical use.

THE S. S. WHITE DENTAL MANUFACTURING COMPANY.

Electric Dental Outfits.—The Company invites particular attention to the marked improvements which have been made in electric dental systems. The electric dental outfits shown may for convenience be divided into two classes—electric dental appliances and electric driving machinery and accessories. The first outfit was the Kells electric system, operated by a 110 volt current. In this outfit is a regulating resistance and a motor, each enclosed in a case mounted on a single cabinet designed to be placed in suitable position against a wall. An additional resistance is provided for operating an electric dental mallet and a mouth lamp on the same 110 volt circuit used for running the motor. The mallet and mouth lamp were connected to a flexible cord, which, when not in use, was automatically drawn in by a reel located within the motor case. The motor shaft is provided with a magnetic clutch for instantly stopping the driving pulley and the tool actuated thereby, the clutch being under the control of a foot treadle. In the base of the treadle is a switch controlled by a sliding rod, which is operated by a metal upright at either side. With the foot upon the treadle, bearing down upon the heel closes the circuit; pushing out the right-hand upright gives the engine a forward or right-hand motion; pushing out the left-hand upright reverses the motion and runs the engine backward, to the left; pressing down the toe stops it instantly, in whichever direction it is running.

Attached to the treadle is a cable 10 feet long, carrying five connecting wires, terminating in an insulated plate having five lugs or arms. To connect with the motor, slip the lugs into place over the corresponding screw contact points on the plate in front of the motor, and tighten the screws. It is impossible to get this wrong, as the lugs only fit one way.

The reverse motor and treadle can be used with any of these electrical outfits. The motor is wound for the 110-volt direct incandescent electric circuit. The outfit further consists of a Kells adjustable bracket arm with an S. S. White engine head, flexible arm and handpiece. The next outfit was similar to the first, the same double resistance, motor, clutch, treadle, mallet and mouth lamp being used, the cabinet, however, being omitted. The resistance case and the adjustable bracket arm are adapted to be mounted separately upon a wall, while the motor is placed on the floor. Another Kells outfit included all the electrical features of the last

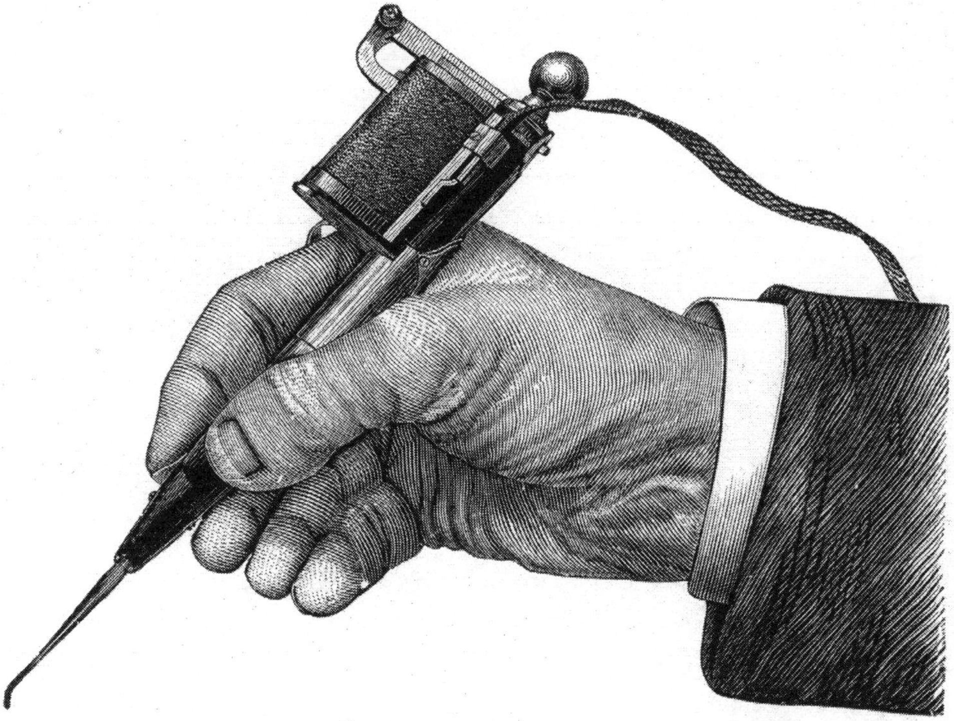

ELECTRIC DENTAL MALLET AS HELD BY OPERATOR WHEN USING.

named with a portable platform with standard substituted for the Kells bracket and taking either the Weber-Perry cord engine, the S. S. White engine head and flexible arm, or the Shaw arm. The company have designed and constructed a special set of improved rheostats, in which they employ the well-known Carpenter form of resistance, the wire being enclosed in an enameled iron plate. These are to replace the wood box resistances used with the Kells outfits and are electrically the same as the wood box resistances, and when substituted for them will control the speed of the motor and operate the electric mouth lamp and mallet equally as well, and, moreover, are absolutely fireproof. In connection with the systems above referred to, any of their brackets, such as the Doriot or the Williamson suspension engine, may be used in place of the Kells bracket arm.

ONE HUNDRED AND TEN-VOLT 6-VOLT MOTOR GENERATOR OUTFIT. —Another of the systems is what is designated as the 110-volt 6-volt motor-generator outfit. It is claimed that this outfit possesses as complete a combination of advantages as is possible in a single apparatus operated by a 110-volt direct circuit, and which uses a wall resistance. The apparatus is designed to be used with the bracket arms and engine heads, the Williamson or Fusee suspension engines, or on a stand on rollers supporting a standard and an S. S. White cord or Shaw arm. In place of the ordinary 110-volt motor, there is substituted a 110-volt 6-volt motor-generator all in one, having two totally distinct sets of coils on the armature, each coil having its own commutator and brushes. On the extension of the shaft are the magnetic stopping clutch and two grooved pulleys of different size, either of which may be used for the belt to drive the dental engine. The 110-volt current passes through the field coils and through one set of coils on the armature and drives the motor; at the the same time the other set of coils on the armature (revolving as they do in a strong field) acts as a generator. These coils are wound to give 6 volts and 12 to 15 amperes. Besides running the dental engine the generator part of the combination will furnish current to operate an electro-mallet, cautery, hot air syringe, etc. This current from the generator has no connection with the street current and can give no shocks. The switch arm on rheostat regulates the amount of current to suit the different appliances, and also, by the use of a patent attachment, using both the 110-volt and 6-volt currents, enables the slowest speed to be obtained with full power. This novel method of obtaining slow speed consists in first throwing a resistance in series with the armature of the motor, which slows it down a certain amount; by the next movement of the switch the current from the generator is sent through a resistance, which resistance is reduced by turning the switch further, allowing more and more current to pass through the generator and slowing the motor down in proportion; but on adding external work to the motor, the current in the generator decreases and gives up power to the external work, thus giving a strong pull on the belt at a very slow speed.

ELECTRICAL HOT AIR SYRINGE AND CAUTERY.—Many attempts have been made to operate a hot air syringe and cautery from the 110-volt current by the use of a resistance, but besides the danger of shocks, especially where connected fountain spittoons give a perfect ground connection, there is great danger in operating a cautery in this manner. If the platinum wire burns in two an arc is set up at the break, with the whole force of the 110-volt current, which may seriously endanger the safety of the patient. The motor generator made for this purpose by this company, when operating the cautery at its greatest heat, does not take

Interior of Electricity Building.—Looking South.

from the 110-volt circuit a current more than equal to that required to illuminate 2½ 16-candle power incandescent lamps, while resistance arrangements for the same work take a current sufficient to maintain 18 to 20 16-c.p. lamps and waste over 95 per cent. of this current. This outfit complete consists of a special iron rheostat with patent slow speed attachment and binding posts to connect the mallet, cautery and hot air syringe, motor generator, furnished with a magnetic stopping clutch and connected to the rheostat, and a foot treadle for stopping, starting and reversing the motor. Although this outfit will operate a mouth illuminator, it is preferable to use the lamp resistance; otherwise the changes in speed of the motor to suit the demands of the various dental operations will vary the brightness of the lamp.

THE NO. 2 MOTOR STAND OUTFIT.—The new S. S. White motor and stand outfit is a combination of a motor on a stand and an iron box containing a fire-proof resistance, fastened to the legs under the stand. Connected to this resistance by a cable and an ingenious removable contact plate, which prevents any possibility of improper connections, is the new starting, stopping, reversing and speed regulating treadle. This facility in reversing will be found of great convenience in finishing a filling with any polishing appliance, as it enables the operator to work continuously against the margin. It is also of use in changing disks or other appliances on screw or chuck mandrels. The stand is mounted on rollers and has a patent magnetic clutch fastened to one side of the iron resistance box, in line with, and directly under, the pulley on the motor. A pulley on the clutch is belted to the pulley on the motor, and a second pulley, which may be varied in size to suit different dental combinations, is mounted on the outer end of the clutch-shaft and connected by belt with the dental appliances to be driven. The clutch has a vertical motion, regulated by a screw at the bottom, to tighten the belt to the motor, while the belt to the dental apparatus is tightened by rack and pinion on the standard. By means of a socket, which screws on the clutch-frame, any of the standards and engine heads, such as the Weber-Perry, S. S. White, or Shaw, can be fastened directly to the stand, which then makes a convenient substitute for the foot engine. By removing the standard and screwing in its place a belt guide, the apparatus can be moved up close to the wall and used to drive a Kells or Doriot bracket, or a Williamson or Elliott suspension engine. The only appliances on the wall are the starting switch and cut-outs. In connection with this apparatus they have also provided means for running an improved dental mallet, a mouth lamp, a cautery, hot air syringe, etc. To accomplish the above results, they place upon the commutator of the motor two collector rings, connecting them with opposite segments of the commutator, and by a proper pair of brushes take from them an alternating current. A suitable transformer of ordinary construction enables a secondary current of any desired electro-motive force to be obtained for the above purpose. In the case of the mallet above referred to, by including more or less wire in the secondary coil of the transformer, the electro-motive force may be varied, and consequently the

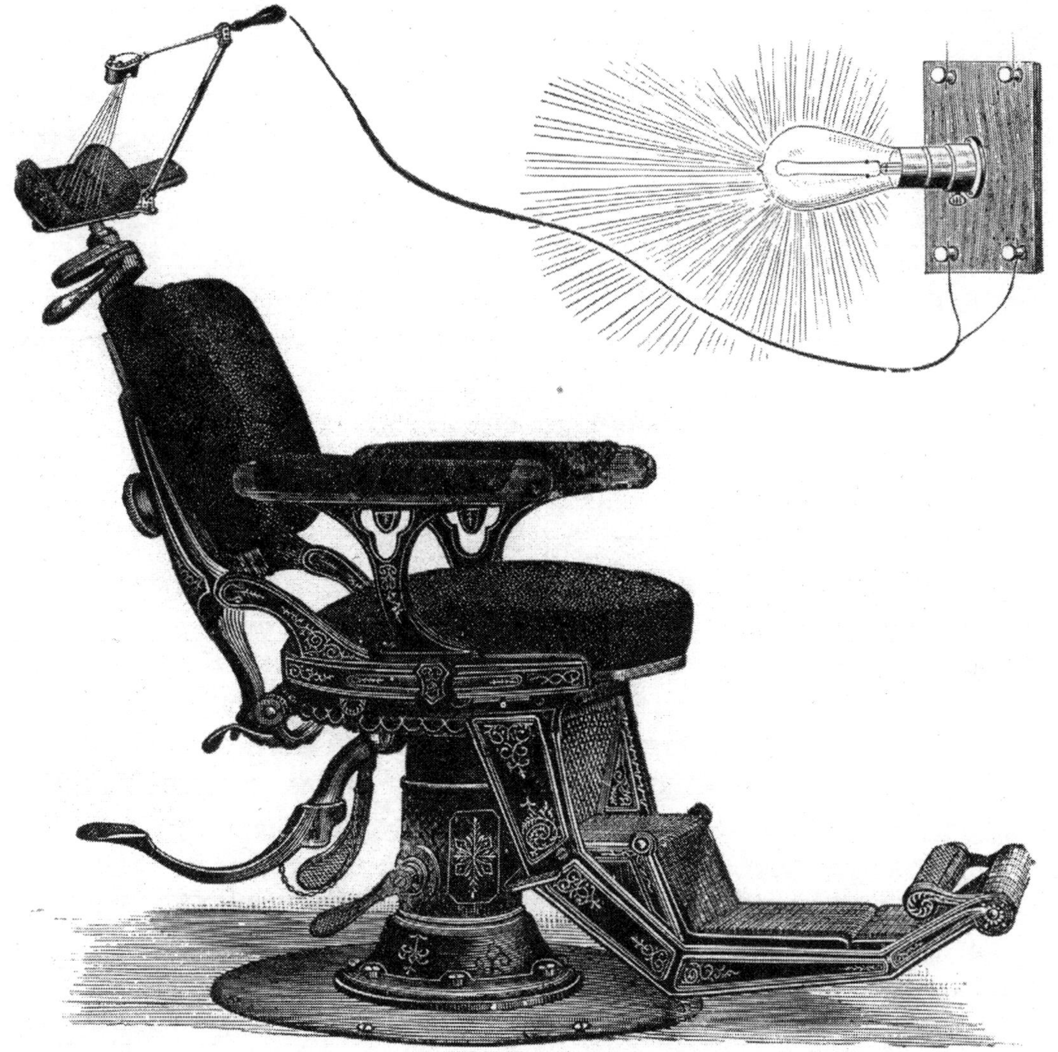

ELECTRIC DENTAL MOUTH ILLUMINATOR ON OPERATING CHAIR, AND LAMP RESISTANCE AS OPERATED ON A 110-VOLT CIRCUIT.

strength of blow, while, by regulating the speed of the motor, the rate of alternation of the current is controlled and with it the number of blows per minute of the mallet. The electric systems operated by currents from high potential street railway circuits are especially deserving of careful consideration. As is well known, the 110-volt direct current, with which these systems have been designed to operate, is in many instances unavailable for the dentist, and for this reason many dentists are prevented from using electricity in their offices. The dangerous 500-volt street railway circuit, on the other hand, often passes directly in front of dental offices, and numerous experiments have been made in order to see if such a circuit could not be brought with perfect safety within the office and made to operate the dental system. The result has been the designing of a number of outfits, differing only in detail, and which may be operated by the 500-volt current, and that, too, without the least danger to the operator or patient. There is no doubt that the 500-volt current, while not necessarily fatal, gives exceedingly unpleasant shocks and has resulted in loss of life. They have, therefore, proceeded on the plan that no part of the 500-volt circuit should come within reach of the operator or patient, nor pass through any switches, rheostats or treadles used to control the dental appliances. They use the 500-volt current merely as a source of power, and transform this power into another current of less potential (110 volts or 6 volts, for instance), which current is used on an outfit similar to the regular 110-volt outfits. One of their systems complete consists of a 500-volt motor which has rigidly connected to its shaft the shaft of a 6-volt dynamo or generator which has no electrical connection of any kind with the motor. On running this motor generator by turning on the 500-volt current, the generator furnishes a perfectly safe 6-volt current, which is connected to an outfit in every way similar to the 110-volt system, except that the rheostat and motor are specially wound to operate properly at 6 volts, and the connecting cords are heavier to carry the increased current. As there is considerable volume to the 6-volt current, binding posts properly connected provide for attachment for any of the electric accessories, such as cauteries, hot air syringes, electric mallets, mouth illuminators, etc. The current to these is regulated by switches. It is intended that the motor generator, with its containing case, should be placed on a shelf out of the way. In this exhibit was shown a 500-100-volt transformer to use with the system where only the 500-volt direct current can be obtained. It is similar to the 500-6-volt system already described, except that the field of the generator is excited by a shunt to the motor field circuit in such a manner that any reduction in the speed of the armature shaft occasions at once an increased current in the generator field coils, thus preventing the drop in electro-motive force which would otherwise greatly interfere with the utility of the apparatus. In the apparatus exhibited, an alternating current was taken from the generator commutator, by means of collector rings and brushes, as previously described in connection with the dental mallet, and, after passing through a choking coil, by means of which its electromotive force was reduced, it was made of use in connection with a mouth illuminator. Among other handsome pieces of apparatus shown were :

electric magnetic mallet, mouth illuminator, mouth lamp, cautery, electric laboratory lathe, hot air syringe, and laryngoscope and octoscope for surgical purposes. The improved hot air syringe is both novel and useful. In general form it is similar to the well-known carbon hot air syringe; the carbon cylinder and containing chamber are of course omitted, and a very

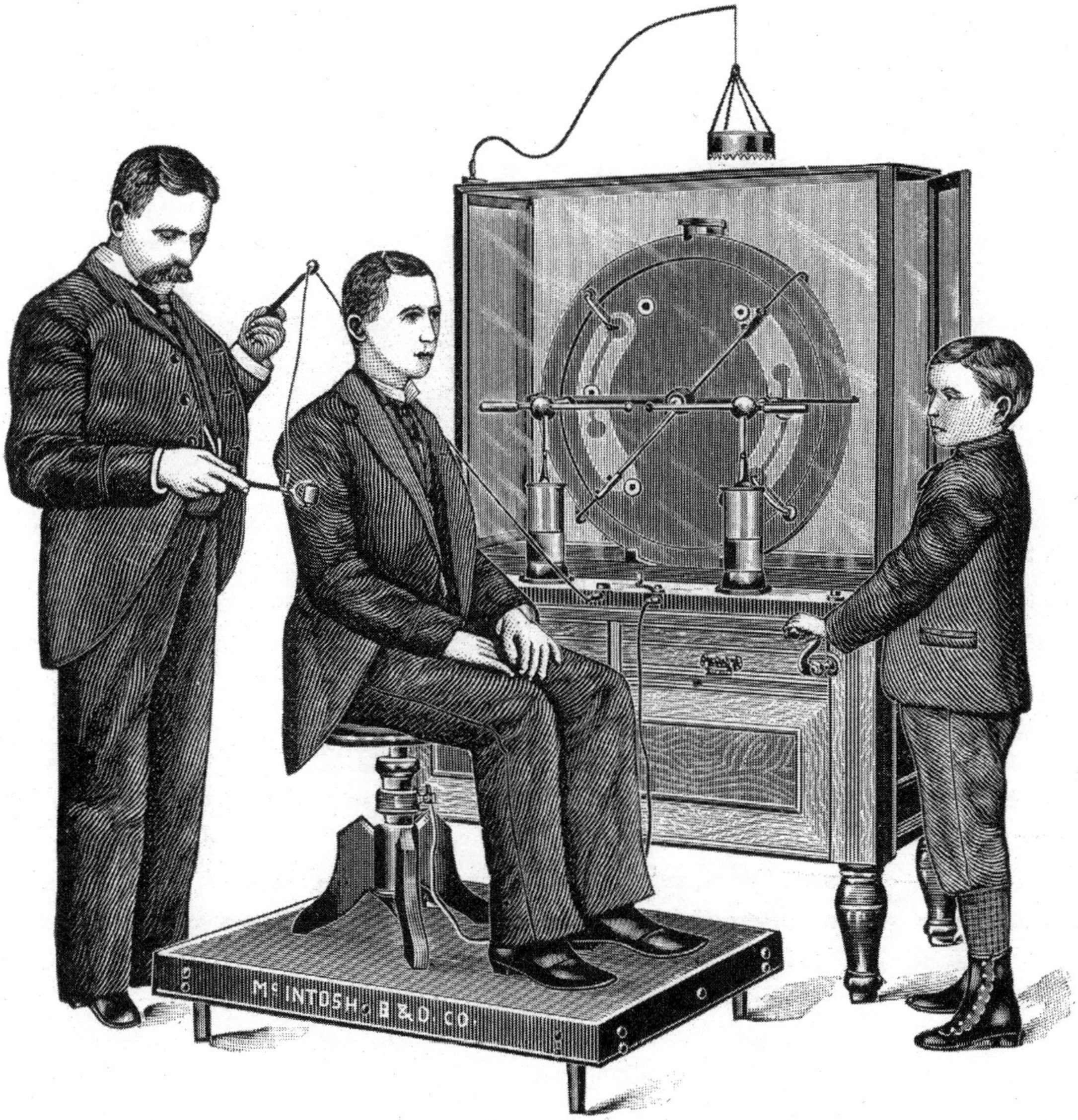

McIntosh Static Machine, with Bath Apparatus.

important improvement consists of a glass nozzle in which is a spiral of platinum wire heated by the current. Glass is a poor conductor of heat, and keeps the air hot while the instrument remains cool. Different size nozzles can be easily and quickly screwed on, giving fine or diffused blasts at will. The current comes in at the back end, and under the thumb is a switch for turning it on.

THE McINTOSH BATTERY & OPTICAL COMPANY.

The exhibit of batteries and electro-therapeutical specialties, illustrated the very best forms of batteries, galvanic, faradic, galvanic and faradic combined, and galvano cautery, in forms portable and for office use. The McIntosh portable batteries embody the most desirable features of a fluid, portable battery, so that, notwithstanding the advancement constantly being made in every form of battery, the McIntosh portable batteries hold the same high rank to-day which they held years ago. The office cabinet batteries are designed for use with fifty or more Leclanche cells, these cells being connected with a galvanic circle of fifty buttons, the cells being brought into circuit through the operation of the McIntosh form of double selector switch. The cabinet is furnished with the customary binding posts, pole changer and selector switches. Resistance is furnished by means of a wire rheostat, introducing from one hundred to fifty thousand ohms. The current is measured by the McIntosh milli-ampere meter. The galvanic current may be interrupted by bringing into circuit the graduated automatic rheotome, this rheotome being constructed for this particular cabinet, so that it is operated either by the usual clock-work machinery, by which the series of interruption can be varied from six (6) to six hundred per minute, and, by the turn of the switch, an electric interrupter is brought into circuit, introducing interruptions from one thousand to ten thousand per minute. The wiring of this cabinet is such that, by the use of particular switches, either the galvanic current or the faradic current in either its primary or secondary forms, or the combined primary and secondary, or the combined galvanic and faradic current, may be obtained at the same binding posts. By the movement of another switch, the current from a battery of 150 ampere storage cells, controlled by a cautery storage battery rheostat, is brought to the same binding posts to enable the operator to use the galvano cautery current. A similar arrangement of switches for combining the various currents is used on the desk cabinet. The need for electric treatment in connection with the bath has caused this company to manufacture several very complete forms of electric bath apparatus. The form shown, connected ready for use, was the combined office and bath apparatus, intended to be used in connection with sixty Leclanche cells. This outfit embraces a complete combined office and bath apparatus, switchboard comprising galvanic circle with double selector switches, faradic coil, twenty-five buttons, wire coil rheostat introducing resistance from one hundred to fifteen thousand ohms, milli-ampere meter and graduated automatic rheotome, large bath coil with double series of switches designed to act as pole changers, as well as to direct the current to any desired

German Historical Exhibit. — Electricity Building.

portions of the tub. These switches are lettered to correspond with the electrodes arranged at the head and foot and along both sides of the tub. There are many cheap forms of bath apparatus in the market, but there are none manufactured by any other company more complete in their appointments, or more certainly calculated to perform the desired purpose. The increased demand for electric massage and for Turkish bath treatment has made necessary the electro-thermal bath cabinet exhibited by the company, an apparatus which offers in a very compact form the opportunity of giving in the physician's office the most complete and thorough electrical treatment in connection with either hot dry air or hot vapor. By the simple manipulation of the switchboard, the current may be directed through electrodes to any portion of the body of the patient within the cabinet. For a somewhat similar purpose, the McIntosh electric chair has been designed, and this appliance has found quite a ready sale for use in sanitariums. No manufacturer of electro-therapeutical instruments can keep pace with the present progressive march in this science without considering the application of the various commercial currents in the physician's practice. In all these applications, safety must always be the first consideration. The McIntosh electric light current switchboard offers a wide range of usefulness in connection with absolute safety. This switchboard embodies the faradic coil, milli-ampere meter, hydro-platinum rheostat, together with the various necessary switches and binding posts. The current is brought to the board through lamp resistance, and being controlled by a half ampere fuse and the McIntosh hydro-platinum rheostat, it is not possible for the patient under any circumstances to obtain a higher current than one-half ampere. In recent years, static electricity has forged its way to the front, as perhaps the most desirable form of electricity in the hands of the neurologist for the treatment of every nervous ailment, together with every form of neuralgic or rheumatic complaint. The McIntosh Atkinson-Toepler electric machine, shown in various sizes in the exhibit, has come to be recognized by the profession as the most reliable of instruments under the most adverse atmospheric conditions. A very elegantly finished machine was shown, constructed with four 36-inch stationary plates, and four 33-inch revolving plates. The most desirable machine, however, for the physician's practice, is the machine constructed of two 28-inch stationary plates and two 25-inch revolving plates. The work on these machines, glass, polished hard rubber and nickel plate work, finished with elegance, made a very showy appearance.

THE EDISON MANUFACTURING COMPANY.

This company exhibited the Edison galvanic cabinet for office work, the Edison physicians' faradic battery and family faradic battery, and the Edison dental motor outfit. The galvanic cabinet consisted of handsomely carved quartered oak roll top cabinet, in the lower part of which were two sliding lead lined trays to hold 50 type "J" Edison-Lalande cells connected with the switchboard on top of the cabinet. The cabinet had two drawers for electrodes, etc., and was fitted with Yale locks. The exhibit also contained a six and eight cell cautery apparatus, also a three cell cautery battery of the same type, also an Edison cautery instrument case of excellent parts and finely finished instruments. An important part of the exhibit was a Delevan condenser for throat illumination. This condenser has been designed to take the place of the old McKenzie condenser where street current can be obtained. The incandescent lamp is so constructed that there is absolutely no shadow thrown on the object examined. It is fitted with a ball and socket attachment which allows the lamp to be adjusted to any angle, making it much more convenient to the operator. Another interesting exhibit was a galvanic switchboard of hard rubber, black walnut case, containing 50 buttons, selector switch, pole changer, faradic coil, with interchangeable fast and slow vibrator, sponge electrodes, battery cord and table.

T. H. PATEE.

Portable Static Machine.—The special features of this machine are its strength and compactness, which render it conveniently portable, and its remarkable electric energy. The plates are firmly held, strengthened and supported in grooved tires or bands of hard rubber to prevent the breakage of glass plates or the warping of hard rubber plates. The band which supports the stationary plate also sustains and strengthens a third plate, usually made of hard rubber and so arranged behind the stationary induction plate as to shield it from injury, to prevent electric leakage, and by its additional surface and induction to increase the flow of current. The revolving plate is adjustable to and from the stationary plate, to which latter it can be firmly clamped for additional support and strength during transportation. The Leyden jars and the conductors leading to them are adjustable and can be folded back against the supporting band of the stationary plate, thus greatly diminishing the size of the carrying case. The axle is adapted to hold and rotate Geissler tubes and other display devices. The base of the machine is also provided with various display attachments so controlled by switches as to produce a variety of beautiful and instructive effects of great value to the lecturer or teacher, enabling him to easily and plainly illustrate the principles of induction, attraction, repulsion, and the laws of electric action. The machine is self generating and remarkable in its electric energy. The character and strength of the current is changed or modified by the use of sliding electrodes and switches, and can be varied from the fierce, lightning-like spark several inches in length, to the gentle, even flow necessary for medical purposes.

UNION ELECTRIC COMPANY.

The medical batteries shown by this company consisted of several types of cells placed in neat mahogany and oak boxes with terminals and carrying strap. They were light and applicable to lighting small lamps, also operating cautery apparatus and small motors. For use with these or other batteries the Union Electric Company showed an aluminum head lamp the weight of which was 3¼ ounces. There was a universal joint focussing arrangement and a light, flexible cord running back of the ear to a storage battery near by. It is found extremely useful by surgeons in operating on the eye, ear and throat and in the cavities of the body.

THE ELECTRO-MEDICAL BATTERY COMPANY.

Improved American Pocket Batteries.—The points of superiority claimed for this electrical instrument over all other pocket batteries are as follows: The patent hard rubber cell, 3 inches by 1½, cylindrical in form, containing the zinc and carbon elements; the solution used to charge the cell is composed of a small quantity of bi-sulph. mercury and water. A screw cap to the cell prevents leaking, so that the battery can be carried in the pocket, fully charged. The connections between cell and coil are made by nickel plated strips securely fastened on each side case, always in sight, and doing away with the old style of small batteries wire connections running along between the false bottom of the case out of sight, liable to rust, and corrode, necessitating frequent and expensive repairs. Portability, power, durability, compactness, and strength of current are among its essential elements of value.

CHAPTER XXI.

ELECTRICAL APPARATUS FOR WAR, MARINE AND NAVAL SERVICE.

SPEAKING tube and cord log practices for vessel service are gone, never to return. All the new vessels now, either merchant or naval, are provided with electric light, electric bells, electric signal lights, and electric signals operated between the different parts of vessels. In naval practice, search lights for various purposes, electric torpedo boats, electrically exploded guns, and an infinite variety of electrical apparatus, have replaced older methods. Speed indicators, logs, rudder, gear, regulator, etc., are electrical. Moreover, it is hardly to be doubted that, within a generation, the propelling power of vessels will be electricity generated en route from either air or water by means of chemical reaction of elements.

THE BATTLE-SHIP ILLINOIS.

ELECTRICAL EQUIPMENT.—In the whole exposition there were few more interesting exhibits than that made by the Navy Department on the model battle-ship Illinois. To the electrical man the lighting features possessed an especial attraction, but, before passing to the consideration of the electrical plant, a few words descriptive of this unique piece of naval architecture may be pardoned. The structure was a full sized model, above the water line, of the Oregon type of battle-ships now building for the United States navy. These vessels have a length on load-water-line of 348 feet; extreme breadth, 69¼ feet; displacement, 10,231 tons; maximum indicated horse power, 9000; estimated sea speed, 15 knots. The main battery consisted of four 13-inch rifles, mounted in pairs in turrets forward and aft on the main deck, together with eight 8-inch rifles mounted in pairs in turrets on each side of the superstructure deck, forward and aft, and four 6-inch rifles. The secondary battery was efficient and provision was made for the discharge of torpedoes and for ramming. The ship was protected by heavy armor, ranging from 18 to 5 inches in thickness.

The Illinois was built in Lake Michigan, opposite Fifty-ninth street, Jackson Park, on a substantial foundation of piling and heavy timbers. The sides of the hull, up to the line of the main deck, were built of brick, and shaped accurately to the lines of the vessel, finished inside and outside with a heavy coating of cement. The sides of the superstructure, the redoubts, and the 13-inch and 8-inch barbettes, turrets and guns were constructed of wood framing, covered with cement worked over metal lathing. Other parts were, in most cases, of similar material to that used in the construction of the coast-line battle-ships. Nothing was shown below the berth deck, which rested on the timber framing over pile heads, rendering it impossible to exhibit the boilers and machinery. The berth deck was the exhibit space proper, and on it were to be found the greater part of the exhibits. All the principal fittings of a modern war vessel were shown in their places, and although many alterations in the plans of the originals had been made (between decks especially), due to the double function of the structure, as an exhibit of itself and to house the exhibit of the Navy Department, the Illinois was a faithful reproduction, in appearance, of the coast-line battle-ships.

The electrical plant was located on the berth deck. It provided current for 350 16-c.p. incandescent lamps, two 38,000-c.p. search lights, and a few motors. In the ships of which the Illinois was a model, the plant consists of two 16-kw. marine generating sets, the dynamos being multipolar and directly coupled to engines. Two of these sets were installed on the brick steamship. In the absence of steam on the vessel, however, these engines and dynamos performed no actual service, but were merely shown as exhibits. Occasionally the dynamo of one set was run as a motor to give visitors an idea of its operation. The real work was done by a motor generator located just forward of the marine sets. This consisted of a 500-volt, multipolar, 50-h.p. motor, directly coupled to an 80-volt, 32-kw. multipolar dynamo, the whole constituting a direct current transformer by which the 500-volt current, taken from the Intramural railway circuits at the end of the pier, was reduced to a pressure of 80 volts, the regulation navy standard. In order to overcome the fluctuations of current occurring on an electric railroad line it was found necessary to employ a wall controller especially designed for the purpose. This device was operated automatically, and cut resistance in and out of the shunt winding motor. All the wiring was done in strict accordance with the requirements of permanent naval work, and the whole plant, indeed, might be regarded as the most thorough specimen of marine electrical

Arch of the Peristyle.—Lighted From Beyond.

work thus far undertaken at a point so far inland as Chicago. The wiring was on the water-tight system, and all the fixtures were water-tight. In the junction boxes, for instance, the wires passed through rubber glands and the top was secured down on a rubber gasket. There was shown a battle lantern with slide, so that the light might be completely hidden during an engagement, when these lanterns would be brought into plan in manipulating the guns. There was also a bulkhead fixture, with a portable lamp just below. A big fixture contained one of the running lights—the lights placed on the bows of all vessels in navigable waters to indicate, by their color and position, the course of the vessel at night. There was also a magazine light, by means of which the ammunition magazines were lighted. Rules against admitting fire into powder magazines are extremely rigorous, and so these rooms were lighted by square fixtures, placed in niches in the outer walls of the magazines, the light shining into the rooms through round bull's-eye windows. There were two search-lights, one on the pilot-house, port side, and the other on top of the hammock berthing, superstructure aft, starboard side. Each search-light required 80 amperes of current at 80 volts pressure, and was electrically controlled. In the base of each stand was a motor, and the controlling apparatus was placed in some protected part of the ship. The operator could, by means of switches, elevate or lower the projector, turn it to right or left, or it could be set to work automatically, sweeping the horizon in both directions at any arc the operator might desire. These search-lights were operated on every open evening, and attracted much attention. The ship was equipped with a complete set of Ardois electric signalling lanterns. There were five double lanterns, operated by means of a key-board in the pilot-house. The lanterns were suspended by a double jack-stay on the port side of the military mast, extending downward from the upper fighting top at intervals of ten feet. Any ordinary message could be spelled out by the red and white lights. In addition to the dynamo plant a battery of 100 storage cells, arranged in series of twenty-five each, was provided. The cells could be discharged in multiple with the generator, or, in case of accident, the lights supplied from the batteries alone. Two Sturtevant blowers operated by ¼-h.p. General Electric motors were on exhibition in the dynamo and storage battery rooms, and a two-horse power motor was used in the machine shop to drive lathes, planers, shapers, etc.

GENERAL ELECTRIC COMPANY.

MARINE AND NAVAL APPARATUS.—In the marine department of this exhibit were shown all the electric light fittings peculiar to man-of-war or merchant marine service. A generator of 15 kw. capacity, direct connected to a Case engine, and switchboard, showing lamps, sockets, switches, lanterns and methods of wiring used for ship lighting, were side by side, and around them search-lights of several sizes, one being the first search-light used aboard a United States war vessel, it having been recovered from the U. S. S. Trenton, which was wrecked off the coast of Samoa. Near by stood a 4-winch hoist, such as is used for loading and unloading vessels and for wharf use, mounted on wheels and easily moved about. The motor driving the winches was of 40-h.p. capacity and of the water-proof type. Suspended from the roof over the exhibit was a set of 10 lamps, 5 red and 5 white, connected to a signaling box so that any possible combination might be made. This system of signaling has been adopted by the United States government, and all modern cruisers are equipped with it. In addition to the search-lights mentioned above, which were controlled either by hand or electrically from a distance, there was exhibited a focusing lamp for photographic work, having a dead white reflector instead of a mirror.

N. G. WARTH.

Torpedo Gun.—This electrical torpedo gun was characterized by the novelty of its motive power, electro-magnetism, which was here for the first time used for the projection of missiles; second, by a force, the safest, most flexible, beautiful and economical of any used for the projection of high explosives—the effect being similar to that of throwing a base ball—uniform acceleration without shock, sudden or violent recoil. This invention has for its object the provision of a method, and apparatus for carrying out the same, by which a projectile may be impelled from the gun to a given point, avoiding the use of dangerously high explosives, the necessity of employing complicated machinery and conductors of great strength, or compressed imponderable agent, and dispensing entirely with the consequent noise, noxious fumes, smoke, etc. The object is also to provide a method and apparatus employed in conjunction with a gun propelling the projectile, by which each distance of delivery or length of range may be exactly reproduced, thus constantly and successively maintaining a predetermined range of delivery, increased or diminished accurately and conveniently at the will of the operator, the energy by which the projectile is impelled being absolutely within the control of the operator. It is apparent in this system that it is necessary to apply the current energizing the electro-magnets acting upon the projectile merely for the length of time that said body remains within the influence of each of the magnets, and which application to each magnet, and its discharge therefrom, is instantaneous, so that a current of great magnitude or volume may be applied and caused to circulate in the coils of the magnet without injury thereto, although the cross-sectional area of the winding wire be such that if the same current were applied continuously as in the case, for instance, of an electric motor, the coils would be burned out and destroyed. The efficiency in point of work of an electro-magnet depends upon the sectional area and character of the core, since it is known (the section and character of the core remaining the same) that any increase in ampere turns or increase in the volume, strength or magnitude of the current beyond a certain degree, will produce no further or additional results of work. The inventor ascertained by experiment that when a current is applied momentarily, intermittently and impulsively, with appreciable intervals of time between each intermittent pulsation, upon increasing the volume or magnitude of said current in an electro-magnet having a core of a given section of material, further effects may be obtained from said electro-magnet in addition or excess of the circumscribed and limited results of the electro-magnet, produced with the continuously applied and prolonged current. Therefore it is apparent that with this method of applying electrical energy, the inventor obtains results which are not circumscribed and which are of especial advantage where it is necessary that the size and weight of the core should remain constant and at the same time demanding an exerted energy of great magnitude.

CHAPTER XXII.

ANNUNCIATORS AND ELECTRICAL CALLS.

ANNUNCIATORS are so common and are made for such a variety of purposes that it is almost useless to go into any critical description of them. Their principal uses are for hotel work, guest calls, time keepers for manufacturing concerns, etc. New forms of annunciators are made for recording degrees of temperature, for time signaling service, on street and steam railways, etc. Another comparatively new commercial use of the annunciator is for keeping and recording engagements, especially designed for hotel and livery stable service.

THE WESTERN ELECTRIC COMPANY.

A high partition in the rear of the exhibit office of the company served as a means of displaying a large line of hotel calls, annunciators and burglar alarms. The hotel calls were made in various sizes, the especial feature being the facility with which access could be had to all the working parts of each. The entire front of each was hinged on the side so as to be swung open, thus opening up all the contacts, as well as all the magnets, without disconnecting any of the wires leading thereto. The magnets and drops were designed with special reference to sensitiveness, without in any way sacrificing the durability of the moving parts. These annunciators were usually accompanied by an alarm by means of which a bell in every room in the hotel might be rung automatically in case of fire. There were also shown house annunciators and elevator annunciators, all using the same style of drop and the same style of indicator and moving mechanism as described for hotel annunciators. Burglar alarms were made in several sizes and provided with switches, so that during the day, when the indicators were not desired to be thrown, the switches might be opened and the doors and windows opened and closed at pleasure without affecting the annunciator. The annunciators were all framed in wooden cases of desirable style and finish.

CHARLES E. LEE.

ANNUNCIATOR.—This annunciator was made in a compact form. It was also cheap in cost of manufacture, as only one magnet was required for any number of indications. There was a shaft carrying at its extremity a needle which moved over a dial bearing the indications required, as, for instance, the different apartments of a flat building. Mounted on this shaft, but concealed by the dial, was a segment bearing teeth on its edge which corresponds with the number of indications on the dial. A detent or catch engaged in the teeth of the segment and held it at the zero point until it was released by the electro-magnet ; it then fell until the detent was released by the interruption of the current. There were a number of spring contacts resting on a piece of metal which move with the segment, in electrical connection with the magnet controlling the detent. These springs corresponded in number with the indications of the dial and also with the notches of the segment, and were in electrical connection with push buttons in the various rooms, and so arranged that, when the needle reached the indication corresponding with the particular spring whose button was being pushed, the spring passed off from connection with the magnet, which in turn released the detent and held the segment and needle at the proper point.

NATIONAL CASH REGISTER COMPANY.

This exhibit showed a novel application of electricity in the operation of cash registers. They were operated by electrical instead of manual power. Two registers were in continual operation. One was known as a No. 79, total adding, check printing register. By depressing one or more keys and turning a lever, seven surprising results were obtained, namely: the indicators showing the previous sale were hidden, a new indication of the sale was displayed in dollars and cents columns, the amount registered was added to previous amounts, a check was issued showing the date, consecutive number, the advertisement of the merchant and the amount of the sale, the amount was printed on a strip of paper automatically wound up, a bell was rung, and the cash drawer automatically thrown open. At the end of the day the total of sales on the detail strip must correspond with the total shown on the adding wheels of the register, else there was a discrepancy, which, of course, could be easily detected by means of the checks. The other one exhibited was known as a No. 33 total adding cash register. By simply depressing one or more keys, six results were simultaneously accomplished: the indicators showing the last sale were dropped; one or more indicators raised, showing the new sale; the cash drawer was unlocked and thrown open ; a bell

The Electric Fountains and Row of Cornice Lights.

was rung to show that a sale had been made, and the amount of the sale registered inside.

E. M. EDGERTON.

CLOCK, GUEST CALL AND ANNUNCIATOR.—The automatic guest call was an arrangement whereby any number of hotel guests might be called at any desired time. For example, a guest gives orders to be called at a specified time. The clerk at once sets the index, which is connected with the bell in the room occupied by the party. No writing is necessary. The index being set, the clerk is relieved of further care and responsibility, as the running of a single clock will make the connection and ring the bell in the room at the desired time. The bell will continue to ring for a space of ten minutes unless the guest awakes and pushes the button located in his room. This act automatically stops his bell and at the same time throws the annunciator drop and rings a bell in the office, which indicates to the clerk that the party is up.

E. S. GREELEY & COMPANY.

ELECTRIC PIANO.—This novelty attracted much attention. Fifteen electric bells were mounted with gongs tuned to the automatic reel. The bells were single stroke; connected to these, to be used as the key-board, were fifteen Victor keys. Any tune might be readily played with this instrument.

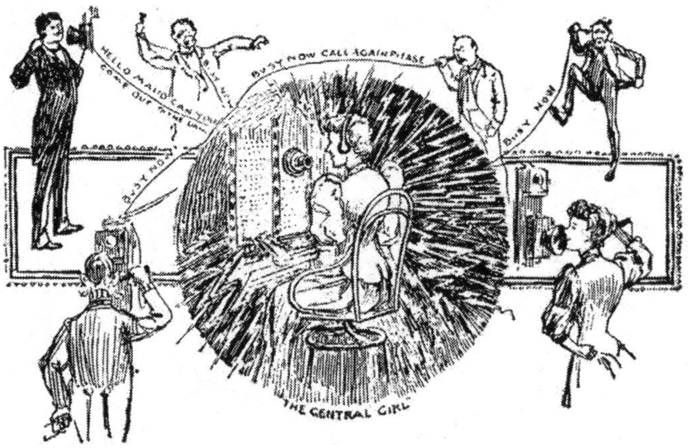

CHAPTER XXIII.

ELECTRIC CLOCKS.

TELEGRAPHY is employed in clock-work to secure a uniform system of time between points remote from each other. This is in fact only a part of their office, but the word electric is a misnomer in that the clocks themselves are mechanical purely; most of them are operated in connection with a master clock at some central point, connection between them and the master clock being maintained by telegraph attachments.

Watchman's time detectors are another form of this same general principle, in this latter class, the master clock being located in a central office or a police headquarters, and employing an auxiliary annunciator for recording the rounds of the watchman, when he performs his duty of pulling the key or operating the recording instrument, of whatever form it may be. There has been a good deal of improvement in the direction of this class of electrical service during the past few years—indeed, the service is of very recent birth and development, and has shown its commercial necessity by going into general use. The governments of the various countries, especially that of the United States, have been among the earliest patrons of this service.

SELF-WINDING CLOCK COMPANY.

Gardner System of Observatory Time.—A standard clock, a mean time clock, and a transmitting clock are used in connection with this time service. Transits of stars are observed every clear night, and the sidereal clock times of the transits are recorded by an electric chronograph. The standard sidereal mean time clocks are allowed to run on and accumulate their errors without interference, a separate clock called the transmitter being used for the purpose of sending the time signals. This transmitter is compared at 11:40 A.M. daily with the standard mean time clock, which has itself been compared with the standard sidereal clock at 9 A.M. as well as on the evening before if observations have been made. The transmitter can be set exactly right by gently touching the pendulum with the finger, making the clock gain or lose as is necessary, by accelerating or retarding the pendulum. At 11:56:45 A.M. the transmitter is switched in and the seconds beats of this clock are transmitted to all parts of the country, being heard in every telegraph office through which they pass. The small-toothed wheel of the transmitting clock, which omits the 29th, 55th, 56th, 57th, 58th and 59th seconds of each minute, is in operation until 11:59:50 A.M., which is the last signal sent out by it. Then to give opportunity to switch in time-ball and clock circuits, the transmitting clock switch is moved to that point which gives a minute contact only so that no more signals go out until the instant of noon, when a signal lasting for about a full second is sent. This long signal is necessary to insure the working of the electro-magnets which drop the time-ball and correct the clocks. While the signals are leaving the observatory both the transmitter and the standard mean time clocks are recording their beats upon a chronograph, giving a record which is preserved so that at any future time the error of the signal may be measured. Thus every day at noon the clocks in the Government offices are set to the correct time, time-balls are dropped, the numerous offices of the Western Union Telegraph Company are furnished correct time, and thousands of clocks on that company's lines are corrected, many thousand miles of railways are given the time signals over their lines, and the number of persons served daily by this system can hardly be estimated. The whole operation is done automatically except the closing of the circuits, which is performed by the officer in charge of the service.

GARDNER TRANSMITTING CLOCK.—As its name implies, this clock transmits its beats over circuits in electric connection with it. In this clock a small toothed wheel on the second's arbor opens an electric circuit as each tooth passes a small spring, which keeps it closed. The teeth corresponding to the 29th, 55th, 56th, 57th, 58th, and 59th seconds are omitted, so that as the clock makes these beats the circuit remains closed and no signal goes out. By moving a switch the current may be made to pass through a second spring which is pushed aside only at the end of each minute. The armature of an electro-magnet, which is in the circuit, is opened and by the action of a spring when so released it closes two other circuits. One of these circuits is used for the time-ball, the other goes to the transmitting table, where it actuates an eight-point telegraphic repeater. The points of this repeater in turn close the main-line circuits of the Western Union Telegraph Company, the fire alarm circuit and the clock line. The first named circuit gives the correct time to all the company's offices, and on the last are located the controlled clocks corrected daily at noon. A relay for the fire alarm circuit furnishes correct time to the fire department of the District of Columbia, three times a day.

TIME-BALL.—The ball, 4 feet 6 inches in diameter, part of the Gardner system of observatory time, was displayed on the dome of the government building. It is made of light steel, canvas-covered, and air-cushioned by two pneumatic plungers. Hoisted by a small windlass, it is held by a lever until released by the action of an electro-magnet on the completion of the electric circuit at noon of the 90th meridian, i.e., central standard time, the telegraphic signal coming from Washington. Similar time-balls are now dropped daily at noon of the 75th meridian, at Boston, Newport, Wood's Hall, New York, Philadelphia, Baltimore, Washington, D. C., Fort Monroe, Savannah, and New Orleans, all by the telegraphic noon signal sent out by the Naval Observatory, over the Western Union lines from Washington, D. C. Under the auspices of the Spanish authorities a time-ball has been dropped at Havana by the aid of signals received from Washington, via the cable between Key West and Havana. A time-ball will also be dropped from the roof of the Masonic Temple in Chicago, under the auspices of the branch hydrographic office. A time-ball is also dropped from the branch observatory at the Navy yard, Mare Island, on Telegraph Hill, San Francisco, through the branch hydrographic office.

The ball exhibited was shown as having an air cushion formed by a cylindrical tube surrounding the base of the staff upon which the ball was hoisted. An equally effective method of cushioning the fall of the ball has been provided at several places by brass pneumatic cylinders, in the ball itself, dropping on two vertical stationary pistons. The form of cushion varies only with the architectural requirements.

GARDNER CORRECTING CLOCK.—By the addition of a simple mechanical device, actuated by an electric current, the hands of this clock can be set to the true time. By this means an error of the clock of 25 minutes can be corrected as easily as one of 25 seconds, but in order to avoid the effect of a false signal the limit within which correction may be made may be decreased to any desired extent by a special device.

TRANSMITTING TABLE.—This table has the various telegraphic transmitting devices used to send out time signals, drop time-balls, correct clocks, indicate the time of noon by the firing of guns or torpedoes, or by the striking of gongs and sounders, capable of sending a signal over large areas of territory.

THE E. S. GREELEY COMPANY.

Watchman's Clock.—In a show case mounted on the wall was shown a watchman's time detector; this instrument was operated by electro-magnets, one for each station. The magnets attracted an armature on which was a lever with a pin. This pin punctured a paper dial placed on the clock in place of a regular dial. This dial was subdivided in hours and minutes and divided into circles, one for each station. When the watchman marked a contact by means of a push button, he closed the circuit, attracted the armature and the pin point and punctured a hole in the dial, thus indicating the minute and station.

CLEVELAND ELECTRICAL MANUFACTURING COMPANY.

This exhibit consisted of eight different styles and forms of the American (electrical) watchman's time detector, combined with electric time; electric fire alarm switches, and the patrol and fire alarm system of the Chicago Telephone Company; also exhibit of the American & Columbian Employe's Time Recorders. In watchman's time detectors were shown instruments ranging in capacity from 1 to 54 stations, with recording dials revolving every 12 hours, 24 hours, 48 hours, and 7 days; these instruments were shown in standard cases, and combined with office clocks and regulators; they also show these instruments operated by both magneto generators and battery. The American watchman's time detector is designed to keep an exact record of the doings of the night watchman in any establishment where it is in use. The instrument, closed with a Yale lock, is placed in the office and connected by insulated electric wires to key stations placed in different parts of the premises where the watchmen are required to pass. As the watchman passes on his round he operates each station in turn, and this is instantly recorded in the detector in the office, showing where each watchman is, and when he is at each station. The record is made on the paper dial, which revolves on time by a clock movement, the dial being perforated by needles showing the exact time at which each station is visited, as well as the number of the same; if the station is operated a record is made; if it is not, no record is made. It makes no difference how many watchmen are employed, there is a separate record for each, and one watchman cannot make a record for the others without it being recorded on the dial. There is no possible way of making a record except by operating the station, and it is impossible to tamper with it in any way without certain detection. The Agricultural and Electrical buildings were both equipped with this system, the same being used by the Columbian guards, a number of guards registering on each instrument, thus insuring the most per-

Newspaper Row.—Electricity Building.

fect watch service. The electric time system shown was a simple device for driving any number of secondary dials from a master regulator by means of a circuit closer placed on regulator movement. The employes' records are designed to automatically keep the time of employes, any number from 1 to 500 being able to register their time on one instrument at the rate of 100 per minute, thus doing away with the annoyance and expense of taking each employe's time in the old way, saving time for the employe and making him his own time-keeper. The fire-alarm system shown is combined with the watchman's time detectors, and is a simple and inexpensive device which enables the watchman on his rounds to send in an alarm electrically from any station, giving the location of the same.

WESTERN ELECTRIC COMPANY.

ELECTRICAL REMINDER.—To the eye the electrical reminder, manufactured for the Electrical Reminder Company of Chicago, was a highly finished, veneered, and ornamented oak case, 36 inches high and 25 inches wide. At the top, and as the basis of the mechanism, was a serviceable clock of the best known make, which fulfills the double duty of keeping correct time and making the electrical connections to be described. The veneered face of the instrument was lettered for each fifteen minutes of the day and night. Copper contact pins were provided, to be inserted under any of the fingers on the face of the reminder as occasion might require. An electric alarm bell was a part of the hidden mechanism, and was operated by a battery, also out of view. The whole instrument was wired internally, so that the placing of one of the pins into any one of the holes, representing each fifteen minutes of the twenty-four hours, would cause the closing of the circuit and the ringing of the bell upon the instant that the clock recorded the time desired to be announced. Any number of calls could be recorded, and engagements kept for every fifteen minutes if desired. The mechanism of the instrument consists of an eight-day lever clock, to which is attached an extra gear wheel for the purpose of carrying a contact hand at the back of the clock, which makes a revolution every twenty-four hours. A roller at the end of this contact hand travels over a system of wire points set into a hard fibre disc. Each one of the wire points is connected with one of the pin holes on the clock front. The instrument is operated by two Leclanche cells, one wire of the battery going to the clock, the other to one pole of the bell. The battery is normally out of circuit, and to be thrown in, a pin is placed in the hole connected to that particular wire point directly in contact with the roller on the contact hand. For instance, if an engagement is to be kept at 3:15 p.m., a pin is placed in that hole in the board. When the contact hand reaches the wire point connected to the 3:15 p. m. hole, the circuit is complete, and the bell rings. A still further development of the machine has another gear wheel in addition to the one above described, which makes a revolution every seven days. Seven holes in the board represent the seven days of the week, and normally all of these are out of circuit. When an engagement is to be kept, a pin is placed in the hole representing the day, and a second pin is placed in the hole indicating the minute and hour of the day. When the day arrives, the circuit for that day only is closed.

A. NEWMAN.

WATCHMAN'S CLOCK.—The features of this electrical detector are novel so far as its combination is concerned. At every station is placed a box similar to the ordinary district telegraph box, except as to the number of teeth in the break-wheel, which is arranged to suit the station, as for instance: station No. 1 makes one break, No. 2, two breaks, and so on. As soon as the breaks are made, another contact piece is connected, and still another immediately thereafter. The instrument in the office is arranged with a ratchet wheel, a pawl holding it in its place, another pawl connected to an armature, a rod lying above the pawls and fastened to another armature, a needle lying in a barrel having a small spring wound loosely around it, an armature behind the needle, a small chain fastened at one end to the barrel, and the other end to a drum. This drum is rigidly attached to the ratchet wheel. When the watchman turns in, say station No. 10, the armature that holds the pawl is attracted ten times, causing the pawl to shove the wheel ten teeth, which raises the chain around the drum, thus bringing the needle ten steps and lying over the tenth circle of a paper dial. The second contact made in the box attracts the armature over the needle and forces it into the paper. The spring of the needle draws it back and the third contact causes the armature that holds the rod over the pawl to draw the pawls from engagement with the wheel, and the weight of the barrel brings back the needle to its normal position at the bottom of the dial. This clock is arranged in such a manner that the installation is simple, and it cannot be tampered with even by an expert. The three wires, with a return, are all that are necessary, even though there were thirty stations. Station boxes can be put on this line at any time after first installation. On this clock is also placed a 48-hour dial and the whole instrument for 30 stations is no larger than an ordinary man's clenched hand. The case is locked with an escutcheon plate, the removal of which records the tampering by pricking the paper at the bottom line, denoting the time of day or night that the clock was opened or closed. Thus is completed a device, simple in construction, absolutely tamper-proof, and arranged so that the needle cannot stick and hold the paper dial from making its revolution. At the exhibit were shown detectors in 15 different styles of cases, also an historical exhibit demonstrating progress in the art.

ALLGEMEINE ELECTRICITATS GESELLSCHAFT.

Electric Clocks.—The most modern form of electric clocks have become of great importance in connection with central plants for lighting. As is well-known, the attempt has often been made to work clocks for public use from a central station. These efforts date back to the very first and most modest beginnings of the technical application of electricity. There are two principal reasons why electrical clocks have not obtained success. The old systems required their own expensive wiring, and a central station with its own batteries and attendance. They became too expensive, both in cost of plant and of working, to be paid for by the consumer. Through the cutting down of the cost of the mains and curtailing the attendance, the reliability of these plants suffered so much that they came to be looked upon with disfavor. These objectionable circumstances are excluded in the clocks exhibited, for they can be attached to wires already in use, in the same way as an incandescent lamp. Every single clock is wound up, kept going by the current; notwithstanding any interruption in supply, caused for instance by altering joints, the clocks on the circuit go on undisturbed, provided the interruption does not exceed longer than about 12 hours. Besides this, all clocks on the circuit are set to the right time at the same hour, once a day, all differences being corrected. This regulating takes place from the central station through a slight and momentary diminution of tension throughout the whole system. At the pillar in the middle of the exhibition several of these clocks showed the time of different towns and were daily corrected at 1 o'clock P. M. (Chicago time), by an apparatus specially constructed for this purpose. The going of the clocks remained unaltered by any accidental variation of the voltage at any other time. A similar clock was also placed on the railway signaling apparatus in the exhibit. This device has lately been devised for stations with large traffic and is of great assistance to travelers. The clock has an automatically working plate, with the inscriptions, departure, direction, and a place for the hour and minute.

CHAPTER XXIV.

CARBONS FOR ELECTRICAL PURPOSES.

THERE is little to be said by way of introduction to the subject of carbons. Improvements in the manufacture of this class of goods have been made pari passu with the general progress of commercial electricity. There was a time, only a short while ago, when there seemed to be a necessity for an arc light carbon of purity and durability sufficient to enable it to burn a longer number of hours, so that single carbon lamps could run all night without renewing. Of late, however, improvements have been made in the feeding mechanism of the lamps themselves that seem to provide a solution of the question, and there is no doubt that there are single carbon lamps on the market to-day that will burn for all night service, using a carbon that would have been consumed by midnight under the feeding of the old lamps. One of these new lamps is that made by the Western Electric Company, and many of them are being placed in service for the city of Chicago. In the manufacture of the carbons themselves improvements have also been largely in overcoming the tendency to disintegrate and spark. Incandescent carbons are more intimately associated with their lamps than in the case of arc carbons, and will be more properly described under the lamp chapter.

THE NATIONAL CARBON COMPANY.

This firm made an exhibit as follows: arc light carbons made by forced process, in all sizes and lengths from ⅛ inch in diameter, 12 inches long, to 1½ inches in diameter; plain or copper coated in various thicknesses; both solid and cored, soft, hard and medium, and of various grades; made to suit all systems of lighting, with currents from the lowest to the highest amperes, either high or low tension; arc light carbons made by moulded process, in all sizes and lengths, plain or copper coated in various thicknesses; soft, hard and medium and of various grades, made to suit all systems of lighting, with currents from the lowest to the highest amperes, either high or low tension; cored carbons of highest quality in all sizes and lengths, for use in arc lamps on incandescent circuits; search lights, focussing lamps, and other special lamps and electrical contrivances where "fancy" lighting is desired; smelting carbons of all sizes and lengths from 5/8 inch in diameter upwards, for use in smelting aluminum, gold, silver, copper and other metals; motor and dynamo carbon brushes of all sizes, either plain or copper coated, with various thicknesses of copper; carbons for chemical purposes, in all sizes and lengths used in connection with electric currents, for the separation and purification of various liquid elements; battery carbons, cylinders, cups and plates of all sizes, thicknesses and dimensions; a most complete assortment of all classes of electric carbon products; the highest quality in purity of contents, uniformity of resistance and the lowest average resistance; the highest in point of efficiency and brilliancy, requiring less power to consume than other carbons; the highest in mechanical excellence; the highest in point of copper plating, in various thicknesses and in uniformity of plating; free from impurity of acids and not tarnishing.

GENERAL ELECTRIC COMPANY (LIMITED), LONDON.

ARC LIGHT CARBONS.—For outdoor as well as indoor lighting, a great variety of solid and cored arc lamp carbons were shown, of superior make. To show that such carbons, which frequently are not straight at the commercial sizes of one foot, can be made straight without trouble, the sizes varying from 3-32 inch upwards were shown in lengths of three feet and disclosed the excellence of make and mode of manufacture. The carbons have a very close and homogeneous grain and have all the qualities desired for obtaining a pure, steady, white light.

WASHINGTON CARBON COMPANY.

This firm showed several boards upon which were carbons of their own manufacture, arranged in attractive forms. The carbons were for use in arç light and battery service; there were also plate and brush carbons, as well as samples of the unfinished product.

HARDTMUTH & COMPANY, VIENNA.

This firm made an exhaustive exhibit of carbons for various purposes. The carbons are undoubtedly of superior make, pure in quality, and, under test made by the electrical engineering department of the Exposition, gave results which place them very high. The arc carbons sell for

THE COURT OF HONOR.—BY NIGHT.

more than those made in the United States, and the duty of 45 per cent makes them still more costly, and it is a question whether, with these odds against them, they are of greater commercial value than our own. They last longer, and give a very superior light, steady in character, mellow and white in quality.

N. P. STEVENS.

SPLICING REFUSE ARC CARBONS.—The pieces of carbon have cut in them a plug and socket on the ends to be joined. The plug and socket are dipped in a preparation and joined together, then placed in the car and run into an oven to bake. The carbons are sufficiently strong, and can be used in either upper or lower holders. There is no apparent change in the light while burning past the splice. The preparation is composed of coal tar and carbon dust, made by cutting the carbon joints. The dust is sifted so as to obtain the finest, and mixed with coal tar while hot to a thick paste, then put in a shallow dish. Heat enough is kept under the dish to keep preparation warm while being used; there is also another small dish containing a sponge saturated with coal tar in which to dip the plug end of the carbon; the dipped plug is inserted in the socket of the piece to be united; it is then revolved so as to moisten and absorb whatever dust there may be in it. This prepares the ends to receive the preparation. Dip the plug end of the carbon in the preparation, taking up sufficient to insure a good joint; wipe the extreme end of the plug so as not to fill the socket too full, then turn the plug in the socket with sufficient pressure to unite them firmly—at the same time see that the carbon is straight; the carbons can be straightened easily by turning them between the thumb and finger. When straightened place them on the car in diagonal layers and run into the oven to bake. If gas is used in baking it will take about two hours for a baking.

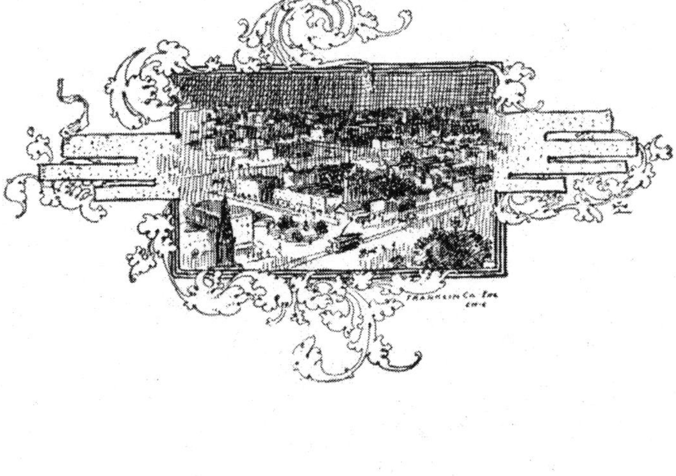

CHAPTER XXV.

ELECTRICAL PARTS AND MISCELLANIES.

WESTERN ELECTRIC COMPANY.

AUTOMATIC SCREW MAKING MACHINE.—This machine is the automatic nut machine, which manufactures nuts from brass strip ⅜-inch in diameter, all polished, drilled, tapped and faced off on one side and chamfered on the other. The machine has a capacity of from 8,000 to 10,000 per day, large quantities of which are used in switchboard apparatus, bells, etc., manufactured by this company. Next to this machine stood the double turreted automatic screw machine made to work upon both sides of a piece before it leaves the machine. The stock is fed in through the lower head, and the work at one end is finished, after which it is cut off and picked up by an automatic finger and placed in the upper head, where the opposite end of the piece is finished at the same time that the machine performs the operation on the first end of the next piece, after which the piece is thrown out into a receptacle made to receive the finished work and separated from all chips. The oil used for both the screw and nut machines is not wasted, but, by means of a pump, is used over and over again. The machine is adapted for a variety of work, such as screws, studs, sleeves, collars, washers, nuts, thumb nuts, binding posts, tips for switchboard use, and many other parts for electric appliances, also a large variety of work for other purposes.

AUTOMATIC DRILL PRESS.—This was a multiple drill press, consisting of twenty-four adjustable spindles. The machine was connected to a line shaft by means of belting, and this line shaft received its motion from a 220-volt slow speed 10-h.p. motor of the Western Electric Company's system. This motor received its current from generators located at a point 2500 feet distant. The product of this machine at the Exposition consisted of a small souvenir of white hollywood, ½ in. thick, 1 3-16 in. wide and 2 in. long. Through these small blocks were drilled twenty-four ⅛ in. holes simultaneously. These holes were drilled at a uniform distance from each other, equal to 3-16 in.; they were drilled in two rows near and around the outer edge, leaving a blank space in the center ; in the blank space, on one side, was stamped the inscription: "Western Electric Co., 1893," and on the other side, "Drilled by one operation." This style of product was chosen for two reasons—first, because a small article was more likely to be kept by the recipients, and second, to demonstrate the general utility as well as the practical features of the machine. Visitors could see at once how it is possible to drill a large number of holes in small pieces, as well as in large ones, by handling the piece but once, and by making but one depression with the feed lever.

THE WM. POWELL COMPANY.

DYNAMO OILER.—The exhibit consisted of patent signal dynamo oilers, Star sight feed cylinder lubricators and regrinding Star globe valves. At a glance the engineer can tell whether the feed is off or on by the position of the signal lever. If it is up, the feed is on—down, it is off. The flow of oil can be regulated to a nicety by the milled head screw, which is secured by the winged jamb nut, and it can be instantly worked or stopped by moving the signal lever up and down, without disturbing the regulation. If necessary to flush the bearings, it can easily be done by canting the signal lever to one side, where it will remain until replaced in position, and this, too, without disturbing the regulation, as the signal lever is independent of feed regulator—an important feature not found in other oilers and well adapted to sudden emergencies sometimes incident to dynamo-motor use. The filling of this oiler is simple and convenient and can be done without stopping or changing the feed. Move the snap lever to the right or left and the filling hole in top cap is exposed. The sight chamber is fitted with one large hole encased with square window panes, which are

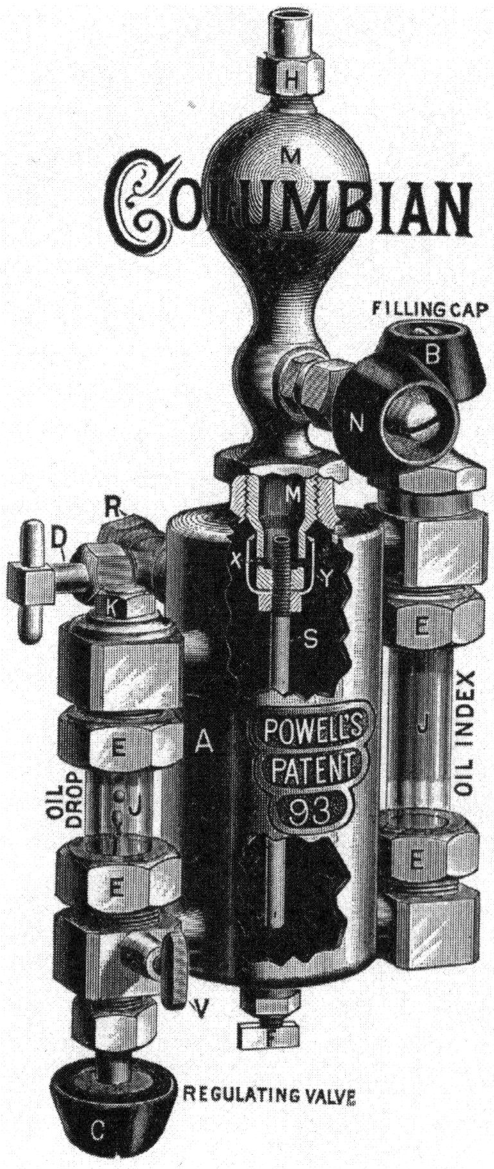

secured by brass clips. They can readily be removed for cleansing or repairs whilst the machine is in motion. The oil feed in this style of sight chamber is exposed to full view from two sides of the oiler. Another valuable feature is the wire cloth filter, which covers the oil passages in the lower end of valve post, which excludes all grit and impurities in the oil from the bearings. The filter can be easily removed for cleansing or renewal. The oiler is positive and reliable in its action, well made, highly finished and ornamental.

STAR SIGHT UP-FEED LUBRICATOR FOR HIGH GRADE DYNAMO ENGINES.—It is extra heavy and intended for high class service and heavy pressure. There are no joints about it to leak, the arms and body being cast in one piece.

The index and sight glass are to the right and left of the oil reservoir, showing the engineer at a glance the drops of oil passing to the cylinder and the quantity of the oil in the chamber. This arrangement also makes it convenient to get at the valves and replace glasses when necessary. Besides the patented condenser water trap, there is attached a vertical delivery tube S, the object being to insure a free delivery of condensed water to the bottom of the oil chamber and at the same time preserve the anti-siphonage feature of the cups. It is impossible to siphon out the contents. The lubricator is of handsome appearance, nicely proportioned and reliable in the regular delivery of oil.

THE UNION ELECTRIC COMPANY, NEW YORK.

Part of the exhibit of this firm was Electricon, so-called, an anti-friction or lubricating metal for use in main rod connections of locomotives, main journal and crank pin bearings, marine engine bearings, and all classes of shaft and spindle bearings. It will not flake or fuse. It requires less oil, lasts longer, heats less and stands heavier pressure than other metals.

PAGE BELTING COMPANY.

BELTING FOR DYNAMOS AND MOTORS.—A beautiful pavilion, a Doric-Grecian temple, was erected on this space. It was constructed of leather, representing the various styles of belting manufactured by this company for electrical purposes. No woodwork appeared above the raised platform upon which this temple was erected. It was brilliantly lighted by means of a motor and a dynamo located in the space, upon which were shown in constant operation, day and evening, an Acme link and a Eureka dynamo belt, the two running tandem. These are their leading styles of belts for electrical work. The spiral columns, made of link belts highly polished, surmounted by a cornice consisting of both link and flat belts, natural color and black, plain and slotted, not only demonstrated the possibilites in leather decoration, but served as a splendid advertisement of the various kinds of belts devoted especially to electrical work, as made by the company.

In addition to belts displayed in the exhibit above mentioned, the company exhibited belts in motion, of which they had ninety-five on electrical machines, most prominent of which were the two tandem belts, 3-ply, one 72 inches wide, 148 feet long, and one 71 inches wide and 185 feet long, running from the Allis engine in Machinery Hall, at a speed of 5600 feet per minute, and transmitting more than 2000 h.p. These constituted a leading exhibit of belts in motion at the Fair. The large Allis engine, the two mammoth generators furnished by the Westinghouse Company, and the two belts supplied by the Page Belting Co., connecting the engine with the generators, combined to make a masterpiece of mechanical work.

THE JEWEL BELTING COMPANY.

BELTING FOR DYNAMOS AND MOTORS.—The exhibit consisted of various grades and styles of belting for the transmission of power. It was all pure oak bark tanned leather and short lap. The Jewel dynamo belt is made from the center cuts, only of the choicest selected stock, pure oak bark tanned leather, and is selected and put together with special reference to the hard work required of it in transmitting power to electrical machinery, great care being taken that it shall be perfectly balanced, as well as of the best stock, in order that the high speed at which this class of belt is run may not cause it to sway, jump or run unevenly. The Jewel extra belt represented the quality and kind of belting in general factory use for running miscellaneous machinery. It, too, was pure oak bark tanned and short lap, and very heavy. The specialty of the exhibit was a 52 inch double belt made of one piece of leather on each side. Not more than one hide in a hundred is suitable to make a belt of this width, and it was exhibited to show what could be done in the direction of width, by a careful selection of the entire product of tanneries for a number of months, and was the widest belt ever made without longitudinal joints. Another interesting feature of the exhibit was a 12 inch belt which had been in use for thirty-six (36) years, and which is computed to have traveled over 1,250,000 miles, and which was still in service at the time it was secured to send to the Exposition. The company had also in the power plant four 24 inch dynamo belts, in use on Phoenix engines and Eddy generators. One of these 24 inch belts, whose capacity under the conditions as they existed was rated at about 250 h.p., was called upon by reason of short circuit, ground on the line,

Interior View Electricity Building.—Looking South.

etc., to stand a strain over 1000 h.p., which it did without the slightest injury to itself, and was afterward used with the best results.

THE CHARLES MUNSON BELTING COMPANY.

BELTING FOR DYNAMOS AND MOTORS.—The exhibit showed the rough oak tanned leather used exclusively in the manufacture of dynamo belting. There were four or five whole oak tanned steer hides hanging on the wall and side of the exhibit, which showed by well marked lines the centers used in the construction of belts. The finished stock was well represented by a full line of Eagle and dynamo belts, single, double and three-ply of different sizes, prominent among which was the large 84-inch double belt hung in the center of the exhibit. The interior decoration and furnishing were quite novel, and entirely made out of leather; the floor consisted of small, triangular pieces of leather of light and dark color, making a beautiful contrast, all being cemented to the wooden floor in handsome squares, resembling an artistic tile pattern. In this floor there were 4974 pieces of leather used, which had to be cut to exact size. The railing in front of the exhibit was constructed entirely out of round pieces of leather firmly cemented together, and represented four columns, connected on each side by rails of the same material. In the construction of this railing 13487 pieces of leather were required. Above the entrance and connecting the two middle columns was a highly ornamental sign of the company cut out of leather.

THE CHICAGO BELTING COMPANY.

BELTING FOR DYNAMOS AND MOTORS.—The exhibit consisted of dynamo belting from three to seventy inches in width, inclusive. This belting is made of the finest material, and the peculiar merits or points of excellence claimed are as follows: perfect uniformity in weight, obtained without filling or skimming, but by the selection of plump hides; solidity and fine grain, combined with great pliability and elasticity; the length of stock used, no piece extending into the shoulder, all being cut 50 inches in length or under from the butts of pure oak tanned leather; its treatment to a preparation that renders them nearly or quite impervious to the action of machine oils or dampness; the excellence of the cement used, no rivets or fastening of any kind being necessary. There was also on exhibition all belting used in the saw-mill exhibit made by the Stearns Manufacturing Company and the E. P. Allis Company. The company furnished the Exposition, under contract, one 32-inch double belt, one 24-inch double belt, one 19-inch double belt, two 12-inch double belts, thirteen 10-inch double belts; these were used in connection with the Buckeye engines and the Fort-Wayne-Jenney-Wood dynamos.

A. GROETZINGER & SONS, ALLEGHENY CITY, PA.

EXHIBIT OF DERMAGLUTINE, BLANKS, PINIONS, GEARS, ETC., FOR ELECTRIC CAR SERVICE AND MECHANICAL PURPOSES.—This article, as the name implies, is rawhide specially prepared and cemented together under a new process. The use of rawhide for mechanical purposes was comparatively unknown prior to the advent of the electric street railway motors; the ordinary cast iron gear wheels used at first in the propulsion of these motors was found to be very unsatisfactory; steel was then resorted to, which was found equally objectionable owing principally to the disagreeable noise and jarring caused by the gear wheels when the car was in motion. All these objectionable features were subsequently completely overcome by the introduction of rawhide, or Dermaglutine, which possesses all the important properties required for this purpose, namely: extreme hardness when properly seasoned; toughness of fibre and power of resisting the great strain in a higher degree than any other material, besides running almost noiselessly. Dermaglutine has, through a series of costly experiments, been brought to perfection and has practically superseded and taken the place of metal pinions and gear wheels formerly used by the electric street railway systems.

THE BILLINGS & SPENCER COMPANY.

TOOLS FOR ELECTRICAL WORKERS AND ELECTRICAL PARTS.—The exhibit consisted principally of Billings patent commutator bars of pure lake copper and Tobin bronze made by the process of drop forging. The particular claims of superiority made for the Billings patent commutator bars are that they are made from guaranteed pure wrought copper, and, being forged instead of cast, they are free from blow holes or porosity; therefore are of the highest possible conductivity. The company displayed also drop-forged eye bolts used in the construction of dynamos and motors; also forged steel wrenches and lineman's vises.

EUREKA TEMPERED COPPER COMPANY.

COPPER FOR ELECTRICAL PURPOSES.—This company, acting upon the theory that pure copper was the only metal that was practical and with sufficient conductivity for electrical work, has assumed to supply the trade with a hardened or tempered copper which they claim to be free from alloys of all kinds, cast solid without blow holes. They had on exhibition a full line of commutators made from their metal for all, or nearly all, of the standard dynamos of the present day, new bearings, and bearings that had seen service and had a record worthy of attention of all those interested in machinery. They also had a full line of commutator segments, both for electric lighting and street car work, made from this metal, and metal brushes of all kinds, styles and dimensions; also a full line of commutator brushes and rolled tempered copper for all the different articles which have heretofore been made of rolled brass or soft-rolled copper. This they have made up into electrical brushes, washers, contact springs, etc. They exhibited tempered copper soldering coppers, with the handles cast solid in them, for which they claim a superiority over others. They also exhibited a full line of tempered copper wire, both bare and insulated, for electrical work, together with a full assortment of bushings, gear pinions, trolley wheels, etc.

L. BAUDREAUX, PARIS.

FOLIATED ROLLED METAL DYNAMO BRUSHES.—These consist of metallic foliated leaves rolled to the least possible thickness, folded, and reduced under pressure to the exact size of the brush holders. The metal of which these brushes are made has copper for its basis, and, being very ductile, possesses all the peculiar advantages of the anti-friction metals—that is to say, is unaffected by rubbing. Every inch in each or any sectional part contains about 1,000 foliated metal strips or leaves, each being from 1-1000th to 1-1200th of an inch only in thickness. This infinitesimal division of the foliated metal produces great softness of friction, thus reducing the wear and tear of the commutator to a minimum, while at the same time giving the brushes a very long life. These brushes, therefore, take the shape of a block made of a sort of foliated metal of the thickness of tissue paper, the main body of which is far more uniform than that of brushes made of wire or metallic gauze. The rubbing surface being continuous and forming an aggregate mass, polishes the commutator to a brilliant surface, whereas the ordinary wire brush presents a finely divided surface, which impairs and eventually destroys the surface of the commutator. Taking into comparison the same sizes, the weight of these new brushes is double that of the wire or gauze brushes, and their conducting power much greater. The resistance and the heat produced by friction are therefore much less, and the capacity of the dynamo is thus correspondingly increased. This increase can easily be ascertained in machines of large current and low tension.

THE PELTON WATER WHEEL COMPANY.

WATER WHEEL FOR OPERATING ELECTRIC MOTORS.—This exhibit, made in conjunction with that of the General Electric Company, consisted of a 32 inch Pelton wheel, running at a speed of 750 revolutions per minute, under a pressure of 182 pounds, equivalent to a vertical head of 420 feet. It was connected direct to the armature shaft of a General

Electric generator, which operated a mining drill, hoist, and other machinery.

Efficiency.—Under this heading reference is made to a brake test at the Idaho mine, showing an efficiency of 87 per cent. Reference is also made to numerous tests of like character, made in the engineering departments of the various universities of the country, which have shown efficiencies ranging from 85 to 92 per cent. Mention may also be made of a test on a Pelton wheel recently reported by Messrs. Gantz & Co., of Budapest, Hungary, showing 90 per cent. efficiency. From an examination of the shape and curves of the buckets, as well as the action of the stream thereon when the wheel is in motion, it is evident that but little, if any, water is carried over. After having spent its force, it is thrown back out of the course of the wheel at an angle of about 10 degrees from the line of impingement.

Maintenance.—The advantages the Pelton wheel offers in this regard are so obvious that they need not be largely commented upon. Reference need only be made to the fact that, there being no metallic contact, there is nothing subject to wear excepting the action of the water upon the buckets, which in practice is so insignificant as not to be worthy of account.

Adaptation to Varying Conditions.—Attention is called to the facility with which the wheel can be adapted to varying conditions of speed and pressure, as well as volume, water wheels being made from 4 inches up to 20 feet in diameter, capable of using from one to six nozzles each, the latter varying from one-fourth of an inch up to twelve inches in diameter—thus compassing a wide range of conditions with but small loss in efficiency. The capacity of any wheel may also be reduced from its maximum, down to fully 25 per cent. of same, without a loss of more than 5 per cent. Except as to the largest nozzle mentioned, the variations above named are constantly met with in practice.

Simplicity and Absence of Wearing Parts.—Under this head but little explanation is needed. The wheel proper consists of but four essential parts, namely—frame or center buckets, gate, and nozzle. The water discharging into open buckets that have no contact surfaces to wear, or water joints to maintain, continuous and reliable service is assured for an indefinite time.

CHAPTER XXVI.

ELECTRICAL JOURNALS AND PERIODICALS.

THE representative electrical journals were awarded space in the gallery of the Electricity building in which to erect booths or pavilions. The representatives of all those awarded space combined to construct headquarters of a uniform style of architecture, and indeed these were no small attraction in the building. Each journal had a small room within an inclosure surrounded by a brass railing. Portieres and drapery were used in profusion to decorate the spaces, and handsome furniture and fixtures further added to the attractiveness of Newspaper Row, as it came to be known. No class of people has done more to further the interests of those in the electrical business, and certainly no class did more to make the electrical department of the Columbian Exposition a success.

THE ELECTRICAL INDUSTRIES PUBLISHING COMPANY.

The exhibit made by this company was confined to the publications which it issues regularly, consisting of a monthly edition and a weekly World's Fair issue, the latter being devoted exclusively to the electrical and allied interests of the World's Columbian Exposition. The company was organized in 1889, for the purpose of publishing a monthly journal devoted to the consideration and advancement of electricity in all its applications. The field which the paper aims to cover is to give a review each month of everything of electrical interest in this country, including illustrations of new electrical and mechanical appliances, descriptions of lighting and power plant installations, together with original articles and papers by the most prominent writers on electrical subjects. In order to make the publication of further use and value to the subscriber, complete directories of the electric lighting central stations, the electric railways and the electrical associations of North America, with a buyer's directory of the manufacturers and dealers, are published in each number. Addi-

Columbian Fountain.—Search Light Effect.

tions and corrections to these directories are made every issue. The field for such a publication was apparent at the start and it was favored immediately with an exceptionally generous patronage at the hands of both subscribers and advertisers, and to-day it ranks as a very strong, healthy and influential paper in the electrical field. It has earned an enviable reputation for honesty, integrity and fairness. The first issue of the *Weekly World's Fair Electrical Industries* appeared in June, 1893. This paper was devoted, as before stated, exclusively to World's Fair matters, the publishers finding such a paper necessary, in order to better serve its readers with news during the Exposition. It contained in each issue illustrated descriptions of exhibits, notes and news of current events at the Fair of particular interest to electrical people, together with a map of the Electricity building, giving the name and location of the different exhibitors. The paper was very popular and its list of subscribers included names from all over the world. It was the only paper published exclusively in behalf of the electrical and allied interests of the Fair, and the enterprise of the publishers was fully recognized.

THE ELECTRICAL REVIEW.

The Electrical Review, of New York, was founded by Mr. George Worthington, in February, 1882, and was the first weekly devoted to electrical interests published in the United States. Mr. Worthington remained as editor of the *Review* until his death in February, 1892, just ten years after the first number appeared. The development of the *Review* has been coincident with that of the electrical field. At the time the journal was started, matters electrical were in a rather chaotic and undeveloped state. It has doubtless been a great mutual benefit to the electrical industry and the *Electrical Review* to have developed side by side, each helping the other. It is estimated that there are $800,000,000 invested at the present time in the electrical industry in this country, and the *Electrical Review* has always presented a faithful and accurate account of this development. As an interesting, influential and independent publication the *Review* has always occupied an enviable position and is recognized as one of the leading class journals of this country. Its headquarters in Electricity building, where files of the paper and pictures of eminent electricians were exhibited, were visited by large numbers of electrical people of this country and Europe. A consistent advocate of the World's Columbian Exposition and zealous in advancing the interests of electrical exhibitors, the *Review* has at all times been of material aid to this department.

THE STREET RAILWAY JOURNAL.

A reception room, handsomely furnished, with chairs, desks, writing material, etc., was in the space of this company, where street railway visitors to the Fair, with their friends, could rest, attend to their correspondence, read technical papers, etc. That a large number of street railway men, both from this country and abroad, availed themselves of this opportunity was testified to by the many signatures on the register and the many personal cards left.

The walls of the office in Electricity building were adorned with numerous engravings and photographs of well-known street railway plants, both cable and electric.

The exhibit proper of the *Street Railway Journal* consisted of a complete file of the copies of the *Street Railway Journal* since the date of its publication, January, 1884. These files contained an unequalled description of the gradual development of the application of mechanical methods of traction to street railway purposes during these years, the most momentous in the history of the industry. Also shown were copies in different styles of binding of the book recently issued from the press of the *Street Railway Journal*, entitled "Street Railways," by C. B. Fairchild. Another exhibit made by the *Street Railway Journal* was a collection of forms for the keeping of accounts and reports made by different departments, order blanks, etc., used in street railway business, together with tickets, transfers, etc. This collection is the largest and most complete in the world, and was completed only after considerable expenditure of time and money. The forms were arranged systematically, so that they could be readily inspected and examined. This collection was undertaken largely by the *Street Railway Journal* in recognition of the popular demand of street railway men for a more uniform system of keeping expense accounts, etc., and its value in advancing this desirable end by collecting the leading systems in use cannot be too highly estimated.

THE WESTERN ELECTRICIAN.

The illustration shows the office and exhibit of the *Western Electrician* in the gallery of Electricity building. This journal, the only weekly electrical newspaper in the West, early recognized the vast importance of the World's Fair to the electrical interests of the country, and, from its location in Chicago and its position as the recognized exponent of that stirring activity in electrical development which distinguishes the central and western portion of the United States, it took a peculiar pride and pleasure in advocating and describing the extensive and varied electrical features of the Exposition. Those who followed the profusely illustrated articles that appeared in the paper from week to week know how well this self-imposed task was executed. It is not too much to say that no other technical journal equalled the *Western Electrician* in quality or quantity of World's Fair matter, or in the number and excellence of accompanying illustrations. As early as 1890 the *Western Electrician* recorded the first movement for electrical representation. From that time until the close of the exhibition, and after, the electrical interests at the Fair found in this journal their warmest champion and most accurate and minute chronicler.

The plans of Electricity building were first made public by reproduced drawings published in the *Western Electrician* early in the summer of 1891. This was soon followed by the first perspective view, which was drawn in the office of the paper from the architects' plans and afterward adopted by the World's Fair authorities as an official picture. The electrical construction work was carefully recorded as it progressed, and the articles on "Electricity at the World's Fair" were made a department of the paper, increasing in size until, during the Fair period, they constituted the leading feature of the journal. At the close of the Exposition every electrical feature of importance had been fully described.

It was the *Western Electrician* which, on behalf of the electrical fraternity, led the opposition to the evening closing rule, which was eventually repealed, and the same journal was the first to publish the eagerly awaited list of electrical awards.

The distinctive World's Fair issue published in the week ending May 6, 1893, and, devoted entirely to World's Fair electrical matters, was undoubtedly one of the finest and most richly illustrated numbers ever issued from the office of an American technical newspaper. It contained much valuable information and elicited handsome compliments even from competitive electrical journals.

From its fortunate location the *Western Electrician* was the first electrical paper to greet the electrical men at the Fair every week, and the coming of the blue-covered journal was eagerly awaited. The exhibit space was decorated with specimens of the fine illustrations for which the paper is noted, and these attracted much attention.

THE ELECTRICAL ENGINEER OF NEW YORK.

This journal made a most interesting historical exhibit of Davenport railway and other motors. A detailed description of this part of the exhibit will be found under the proper heading.

In addition to this exhibit the company maintained a most agreeable headquarters for the electrical people. On view were current files and handsomely bound volumes of *The Electrical Engineer*, covering the period of its existence, from January, 1882. They also showed many photographs and drawings of interesting apparatus and installations.

CHAPTER XXVII.

AWARDS IN THE ELECTRICAL DEPARTMENT.

THE awards system at the Columbian Exposition, if taken as a whole, was not a success; but the scheme adopted by the judges of award for the Department of Electricity would have been in every way satisfactory had it not been for the general disrepute which came upon the system of awards for the Fair as a whole. Out of a total of 60,000 awards, more or less, there were given in the Electrical department only 213 awards. The difference between the awards system adapted for the Electrical department and that in operation for the rest of the Fair grew out of the fact that the judges of award in electricity declined peremptorily immediately upon their appointment, to begin their work under the system promulgated by the Executive Committee on Awards. The fundamental error in the general system, from the point of view of the board of electrical judges, was the entire absence of competition. It was agreed at the outset that this was an error fatal in every way to success and the judges declined to go into their work with failure stamped upon the department at the outset. Another great error was the individual judge system. It was believed that a committee of judges would prove more competent to pass upon the relative merits of exhibits in a given class than a single judge who was asked to pass upon a single exhibit without reference to any other exhibits in that class. The judges therefore declined to act singly in passing upon any of the groups in the Electrical department, but organized themselves into sub-committees of judges and apportioned the work by groups among those sub-committees.

Previous experience of some of the members of the board convinced them that the awards in the department would lose a very large percentage of their value if the judges should pass upon the exhibits and recommend awards and not have the final voice in the matter. All of the members of the board of judges in the Electrical department were highly

Electric Fountain.—Search Lights and Agricultural Building.

educated, scientific and practical electricians, at the head of their profession in the special direction of work for which they were appointed. They declined therefore to proceed until they had come to a distinct agreement in writing with the chairman of the executive committee on awards, providing that their recommendations should carry with them an affirmative action on the part of the awards commission without appeal.

In other words, the board of judges in the Electrical department, immediately upon their organization decided to ignore the instructions of the executive committee on awards in three directions. They provided first, for competition in all classes and an award to the best exhibit in each class. There were only a few cases of deviation from this rule and those were peculiar, and concerned special and meritorious features. The judges provided further, that exhibits should be examined and passed upon by sub-committees instead of by individual judges as provided for under the rules, and the vote of the majority of the judges in any sub-committee was necessary to a recommendation; and the action of the sub-committee was not final until acquiesced in by a majority of the full board of judges. The third provision made by the board of judges was that there should be no appeal from their action before the executive committee on awards. With these three exceptions the rules governing awards for the Fair generally prevailed in the Department of Electricity.

The board of judges met pursuant to instructions from the executive committee on awards, on July 15th, in the offices of the Electrical department. An officer of the Electrical department acted as temporary chairman for the board, pending the election of a presiding officer. A general discussion of the work of the department and preliminary action as outlined above, consumed two or three protracted meetings. On July 17th, Prof. H. S. Carhart, of Ann Arbor, Michigan, was elected permanent president of the board; Dr. E. Rathenau, of Berlin, Germany, was elected first vice president; Prof. W. E. Ayrton, of England, second vice president; and W. E. Anderson, of Virginia, was appointed secretary, though not a member of the board. The board complete was as follows:

Prof. Adolpho Aschoff, Brazil; Prof. W. E. Ayrton, England; Prof. Brown Ayres, Louisiana; Prof. Geo. F. Barker, Pennsylvania; Prof. Henry S. Carhart, Michigan; Prof. A. E. Dolbear, Massachusetts; Dr. Louis Duncan, Maryland; Dr. Chas. E. Emery, New York; Ahmed Fahri Bey, Turkey; Prof. George Forbes, England; Prof. L. C. Hill, Colorado; Dr. W. J. Herdman, Michigan; Prof. D. C. Jackson, Wisconsin; Prof. T. C. Mendenhall, Washington, D. C.; Prof. M. O'Dea, Indiana; Prof. R. B. Owens, Nebraska; Mr. R. W. Pope, New Jersey; Director E. Rathenau, Germany; Lieut. Samuel Reber, U. S. A.; Dr. R. A. Rowland, Maryland; Prof. H. J. Ryan, New York; Prof. William

Shrader, Missouri; Prof. W. M. Stine, Illinois; Prof. B. F. Thomas, Ohio; Mr. E. P. Warner, Illinois; Dr. Lobach, Germany.

The division of the board into sub-committees for active work was as follows:

1. To have charge of groups 122, 123 and indicators, registering meters, ammeters and voltmeters of group 126—Messrs. Ayrton, Mendenhall, Rowland, Owens, Stine, Thomas, Barker, Ryan and Duncan.

2. To have charge of group 124—Messrs. Dolbear, Shrader, Stine, Barker and Hill.

3. To have charge of groups 125, 127 and 128 and j of 138a—Messrs. Carhart, Emery, Forbes, Jackson, Reber, Ryan, Rathenau, Aschoff and Duncan.

4. To have charge of groups 126 and 129—Messrs. Thomas, Ayres, Owens and O'Dea.

5. To have charge of groups 130, 131 and 132, l m of 138a—Messrs. Barker, Ayres, Rathenau and Warner.

6. To have charge of groups 133 and 134—Messrs. Ayres, O'Dea, Pope and Ahmed Fahri Bey.

7. To have charge of groups 135—Messrs. Herdman, Dolbear, Barker and Shrader.

8. To have charge of groups 136, 137, 138 and 138a—Messrs. Barker, Ryan, Pope, Mendenhall and Lobach.

The members of the board were extremely conscientious and energetic in the execution of their work, made efficiency and time tests wherever it was possible, and in their recommendations were fair and careful. In the groups covering incandescent lamps, primary and secondary batteries and insulated wire, the tests were especially exhaustive and were made in the constant presence of at least one member of a sub-committee. The incandescent lamp tests extended over a period of sixty days, beginning with the first of September and ending with the close of the Fair. About 1,000 lamps were burned constantly on a rack provided for the purpose, proper security from interruption or tampering being provided. There were two tests of insulated wires; one commonly called the break-down test, and the other the soak test. The test was made with No. 14, 5-32 rubber insulation B. & S. gauge copper wire. Current used was the Westinghouse alternating 2,000 volt current with step-up transformer. The highest potential reached, during the test, was 38,000 volts made with Kerite compound insulation, the wire being the manufacture of W. R. Brixey, of Seymour, Connecticut. A good deal of difficulty was met with by the judges in the preparation of the soak test for insulated wires, and competent tests were made possible only by an offer from the Electrical department of the city of Chicago to pay all expenses connected therewith. The city being interested to the extent of many thousands of dollars annually in the purchase of wire, it was deemed absolutely essential that some authoritative and entirely trustworthy test should be made which would enable the city to buy the best wire, irrespective of its first cost. For the purpose of the tests, a sub-committee was appointed consisting of Professors D. C. Jackson, R. B. Owen, M. O'Dea and B. F. Thomas. A vault was secured in the basement of the City Hall, at Chicago, fire and burglar proof, in character, with double doors. These doors were locked constantly with Yale locks and sealed with wax by the judges. Measurements by the galvanometer were made once each week over a period extending from December 1st, 1893, to April 1st, 1894. The results of these tests were the giving of the first award to W. R. Brixey, for Kerite wire, and the second to Okonite.

Although the tests for incandescent lamps were closed on November 1st, the record of the result was not at hand at the date of this publication.

The following is a complete list of awards recommended in the department by the board of judges. This list is correct and complete (with the exception of the awards on incandescent lamps and insulated wire), as the written agreement between the board of judges and the executive committee on awards, through its chairman, took it out of the power of the latter to sustain appeals or to make new awards.

American Battery Co., storage batteries.

Albert and J. M. Anderson, Boston, Mass.—(1) Trolleys; (2) railway insulators.

W. R. Brixey, New York.—Underground, aerial and submarine telegraph and telephone cables.

Brush Electric Co., Cleveland, Ohio.—(1) Direct current dynamos for series arc lighting; (2) direct current dynamo for series arc lighting coupled engine, 125 2000-c.p. lamps; (3) Alternating current dynamos constant potential 36 150 k.w.; (4) arc circuit switchboard; (5) direct current dynamos constant potential 20 100 k.w.; (6) arc lamps all types.

Bryant Electric Co., Bridgeport, Conn.—Snap switches.

J. H. Bunnell & Co., New York.—(1) Standard dry batteries; (2) telegraphic apparatus.

Carpenter Enamel Rheostat Co., Bridgeport, Conn.—Rheostats.

C. & C. Electric Motor Co., New York.—(1) Direct current motors, constant potential 3 50-h.p.; (2) electric motor, fan and blower combinations.

Commercial Cable Co., New York.—(1) Ocean telegraphic apparatus operating through Murihead's artificial resistances; (2) Cuttriss improved cable telegraph apparatus.

Copenhagen Fire Alarm Co., Chicago.—Automatic fire alarm.

Crane Electric Co., Chicago, Ill.—Lamp supporting pulley.

Cutter Mfg. Co., Philadelphia, Pa.—Push-switches for electric lights.

Eddy Electric Mfg. Co., Windsor,Conn.—Direct current motors constant potential.

Edison Mfg. Co., New York.—Edison-LeLande batteries.

Electrical Conduit Co., New York.—Underground conduit for electrical wires.

Electrical Engineer.—Historical electrical railway model (Davenport's).

Electrical Forging Co., Boston, Mass.—Electric heating and welding apparatus.

Electric Heat Alarm Co., Boston, Mass.—Thermostat for automatic fire, hot journal and hot grain alarms.

Electric Launch and Navigation Co., New York.—Electric launches.

Electric Selector & Signal Co., New York.—Electrical system for locking and unlocking.

Elektron Mfg. Co., Springfield, Mass.—(1) Direct current motors constant potential multipolar, slow speed; (2) automatic motor starter.

Eureka Tempered Copper Co., North East, Pa.—Tempered copper for use in electrical construction.

Excelsior Electric Co., New York.—(1) Arc lamps for direct current series circuits. (2) direct current dynamos for series arc lighting.

Fort Wayne Electric Co., Fort Wayne, Ind.—(1) Direct current Wood dynamo for series arc lighting; (2) alternating current Wood dynamo constant potential compound wound 150 k.w.; (3) arc lamps for constant current; constant potential dynamos and motors.

I. P. Fink, New York.—Screen reflectors for incandescent lamps.

Gamewell Fire Alarm Co., New York.—Automatic fire alarm telegraph system.

General Electric Co., New York.—(1) Electric locomotives for factory and switching services; (2) Electric elevated railway system; (3) Long distance power transmission, plant in operation, (tri-phase); (4) Arc lamps for direct current series circuits; (5) Search lights and focusing lamps; (6) Transformers, 250-125,000 watts; (7) Engine dynamos; (8) Automatic over-load switch; (9) Electrically illuminated fountains; (10) Thomson eccentric coil ammeters and voltmeters for alternating currents; (11) Pumping machinery driven by electric motor; (12) Electrically driven rock working machinery; (13) Mine locomotive; (14) Haskin's astatic ammeter; (15) Arc lamps for constant potential circuits, direct and alternating (Knowles); (16) Jaw switches, fuses, sockets and branch blocks; (17) Direct current dynamos for series arc lighting; (18) alternating current dynamos constant potential, 30-300 k.w.; (19) System of street railway service; (20) Direct current dynamos constant potential (direct connected excepted) and direct current shunt wound motors constant potential; (21) Edison feeder system for distribution of electricity; (22) Slate switchboard for arc light circuits. (23) Ventilating set, portable, Government standard; (24) Historical apparatus; (25) Edison three-wire system for distribution of electrical energy; (26) Exhibit of incandescent lamps, all styles, ½ to 250 c.p.; (27) Underground systems complete in all details; (28) Hoisting apparatus driven by electric motors; (29) Integrating watt meter.

General Electric Co.—Ornamental lamp posts.

General Electric Co.—Incandescent lamps used for decorating rooms and other structures.

General Incandescent Arc Lamp Co., New York.—Arc lamps for constant potential circuits.

Graves Arc Lamp Co.—Arc lamps.

Elisha Gray—Telautograph.

The E. S. Greeley & Co., New York.—(1) Testing Instruments; (2) Exeter dry battery; (3) Telegraph apparatus.

The Hanson Battery Co., Washington D.C.—Primary batteries.

Hart & Hageman Mfg. Co., Hartford, Conn.—Snap switches.

Helios Electric Co.—Arc lamp for alternating current circuits.

Interior Conduit & Insulation Co., New York.—(1) System of interior insulating conduits; (2) Snap switches.

Jenney Electric Motor Co., Indianapolis, Ind.—Direct current dynamos and motors, constant potential.

H. W. Johns Mfg. Co., New York.—Vulcabeston and molded mica insulating materials worked into all kinds of insulations.

N. S. Keith.—Constant current motors.

LeClanche Battery Co., New York.—LeClanche batteries, especially the Vole and Cylinder cells.

Mather Electric Co., Manchester, Conn.—Direct current dynamo, constant potential, 500 volts. Automatic adjustable circuit breakers. Constant potential dynamos and motors, latest ring type.

McIntosh Battery & Optical Co., Chicago, Ill.—Electro, medical, dental and surgical apparatus.

National Carbon Co., Cleveland, Ohio.—Carbons for arc lamps.

National Engraving Machine Co.—Jewelers' engraving machine.

Nutting Electric Mfg., Co., Chicago.—Nutting arc lamp.

Otis Bros., New York.—(1) Electric pump; (2) Electric motor and controlling devices for elevator and hoisting service.

H. T. Paiste, Philadelphia, Pa.—Snap switches.

A. H. Phelps.—Electro pyro gravure process.

Phoenix Glass Co., Chicago, Ill.—Electric and gas globes and shades, cut, etched and colored.

Police Telegraph & Signal Co., Chicago.—System of police patrol telegraph.

Queen & Co., Philadelphia, Pa.—(1) Electrometer (Ryan); (2) galvanometers; (3) testing sets and resistances; (4) portable medical induction apparatus for physicians' use; (5) commercial ammeters and voltmeters; hot wire voltmeter.

F. A. Ringler & Co., New York.—Half-tone photo-electrotype steel-faced.

J. A. Roeblings Sons Co., Trenton, N. J.—Bare copper and trolley wire.

Self Winding Clock Co., New York.—Special application of an iron-clad solonid magnet.

Short Electric Railway Co., Cleveland, Ohio.—Short electric railway system.

Sperry Electric Railway Co., Cleveland, Ohio.—Electric railway system.

Standard Electric Co., Chicago.—(1) Arc lamps for direct current series circuits; (2) direct current dynamos for series arc lighting.

Standard Paint Co.—Insulating compound, liquid.

Stevenson Hoggson Electric Co., St. Louis, Mo.—Automatic electric time stamp.

Thomson Electric Welding Co., Boston, Mass.—Apparatus for electric welding and forging.

Union Electric Works, Chicago, Ill.—Primary battery.

J. C. Vetter & Co., New York.—(1) Incandescent current adapter; (2) dry LeClanche battery.

Waite & Bartlett Mfg. Co., New York.—(1) Holt induction machine in air-tight case with six 40-inch revolving plates; (2) special faradic apparatus for varying the tension and strength of current (Engleman's apparatus).

Wm. Wallace, Ansonia, Conn.—Historical electric light exhibit.

Electricity Building.—From the South.

Walworth Mfg. Co.—Poles for trolleys and arc lamps.

Washburn & Moen.—Bare copper wire.

Washington Carbon Co., Pittsburgh, Pa.—Carbons for arc lamps, batteries and dynamos and motor brushes.

Western Electric Co., Chicago.—(1) Columbian street lamp post; (2) telegraph apparatus; (3) telephone cables (Paterson); (4) annunciators and signaling apparatus; (5) multiple switchboard for telephone service; (6) direct current dynamos for series arc lighting; (7) application of electric lights for the production of scenic effects in theaters and for the decorations of rooms, etc.; (8) lamps for constant potential circuits; (10) direct current dynamos and motors constant potential.

Western Union Telegraph Co., New York.—Instruments used in quadruplex telegraph, latest design.

Westinghouse Electric & Mfg. Co., Pittsburg, Pa.—(1) Engine dynamos; (2) transformers, 250—125, 90-kw.; (3) direct current dynamos and motors constant potential bipolar and multipolar (except direct connected dynamos); (4) alternating current dynamos for series arc lighting; (7) long distance power transmission, plant in operation; (8) two phase alternating current motors (Tesla); (9) incandescent system of street lighting; (10) switches; (11) complete switchboard for controlling 17 dynamos and 40 circuits; (12) lightning arresters; high tension experimental apparatus; electric meter, (Shallenberger); constant potential, alternating current arc lamps; (1) regulator (Stillwell); (2) continuous current, constant E.M.F. "letter" type dynamos and motors.

Weston Electrical Instrument Co., Newark, N. J.—(1) Alternating current instrument including wattmeters; (2) standard resistances and bridges; (3) electrical measuring instruments for physicians' use; (4) switchboard instruments; (5) direct current ammeters and voltmeters, standard and portable.

S. S. White Dental Mfg. Co., Philadelphia, Pa.—Acid gravity batteries (Partz); application of electricity as a motive power for electric drills.

Zucker & Levett Chemical Co., New York.—Collection of chemicals and appliances for electroplating.

FOREIGN AWARDS.

AUSTRIA.

F. Hardmuth & Co., Vienna.—Carbons for arc lamps, etc.

Schindler & Jenny.—Electrical cooking apparatus.

BRAZIL.

Directoria General dos Telegraphos, Rio de Janerio.—Telegraphic apparatus.

GREAT BRITAIN.

British Government Postal Telegraph Department.—(1) Modern telegraph apparatus in operation; (2) Historical telegraph apparatus.

Corporation of Birmingham.—Original Woolwich dynamo.

Epstein Accumulator Co.—Storage batteries.

General Electric Co. (Limited), London.—H. I. switches and other incandescent house fittings; carbons.

James White, Glasgow.—Electro magnetic balances (Kelvin).

AWARDS IN THE ELECTRICAL DEPARTMENT.

GERMANY.

Prof. Aron, Berlin.—Electric meters.

J. Berliner, Hanover.—Universal transmitter, long distance.

Geo. Carette & Co., Nuremburg.—Optical, physical, and mechanical instruments and toys.

Dr. Edelmann, Munich.—Electro medical apparatus.

Felton & Guilleaume, Muelhausen on Rhine.—(1) Electric cables with special armor; (2) Electric cables.

Hartmann & Braun, Frankfort.—(1) Differential arc lamps; (2) Galvanometers; (3) Electrical measuring instruments including instruments of precision; (4) Photometer, large universal; (5) Apparatus for testing iron and steel with respect to magnetic permeability; (6) Reading telescopes, mirrors, and scales; (7) Portable measuring apparatus used in laying cables.

Gasmotoren Fabrick "Deutz."—Dynamo, directly connected to gas engine.

Hartmann & Braun.—Galvanometers of special form.

W. A. Hirshmann.—Electro medical apparatus.

Imperial German Postal Telegraph Department, Berlin.—(1) Maps and drawings showing system of government telegraph lines using armored underground cables; (2) Telegraphic apparatus of the German telegraph service; (3) Historical telegraphic apparatus.

Korting & Matthissen, Leipsig.—Arc lamps for constant potential circuits.

Chas. Pollak.—Storage batteries.

Reininger, Gebhardt & Schall.—Electro medical apparatus.

Siemens Bros. & Co.—Cored carbons.

Siemens & Halske, Berlin.—(1) Arc lamps for constant potential circuits; (2) Direct current dynamos, constant potential 750 kw.; (3) historical apparatus.

Schmelzer, Nuremburg.—Carbons for arc lights.

Schmidt & Haench, Berlin.—Lummer-Broadham photometer with accessories.

H. Schomburg & Sons, Berlin.—(1) Insulators and insulator material; (2) Dry batteries.

Schuckert & Co., Nuremburg.—Registering watt meters. (1) Search lights with parabolic glass mirrors; (2) Annunciators for vessels; (3) Arc lamps for constant potential circuits.

J Zacharias, Berlin.—Dry batteries.

ITALY.

Prof. Galileo Ferraris, Turin.—Historical alternating current motors.

Societa Ceramina Richard, Milan.—Insulators.

JAPAN.

Department of Engineering, Imperial University, Tokio.—Automatic electric current recorder.

Eizaemon Fukagawa, Saga.—China insulator.

Imperial University, Tokio.—(1) Seismographs and accessory apparatus; (2) Model of an earthquake.

RUSSIA.

Imperial Russian State Paper Manufactory, St. Petersburg.—Collection of electrotypes.

Imperial Artillery Arsenal, St. Petersburg.—Electrical registering attachment for testing machine (Prince Bagarian).

TURKEY.

Imperial Ottoman Government.—Telegraphic apparatus.

CHAPTER XXVIII.

THE ELECTRICITY BUILDING.

WHEN completed after considerable re-designing, the Electricity building was 350 feet wide and 767 feet long, the major axis running north and south. The south front was on the great Quadrangle or Court; the north front faced the Lagoon; the east front was opposite the Manufactures building, and the west faced the Mines building.

The general plan was based upon a longitudinal nave 115 feet wide and 114 feet high, crossed in the middle by a transept of the same width and height. The nave and the transept had a pitched roof with a range of skylight at the bottom of the pitch and clear-story windows. The rest of the building was covered with a flat roof, averaging 62 feet in height and provided with skylights.

The second story was composed of a series of galleries connected across the nave by eight bridges, with access by eight grand staircases. The area of the galleries in the second story was 118,546 square feet.

The exterior walls of this building were composed of a continuous Corinthian order of pilasters 3 feet 6 inches wide and 42 feet high, supporting a full entablature, and resting upon a stylobate 8 feet 6 inches. The total height of the walls from the grade outside was 68 feet 6 inches.

The north pavilion was placed between the two great apsidal or semi-circular projections of the building; it was flanked by two towers 195 feet high. The central feature was a great semi-circular window, above which, 102 feet from the ground, was a colonnade forming an open loggia or gallery, commanding a view over the Lagoon and all the north portion of the grounds.

The east and west central pavilions were composed of two towers 168 feet high. In front of these two pavilions was a great portico composed of the Corinthian order with full columns.

The south pavilion was a hemicycle or niche, 78 feet in diameter and

103 feet high. The opening of this niche was framed by a semi-circular arch, crowned by a gable or pediment with smaller gables on the returns, and surmounted by an attic, the whole reaching the height of 142 feet. In the center of this niche, upon a lofty pedestal, was a colossal statue of Franklin whose illustrious name intimately connects the early history of the Republic with one of the most important discoveries in the phenomena of electricity.

At each of the four corners of the building there was a pavilion above which rose a light open spire or tower, 169 feet high. Intermediate between these corner pavilions and the central pavilions on the east and west sides, was a subordinate pavilion bearing a low, square dome upon an open lantern. There were thus ten spires and four domes. The entablature of the great Corinthian order broke around each of the pilasters of the four fronts, and above each pilaster in the Attic order was a pedestal bearing a lofty mast for the display of banners by day and electric lights by night. Of these masts there were in all fifty-four.

The first story of the building had indicated between the great pilasters of the Corinthian order, by a subordinate Ionic order, with full columns and pilasters forming an open screen in front of the windows.

The building had an open portico extending along the whole of the south façade, the lower or Ionic order forming an open screen in front of it. The various subordinate pavilions were treated with windows and balconies. The details of the exterior orders were richly decorated, and the pediments, friezes, panels and spandrils had a decoration of figures in relief, with architectural motifs, the general tendency of which were to illustrate the purposes of the building.

The color of the exterior was ivory but the walls of the hemicycle and of the various porticos and loggia were highly enriched with color, the pilasters in these places being decorated with scagliola and the capitals with metallic effects in bronze.

In the design of this building it was proposed to so devise its details and general outlines that they might be capable of providing an electric illumination by night on a scale hitherto unknown the flag-staffs, the open porticos, and the towers, especially, being arranged with this in view.

It was proposed that the hemicycle or niche which formed the south porch should have either a great chandelier or crown of lights suspended from the center of the half dome, or should be provided with electric lights masked behind the triumphal arch which formed the opening of the niche.

Other books from CGR Publishing at Amazon.com

Ultra Massive Video Game Console Guide Volume 1

Ultra Massive Video Game Console Guide Volume 2

Ultra Massive Video Game Console Guide Volume 3

Ultra Massive Sega Genesis Guide

Tome of Infinity Volume 1: World's Best Obscure Games

Chicago's White City Cookbook

Official Guide to the World's Columbian Exposition

How To Grow Mushrooms: A 19th Century Approach

The Great War Remastered WW1 Standard History Collection Vol. 1

Sinking of the Titanic: The Greatest Disaster at Sea

All Hail the Vectrex Ultimate Collector's Guide

Old Timey Pictures with Silly Captions Volume 1

P.T. Barnum The Greatest Showman on Earth

Retromegatrex: The Art of Mark Bussler 1995-2017

Captain William Kidd and the Pirates and Buccaneers Who Ravaged the Seas

Robot Kitten Factory Issue #1

Electricity at the World's Fair of 1893 Columbian Exposition:
Illustrated Enlarged Special Edition

www.CGRpublishing.com

Made in the USA
Monee, IL
12 November 2026

25f8b37f-8da0-4d17-920d-1da103a6cfb5R01